A Course in Vector Analysis

Ll. G. CHAMBERS
D.Sc. F.I.M.A.
Senior Lecturer in Applied Mathematics
University College of North Wales, Bangor

CHAPMAN AND HALL LTD
11 NEW FETTER LANE LONDON EC4

First published 1969

Printed in Great Britain by
Spottiswoode, Ballantyne & Co. Ltd.
London and Colchester
SBN 412 08930 0

Distribution in the U.S.A.
by Barnes & Noble, Inc.

A Course in Vector Analysis

Contents

	page
Preface	ix
1. Elements of Vector Algebra	1
1.0. **Introduction**	1
1.1. **Formal definitions of vector and scalar**	2
1.11. *Position vector*	4
1.12. *Unit vector and modulus*	5
1.2. **General laws of vector manipulation**	5
1.21. *Division of a line in a given ratio*	7
1.22. *Centroid*	8
1.3. **Components of a vector**	8
1.31. *Orthogonal frames of reference*	9
1.32. *Abstract treatment of vectors*	12
1.4. **Geometrical applications of vectors**	14
1.5. **The vector as a function**	16
1.51. *Derivative of a vector with respect to time*	17
1.52. *Relative velocity and acceleration*	19
1.53. *Force*	20
1.6. **Coordinate systems**	22
Worked examples	25
Exercises	36
2. Products of Vectors	39
2.0. **Introduction**	39
2.1. **Scalar product**	39
2.11. *The scalar product in rectangular cartesian coordinates*	41
2.12. *The concept of work*	41
2.13. *Geometrical applications of the scalar product*	42
2.14. *Differentiation of the scalar product*	43

2.2. **Vector product** 44
2.21. *The vector product in rectangular cartesian coordinates* 46
2.22. *Mechanical applications* 46
2.23. *Vector product as an area* 47
2.3. **Scalar triple product** 47
2.31. *Reciprocal vectors* 49
2.4. **The vector triple product** 50
2.5. **Quadruple products of vectors** 51
2.6. **Solution of vector equations** 52
2.7. **Representation of a rotation of a rigid body by a vector** 53
2.71. *Elementary rotations* 56
2.72. *Angular velocity* 57
2.73. *Rotating frame of reference* 57
2.8. **Axial and polar vectors** 58
2.81. *Scalars and pseudo-scalars* 60
2.9. **Orthogonal curvilinear coordinates** 60
2.10. **Complex numbers** 62
Worked examples 64
Exercises 77

3. The Vector Operator ∇ 81
3.0. **Point functions** 81
3.01. *Differentiation of point functions* 81
3.1. **Gradient** 83
3.11. *Special gradients* 86
3.12. *The operator ∇'* 86
3.13. *Differentiation with respect to a vector* 87
3.14. *Differentiation following the fluid* 88
3.2. **Divergence of a vector** 90
3.21. *Special divergences* 91
3.22. *Laplacian operator* 91
3.3. **Curl of a vector** 91
3.31. *Special curls* 92
3.32. *Second-order differential formulae* 93
3.33. *Expansion formulae* 94
3.34. *Beltrami fields* 96

3.4. **Expressions involving ∇ in orthogonal curvilinear coordinates** 96
3.41. **∇** *in cylindrical and spherical coordinate systems* 100
Worked examples 102
Exercises 112

4. Line Surface and Volume Integrals 116
4.0. **General** 116
4.1. **Integral along a curve** 116
4.11. *Tangential line integral of a vector* 117
4.12. *Vector integrals along a curve* 120
4.2. **Surface integrals** 122
4.21. *Integral over a plane domain* 123
4.22. *General surface integrals* 124
4.23. *Solid angle* 126
4.3. **Volume integral** 127
Worked examples 130
Exercises 141

5. Stokes's and Gauss's Theorems and their Applications 145
5.1. **Stokes's theorem** 145
5.11. *Extensions of Stokes's theorem* 147
5.2. **The Gauss divergence theorem** 148
5.21. *Extensions of the divergence theorem* 150
5.22. *Green's theorems* 151
5.23. *Green's formula* 153
5.3. **Green's function for Laplace's equation** 156
5.31. *Uniqueness theorem* 159
5.32. *Helmholtz's theorem* 160
5.4. **Bounds for the Dirichlet integral** 161
Worked examples 163
Exercises 177

6. Miscellaneous Topics 181
6.1. **Linear vector functions and dyadics** 181
6.11. *Scalar and vector of a dyadic* 185
6.12. *Dyadics in rectangular cartesian coordinates* 186
6.13. *Dyadic invariants* 187
6.14. *Dyadics and the vector operator* **∇** 189
6.15. *Integral theorems involving dyadics* 190

6.2. **Vector spaces** 191
6.21. *n dimensional Euclidean space* 196
6.22. *The space* L^2 198
6.3. **Orthogonal transformations and tensors** 200
6.31. *Stokes's theorem and the Gauss divergence theorem* 203
6.32. *Covariance and contravariance* 204
6.4. **Relations with matrices** 205
6.41. *Vectors and matrices* 205
6.42. *Complex numbers and matrices* 206
6.43. *Matrices and dyads* 207
6.44. *Tensors and matrices* 208
Worked examples 208
Exercises 221

Appendix: The Delta Function 225

Index 229

Preface

This book has been written in the hope of providing the basic work on vectors necessary for a degree in Mathematics, Physics or Engineering. I have aimed at making a book which will be useful in applying vectors, rather than one which discusses vectors from the point of view of the pure mathematician. It is for this reason that I do not mention, for example, conditions under which the various vector integrals hold. Everything is assumed to be sufficiently well behaved. Similarly, the treatment of abstract vector spaces is intended to indicate the possibility of extending the concept of a vector and is not intended as a rigorous treatment. In certain places, however, I have indicated the possibility of a more formal approach. For example, I have given a fairly long discussion of the definitions of the scalar and vector product. In this way, I hope, the question frequently asked by students when they meet the products for the first time as to why they are defined in this way, will be answered.

It will be observed that there is no reference to Differential Geometry. This is a subject which, I feel, is best dealt with by Geometers. Whilst I have indicated physical applications of vectors in many places, I have had in mind that this is a book on vector analysis and not on theoretical physics. In the discussion of vector and scalar fields, I have laid particular emphasis on the fact that the field at a point depends not only on the position of the point at which the field is measured, but also on the configuration, relative to this point, of the sources causing the field.

The worked examples and exercises are miscellaneous in character. Some are supplementary pieces of bookwork. Some have been familiar for many years, and I have made up others. I am, however, grateful for permission to use some and the sources of these examples are indicated.

I should like to thank Professor H. N. V. Temperley for many helpful comments on the first and second drafts, Dr R. Brown for some comments on the second draft and drawing my attention to the result given in exercise 20 of Chapter 5 and Mr. R. Buckley for drawing my attention to the result given in worked example 22 of Chapter 5.

I am also grateful to Miss V. Hooper for the typing of a troublesome manuscript.

LL. G. CHAMBERS

1 Elements of Vector Algebra

1.0. Introduction

In Natural Philosophy, various types of entities are discussed. Some entities such as temperature and mass are completely defined by a reading on a scale, or a single number. There are, however, other entities which have associated with them not only a magnitude, but also a direction in space. Examples of these are displacement, velocity and force, and it is found that there are certain definite rules for compounding these entities. For example, a displacement of a particle from a point A to a point B, followed by a displacement from the point B to a point C is equivalent to a displacement from the point A to the point C.

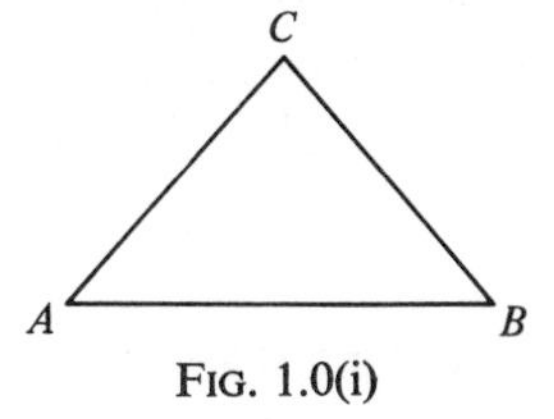

FIG. 1.0(i)

If a further displacement be made from the point C to the point A, the net result of all three displacements is the zero displacement. Clearly the order of the displacements is immaterial and a displacement from the point A to the point C, followed by a displacement from the point C to the point B is equivalent to the displacement from the point A to the point B.

As a further example, it is found experimentally that the condition for a particle to be in equilibrium under three forces is related to the property of displacements mentioned above. Suppose that an experiment is set up in which a mass on a smooth table is subject to pulls from, say, three fixed spring balances through three massless strings.

If the readings on the spring balances and the angles between the strings are as indicated, the relation between them is as follows

$$\frac{P}{\sin\alpha} = \frac{Q}{\sin\beta} = \frac{R}{\sin\gamma} \tag{1.0.1}$$

(Lami's Theorem).

The relation (1.0.1) is, however, equally true for the sides of a triangle. Thus three forces in equilibrium may be represented by three sides of a triangle (the triangle of forces). P is represented by the

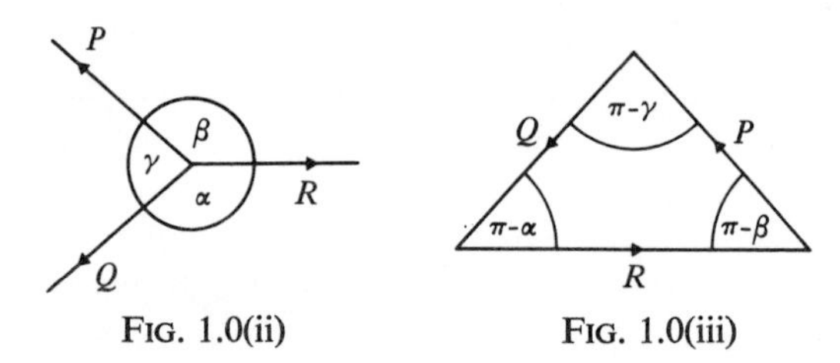

FIG. 1.0(ii) FIG. 1.0(iii)

side BC, Q by the side CA and R by the side AB. If the three forces be represented by the sides of a triangle as indicated, the property of three forces being in equilibrium is equivalent to the result of successive displacements from A to B, B to C and C to A being a zero displacement. It is this equivalence which leads to the concept of a vector which will now be defined formally. (An abstract definition will be discussed in section 1.32.)

1.1. Formal definitions of scalar and vector

(*a*) **Scalar.** A scalar is a quantity which has magnitude only. To specify a scalar, it is necessary to have a unit quantity of the same type and the ratio which the given quantity bears to the unit. The ratio may be positive or negative or zero or complex. (In this book quantities will be real unless the contrary is explicitly stated.)

(*b*) **Vector.** A vector is a quantity which (*i*) has magnitude, (*ii*) has direction, (*iii*) obeys the vector law of addition. That is if the vector $\mathbf{P}$ (in print a vector being usually denoted thus in **bold type** and in writing $\underset{\sim}{P}$ or $\vec{P}$) is represented in magnitude and direction by the displacement from the point A to the point B and the vector $\mathbf{Q}$ by the displacement from the point B to the point C, then the vector $\mathbf{P}+\mathbf{Q}$ is represented in magnitude and direction by the displacement from the point A to the point C. If now the displacement from the point A to the point B is written as $\overrightarrow{AB}$ (obviously a convenient way of writing displacement), the vector law of addition is equivalent to the statement

$$\overrightarrow{AB} + \overrightarrow{BC} = \overrightarrow{AC}. \tag{1.1.1}$$

Displacement is clearly a vector. $\overrightarrow{AB}$ may be termed a representation of **P**.

Two vectors $\overrightarrow{A'B'}$ and $\overrightarrow{AB}$ are equal if (*i*) $A'B'$ is parallel to AB, (*ii*) the length $A'B'$ is equal to the length AB, and (*iii*) the sense $A'B'$ is the same as the sense AB. If $\overrightarrow{AB}$ is a representation of **P**, all such vectors are representations of **P**.

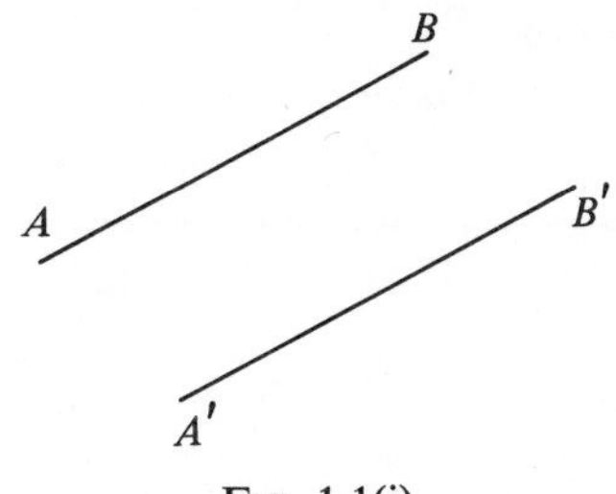

FIG. 1.1(i)

Should B coincide with A,

$$\overrightarrow{AB} = \overrightarrow{AA} = \mathbf{0} \tag{1.1.2}$$

the zero vector or the null vector. This vector may be regarded as parallel to all vectors.

The concept of subtraction follows without difficulty. By the vector law of addition of displacements

$$\overrightarrow{AB} + \overrightarrow{BC} = \overrightarrow{AC} \tag{1.1.3a}$$

$$\overrightarrow{AC} + \overrightarrow{CB} = \overrightarrow{AB}. \tag{1.1.3b}$$

Comparing these it is clear that

$$\overrightarrow{BC} = -\overrightarrow{CB}. \tag{1.1.4}$$

That is two vectors **P**, **Q** obey the relation $\mathbf{P} = -\mathbf{Q}$, if when **P** is represented by $\overrightarrow{AB}$ and **Q** by $\overrightarrow{CD}$, (*i*) CD is parallel to AB, (*ii*) the length CD is equal to the length AB and (*iii*) the sense CD is the opposite to the sense AB.

Thus

$$\overrightarrow{AC} = \overrightarrow{AB} - \overrightarrow{CB}. \tag{1.1.5}$$

There are three different types of vectors:

(*a*) **Free vectors.** A free vector is a vector which is not associated with any particular point. An example is the translation of a rigid body without rotation. This translation is equally well represented by any one of the displacement vectors associated with a point in the rigid body.

(*b*) **Line vectors.** A line vector is a vector which lies along a line of action, but is not associated with any particular point on that line of action. It is sometimes referred to as a sliding vector. An example is a force acting on a rigid body. By the principle of transmissability of force, a force may be applied to a rigid body at any point on its line of action.

(*c*) **Bound vector.** A bound vector is a vector which is associated with one particular point only. An example is the particle velocity of a fluid.

In general it will be clear from the context into which of these categories a vector falls.

It is to be remarked that a quantity must obey the vector law of addition for it to be a vector. For example, a finite rotation of a rigid body has a magnitude (the angle turned through) and a direction (the axis of rotation) associated with it, but it is not possible to compound two finite rotations by means of the vector law of addition and so finite rotations may not be represented by vectors. *Infinitesimal* rotations are however represented by vectors. (This is discussed in detail in section 2.71.)

It should be noted that vectors can only be added when they are of the same nature. For example, a force cannot be added to a displacement.

1.11. *Position vector*

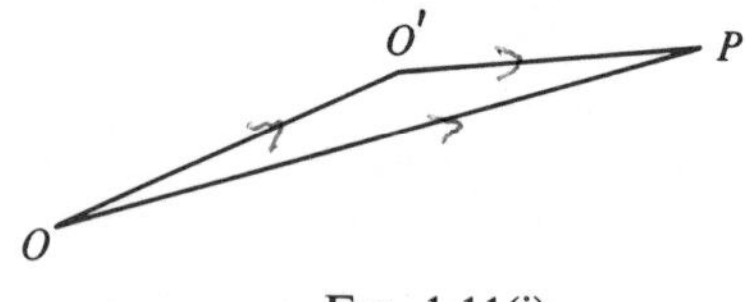

FIG. 1.11(i)

Let P be a point, and O an origin. Then $\overrightarrow{OP}$ (often written as $\mathbf{r}$) is the position vector of P with respect to O. If O' is some other origin then

$$\overrightarrow{O'P} = \overrightarrow{OP} - \overrightarrow{OO'}. \qquad (1.11.1)$$

$\overrightarrow{O'P} = \mathbf{r}'$ is the position vector of P with respect to O'. Thus the position vector of P with respect to O' is the position vector of P with respect to O minus the position vector of O' with respect to O. In this manner it is possible to transform from one origin to another.

1.12. *Unit vector and modulus*

If AB is a displacement vector, its length, which is positive is AB. This is termed the modulus. Furthermore, there exists parallel to $\overrightarrow{AB}$, a vector of unit length. This is denoted by $\hat{\overrightarrow{AB}}$. (The sign ˆ is to be read as unit vector in the direction of.) More generally, any vector has associated with it a modulus P or $|\mathbf{P}|$ and a unit vector $\hat{\mathbf{P}}$. Clearly

$$|\hat{\mathbf{P}}| = 1. \tag{1.12.1}$$

The unit vector is defined uniquely if and only if the vector is non-zero. As mentioned earlier, the zero vector may be thought of as parallel to all directions.

If $\mathbf{P}$ and $\mathbf{Q}$ are parallel, then it is obvious from the definition that

$$\hat{\mathbf{P}} = \hat{\mathbf{Q}}. \tag{1.12.2}$$

If $\mathbf{P}$ and $\mathbf{Q}$ are antiparallel,

$$\hat{\mathbf{P}} = -\hat{\mathbf{Q}}. \tag{1.12.3}$$

Also,

$$|-\mathbf{P}| = |\mathbf{P}|. \tag{1.12.4}$$

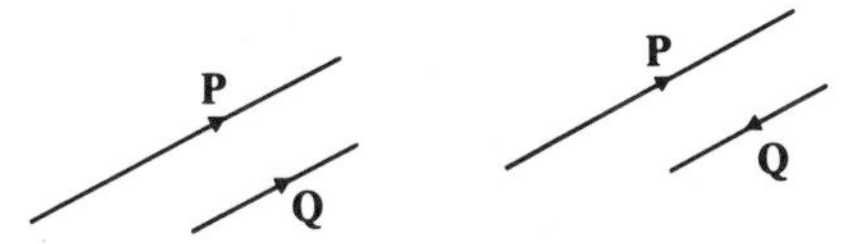

FIG. 1.12(i). **P** and **Q** parallel; **P** and **Q** antiparallel

1.2. General laws of vector manipulation

Many of the laws of vector manipulation are obvious from a diagram.

(*a*) **Product of scalar and vector.** From what has been said previously, the vector

$$(-1)\mathbf{P} = -\mathbf{P} \tag{1.2.1}$$

is the vector with the same modulus as $\mathbf{P}$ and the opposite unit vector.

In figure 1.2(i), $\overrightarrow{CA} = \overrightarrow{AB}$, $\mathbf{P}$ is represented by $\overrightarrow{AB}$ and $-\mathbf{P}$ by $\overrightarrow{AC}$. More generally, if $\mathbf{P}$ is a vector and λ is a scalar, $\lambda\mathbf{P}$ is defined as follows:

(*i*) If λ is positive, $\lambda\mathbf{P}$ is the vector with modulus $\lambda\mathbf{P}$ and unit vector $\hat{\mathbf{P}}$.

(*ii*) If λ is zero, $\lambda\mathbf{P}$ is the null vector.

(*iii*) If λ is negative, $\lambda\mathbf{P}$ is the vector with modulus $-\lambda P$ and unit vector $-\hat{\mathbf{P}}$.

Thus if λ and μ are scalars, $\lambda\mathbf{P}$ and $\mu\mathbf{P}$ are parallel if λ and μ are both of the same sign and antiparallel if they are of opposite sign.

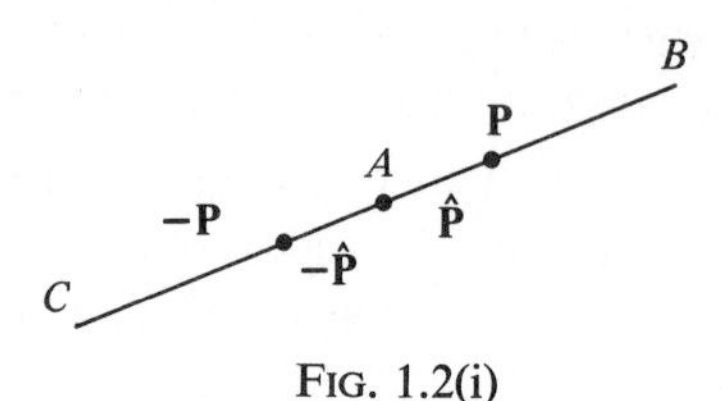

FIG. 1.2(i)

It is convenient to define $\mathbf{P}\lambda$ as equal to $\lambda\mathbf{P}$. From the definition above, it follows that

$$\mathbf{P} = P\hat{\mathbf{P}}. \tag{1.2.2}$$

(*b*) **Addition and subtraction of vectors.** The quantities $\mathbf{P}+\mathbf{Q}$ and $\mathbf{P}-\mathbf{Q}$ are the vectors which are represented by the diagonals of a parallelogram in which the pairs of parallel sides represent $\mathbf{P}$ and $\mathbf{Q}$ respectively.

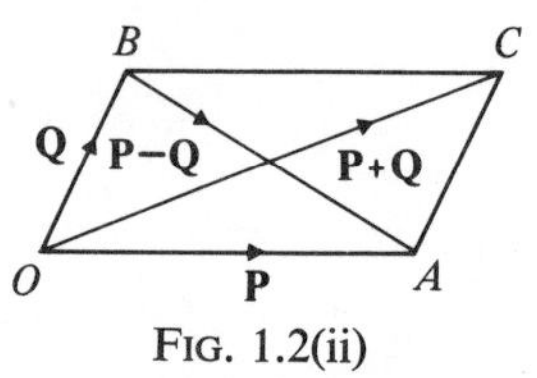

FIG. 1.2(ii)

Let $OACB$ be a parallelogram in which $\overrightarrow{OA}$ (and hence $\overrightarrow{BC}$) represents $\mathbf{P}$, and $\overrightarrow{OB}$ (and hence $\overrightarrow{AC}$) represents $\mathbf{Q}$.

$$\begin{aligned}\mathbf{P}+\mathbf{Q} &= \overrightarrow{OA}+\overrightarrow{OB} = \overrightarrow{OA}+\overrightarrow{AC} = \overrightarrow{OC}\\ &= \overrightarrow{OB}+\overrightarrow{BC} = \overrightarrow{OB}+\overrightarrow{OA} = \mathbf{Q}+\mathbf{P}.\end{aligned} \tag{1.2.3}$$

This incidentally proves that vectors obey the commutative law of addition, and so the order in which they are added is immaterial. Also,

$$\mathbf{P}-\mathbf{Q} = \overrightarrow{OA}-\overrightarrow{OB} = \overrightarrow{OA}+\overrightarrow{BO} = \overrightarrow{BA} \tag{1.2.4}$$

This, in effect, is the definition of the subtraction of two vectors.

The multiplication of scalars and vectors and the addition of vectors obey the so-called affine laws.

Associativeness:

$$(\lambda\mu)\mathbf{P} = \lambda(\mu\mathbf{P}) \tag{1.2.4}$$

$$\mathbf{P} + (\mathbf{Q} + \mathbf{R}) = (\mathbf{P} + \mathbf{Q}) + \mathbf{R} \tag{1.2.5}$$

Commutativeness:

$$(\lambda\mu)\mathbf{P} = (\mu\lambda)\mathbf{P} \tag{1.2.6}$$

$$\mathbf{P} + \mathbf{Q} = \mathbf{Q} + \mathbf{P} \tag{1.2.7}$$

$$\mathbf{P}\lambda = \lambda\mathbf{P} \qquad \text{(by definition)} \tag{1.2.8}$$

Distributiveness:

$$(\lambda + \mu)\mathbf{P} = \lambda\mathbf{P} + \mu\mathbf{P} \tag{1.2.9}$$

$$\lambda(\mathbf{P} + \mathbf{Q}) = \lambda\mathbf{P} + \lambda\mathbf{Q} \tag{1.2.10}$$

These properties may all be proved with little difficulty either by using the affine properties of real numbers, or by drawing a diagram. For example,

$$(\lambda\mu)\mathbf{P} = (\lambda\mu)P\hat{\mathbf{P}} = (\lambda\mu P)\hat{\mathbf{P}} = \lambda(\mu P)\hat{\mathbf{P}} = \lambda(\mu\mathbf{P})$$

thereby proving equation (1.2.4). The proof of equation (1.2.7) has been given above. The proofs of the others follow similar lines.

If equation (1.2.9) be generalized to

$$\left(\sum_{r=1}^{n} \lambda_r\right)\mathbf{P} = \sum_{r=1}^{n} (\lambda_r \mathbf{P}) \tag{1.2.11}$$

it follows that the sum of a number of parallel vectors is a vector which is parallel to them. (In general the term parallel will include antiparallel. The latter term will only be used when it is necessary to distinguish the sense.)

1.21. *Division of a line in a given ratio*

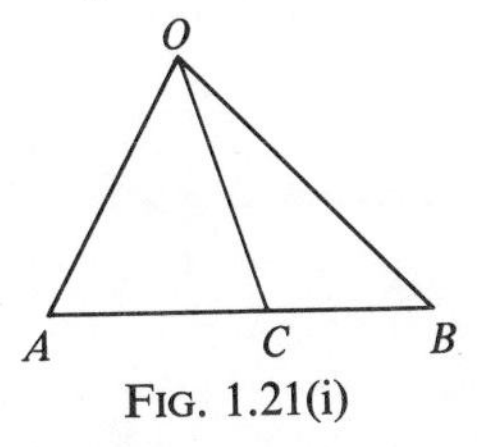

FIG. 1.21(i)

Suppose that C divides the line AB in the ratio $\lambda:\mu$. If λ and μ are both positive, C is between A and B. If λ is negative and μ is positive or vice versa, C is to the right of B when $|\lambda/\mu| < 1$ and to the left of A when $|\mu/\lambda| < 1$.

This property may be expressed as

$$\lambda\overrightarrow{AC} + \mu\overrightarrow{CB} = 0. \tag{1.21.1}$$

Then

$$(\lambda+\mu)\overrightarrow{OC} = \lambda(\overrightarrow{OA}+\overrightarrow{AC})+\mu(\overrightarrow{OB}+\overrightarrow{BC})$$
$$= \lambda\overrightarrow{OA}+\mu\overrightarrow{OB}+\lambda\overrightarrow{AC}+\mu\overrightarrow{CB}$$
$$= \lambda\overrightarrow{OA}+\mu\overrightarrow{OB}, \tag{1.21.2}$$

an alternative form for which is

$$\overrightarrow{OC} = \frac{\lambda}{\lambda+\mu}\overrightarrow{OA}+\frac{\mu}{\lambda+\mu}\overrightarrow{OB}. \tag{1.21.3}$$

In particular, if C is half way between A and B

$$\overrightarrow{OC} = \tfrac{1}{2}(\overrightarrow{OA}+\overrightarrow{OB}). \tag{1.21.4}$$

1.22. *Centroid*

If m_n are a set of scalars associated with the points P_n, the centroid of the system G is defined by

$$\sum_n m_n \overrightarrow{OG} = \sum_n m_n \overrightarrow{OP_n} \tag{1.22.1}$$

provided that $\sum_n m_n$ is not zero. If this condition is not satisfied, the centroid is not defined. When the m_n represent masses, $\sum_n m_n$ is always positive and G is definable. In this case, it is often loosely termed the centre of gravity. When the centroid of a line, or surface, or volume is required, the summation becomes an integral.

1.3. **Components of a vector**

The sum of two arbitrary vectors **P** and **Q** is represented by the diagonal of the parallelogram which has the representatives of these vectors as sides.

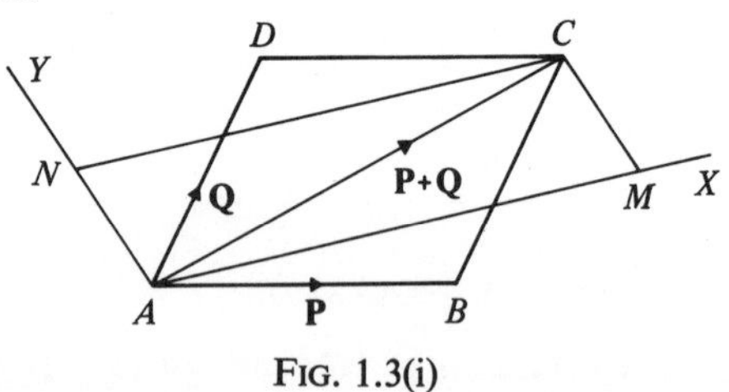

FIG. 1.3(i)

The vectors $\overrightarrow{AB}$ and $\overrightarrow{BC}$ are the components of the vector $\overrightarrow{AC}$ which is said to be resolved into these components. Clearly if AX, AY are two arbitrary lines through A such that the three lines AX, AY, AC are coplanar, it is possible to resolve $\overrightarrow{AC}$ into two components

$\overrightarrow{AM}$, $\overrightarrow{MC}$ parallel to AX, AY respectively, $AMCN$ being a parallelogram.

A set of three or more vectors are said to be coplanar when a plane can be drawn parallel to them. If this is not the case, they are non-coplanar. Any vector may be expressed as the sum of three components, respectively parallel to any three non-coplanar vectors.

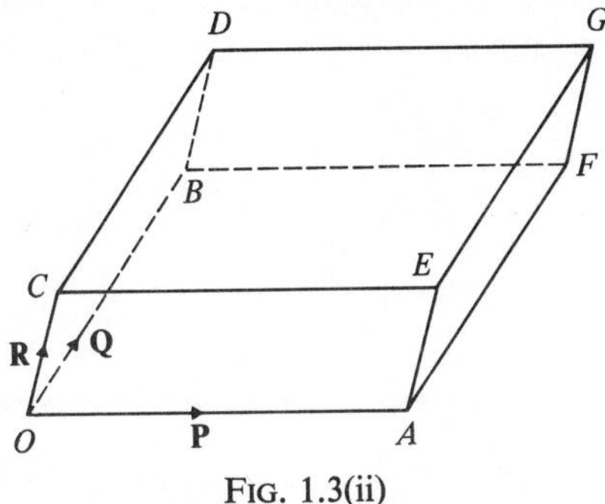

FIG. 1.3(ii)

Suppose that the three non-coplanar vectors are **P**, **Q**, **R** and that the vector whose components are desired is represented by $\overrightarrow{OG}$. Construct the parallelopiped with OG as diagonal and with edges OA, OB, OC parallel to **P**, **Q**, **R** and of length a, b, c respectively. $\hat{\mathbf{P}}$, $\hat{\mathbf{Q}}$, $\hat{\mathbf{R}}$, together with the common origin O may be said to form a frame of reference. $\overrightarrow{OG}$ is the resultant of $\overrightarrow{OA}$, $\overrightarrow{OB}$, $\overrightarrow{OC}$ which are the components of $\overrightarrow{OG}$ in the three directions defined by $\hat{\mathbf{P}}$, $\hat{\mathbf{Q}}$, $\hat{\mathbf{R}}$ respectively. Loosely, a, b, c may be referred to as the components. The resolution is unique as only one such parallelopiped can be constructed

$$\overrightarrow{OG} = a\hat{\mathbf{P}} + b\hat{\mathbf{Q}} + c\hat{\mathbf{R}} \tag{1.3.1}$$

a, b, c may be positive, negative or zero. a is positive if $\overrightarrow{OA}$ is parallel to $\hat{\mathbf{P}}$ and is negative if $\overrightarrow{OA}$ is antiparallel to $\hat{\mathbf{P}}$. It will be shown in section 2.5 how a, b, c may be calculated. It may be remarked that the component in the direction of $\hat{\mathbf{P}}$ depends upon $\hat{\mathbf{Q}}$ and $\hat{\mathbf{R}}$ also. Obviously, two equal vectors have equal components and the components of the sum of two vectors are the respective sums of the components.

1.31. *Orthogonal frames of reference*

The case where the three vectors defining the frame of reference are mutually perpendicular is of great importance. Such a frame of reference is termed an orthogonal frame of reference. It is now necessary to distinguish between left-handed and right-handed frames of reference (the distinction exists also in non-orthogonal frames of reference and the extension is obvious). Suppose that Ox, Oy, Oz, *in that order*, are three mutually perpendicular lines. By definition they form an orthogonal frame. Suppose that the frame be rotated, so that Ox and Oy lie in the plane of the paper as indicated. Then xOy is a right angle and Oz is perpendicular to the plane of the paper.

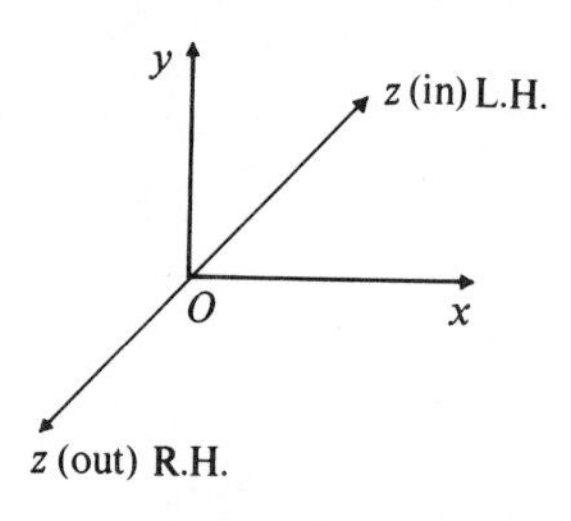

FIG. 1.31(i)

If Oz is in the direction out of the paper, the system of axes is right-handed. If Oz is in the direction into the paper, the system of axes is left-handed. Thus a left-handed system may always be transformed into a right-handed system by a reflection in the xOy plane. The order of Ox, Oy, Oz is important. If Ox, Oy, Oz is right-handed, Oy, Ox, Oz is left-handed. An example of a right-handed system is the following. Ox is eastwards, Oy is northwards and Oz is vertically upwards. The use of left-handed frames of reference is very rare (see S. Banach, *Mechanics*, Hafner, New York (1951) p. 2) and the usual right-handed frames will be employed in this book.

A special notation is used for the unit vectors associated with the right-handed orthogonal system Ox, Oy, Oz. These are normally written **i**, **j**, **k**.

$$\mathbf{A} = A_x\mathbf{i} + A_y\mathbf{j} + A_z\mathbf{k}. \tag{1.31.1}$$

A_x, A_y, A_z are the components of **A** in the x, y, z directions. It is often convenient to write

$$\mathbf{A} = (A_x, A_y, A_z). \tag{1.31.2}$$

Obviously

$$\lambda\mathbf{A} = (\lambda A_x, \lambda A_y, \lambda A_z) \tag{1.31.3}$$

$$\mathbf{A} + \mathbf{B} = (A_x + B_x, A_y + B_y, A_z + B_z). \tag{1.31.4}$$

The projection of the vector **A** on a line or plane is the vector whose initial and terminal points are the projections of the corresponding points of **A**. If the vector is normal to the plane or line, the projection consists of a single point, that is the null vector. Thus the projection of **A** on the z axis is $A_z\mathbf{k}$ or loosely A_z and the projection on the xy plane is $A_x\mathbf{i} + A_y\mathbf{j}$. Projections upon arbitrary lines and planes are discussed in section 2.11.

The position vector is

$$\mathbf{r} = x\mathbf{i} + y\mathbf{j} + z\mathbf{k} = (x, y, z). \tag{1.31.5}$$

This may be written in the alternative form

$$\mathbf{r} = r\cos\alpha\mathbf{i} + r\cos\beta\mathbf{j} + r\cos\gamma\mathbf{k} \tag{1.31.6}$$

where $\cos\alpha$, $\cos\beta$, $\cos\gamma$ are the direction cosines. Equation (1.31.5) may also be written as

$$r\hat{\mathbf{r}} = (r\cos\alpha, r\cos\beta, r\cos\gamma) \tag{1.31.7}$$

where $\hat{\mathbf{r}} = (\cos\alpha, \cos\beta, \cos\gamma)$.

More generally

$$\mathbf{A} = A(\cos\alpha, \cos\beta, \cos\gamma) \tag{1.31.8}$$

where $(\cos\alpha, \cos\beta, \cos\gamma)$ are the direction cosines of $\hat{\mathbf{A}}$.

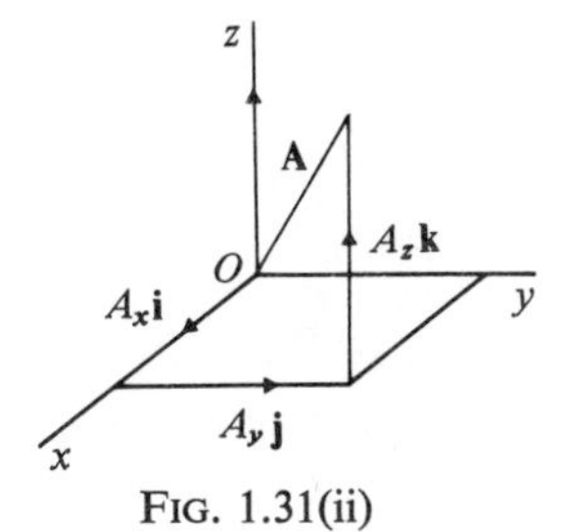

FIG. 1.31(ii)

Now equation (1.31.2) may be rewritten in the form

$$\mathbf{A} = (A_x^2 + A_y^2 + A_z^2)^{1/2}\left\{\frac{A_x}{(A_x^2 + A_y^2 + A_z^2)^{1/2}}, \frac{A_y}{(A_x^2 + A_y^2 + A_z^2)^{1/2}}, \frac{A_z}{(A_x^2 + A_y^2 + A_z^2)^{1/2}}\right\}. \tag{1.31.9}$$

Comparing this with equation (1.31.8), it may be seen that

$$A = (A_x^2 + A_y^2 + A_z^2)^{1/2} \tag{1.31.10}$$

and that α, β, γ are defined by

$$A_x = (A_x^2 + A_y^2 + A_z^2)^{1/2}\cos\alpha, \text{ etc.} \tag{1.31.11}$$

1.32. *Abstract treatment of vectors*

It is possible to give an abstract treatment of vectors by initially defining the vector in terms of its components. This definition is not particularly useful for physical problems, but may be generalized to give definitions of vectors in spaces of more than three dimensions (see section 6.1). A vector $\mathbf{x}$ is defined as an ordered triple of real numbers

$$\mathbf{x} = (x_1, x_2, x_3) \tag{1.32.1}$$

the zero vector $\mathbf{0}$ being defined by

$$\mathbf{0} = (0, 0, 0). \tag{1.32.2}$$

Real numbers are termed scalars. Thus a vector is an ordered triple of scalars. If α is a scalar, the scalar multiple of $\mathbf{x}$ by α is defined by

$$\alpha\mathbf{x} = \mathbf{x}\alpha = (\alpha x_1, \alpha x_2, \alpha x_3) \tag{1.32.3}$$

with the convention that

$$-\mathbf{x} = (-1)\mathbf{x} = (-x_1, -x_2, -x_3). \tag{1.32.4}$$

The sum of two vectors is defined by

$$\mathbf{x} + \mathbf{y} = (x_1 + y_1, x_2 + y_2, x_3 + y_3) \tag{1.32.5}$$

with the convention that

$$\mathbf{x} - \mathbf{y} = \mathbf{x} + (-\mathbf{y}). \tag{1.32.6}$$

From these definitions, the following properties follow immediately

$$\mathbf{x} + \mathbf{y} = \mathbf{y} + \mathbf{x} \tag{1.32.7}$$

$$\mathbf{x} + (\mathbf{y} + \mathbf{z}) = (\mathbf{x} + \mathbf{y}) + \mathbf{z} \tag{1.32.8}$$

$$(\alpha\beta)\mathbf{x} = \alpha(\beta\mathbf{x}) \tag{1.32.9}$$

$$(\alpha + \beta)\mathbf{x} = \alpha\mathbf{x} + \beta\mathbf{x} \tag{1.32.10}$$

$$\alpha(\mathbf{x} + \mathbf{y}) = \alpha\mathbf{x} + \alpha\mathbf{y} \tag{1.32.11}$$

$$1\mathbf{x} = \mathbf{x} \tag{1.32.12}$$

and there exist vectors $\mathbf{0}$, $-\mathbf{x}$, such that

$$\mathbf{x} + \mathbf{0} = \mathbf{x} \tag{1.32.13}$$

$$\mathbf{x} + (-\mathbf{x}) = \mathbf{0}. \tag{1.32.14}$$

The properties of vectors as defined by equations (1.32.7)–(1.32.14) can easily be seen to agree with the properties of vectors which are already known. From the abstract point of view, these properties define what is termed a vector space.

A collection of n vectors $\mathbf{x}^{(i)}$ is said to be linearly dependent if and only if there exist scalars α_i, not all α_i zero, such that

$$\sum_{i=1}^{n} \alpha_i \mathbf{x}^{(i)} = \mathbf{0}. \tag{1.32.15}$$

If this is not the case, they are linearly independent. If two vectors are linearly dependent

$$\alpha_1 \mathbf{x}^{(1)} + \alpha_2 \mathbf{x}^{(2)} = \mathbf{0} \tag{1.32.16}$$

and the two vectors are said to be collinear, one being a scalar multiple of the other. If three vectors are linearly dependent,

$$\alpha_1 \mathbf{x}^{(1)} + \alpha_2 \mathbf{x}^{(2)} + \alpha_3 \mathbf{x}^{(3)} = \mathbf{0} \tag{1.32.17}$$

they are said to be coplanar. Collinear vectors are of course coplanar.

It may be shown that four vectors are always linearly dependent as are three coplanar vectors or two collinear vectors.

Let

$$\alpha_1 \mathbf{x}^{(1)} + \alpha_2 \mathbf{x}^{(2)} + \alpha_3 \mathbf{x}^{(3)} + \alpha_4 \mathbf{x}^{(4)} = \mathbf{0} \tag{1.32.18}$$

where the α_i are arbitrary and non-zero. Then

$$\mathbf{x}^{(4)} = \beta_1 \mathbf{x}^{(1)} + \beta_2 \mathbf{x}^{(2)} + \beta_3 \mathbf{x}^{(3)} \tag{1.32.19}$$

where

$$\beta_r = -\alpha_r/\alpha_4. \tag{1.32.20}$$

Considering the components of the vectors in equation (1.32.19), there are three equations for the three unknowns β_r

$$\sum_{r=1}^{3} \beta_r x_s^{(r)} = x_s^4 \qquad 1 \leqslant s \leqslant 3. \tag{1.32.21}$$

These equations have a solution provided that

$$\begin{vmatrix} x_1^{(1)} & x_1^{(2)} & x_1^{(3)} \\ x_2^{(1)} & x_2^{(2)} & x_2^{(3)} \\ x_3^{(1)} & x_3^{(2)} & x_3^{(3)} \end{vmatrix} \neq 0 \tag{1.32.22}$$

The determinant vanishes only if $\mathbf{x}^{(1)}, \mathbf{x}^{(2)}, \mathbf{x}^{(3)}$ are linearly dependent. Thus the β_r can be found and so the result is proved. Equation (1.32.19) corresponds with the fact that any vector can be expressed as the sum of three vectors respectively parallel to three non-coplanar lines.

The abstract development of vectors will not be considered here. The utility of the abstract approach lies in the extension of the vectorial concepts to spaces of more than three dimensions rather than in applications to the three-dimensional physical world. In any case, the properties of vectors are the same, whether considered abstractly or not, and it is merely a question of which postulates and definitions are adopted in the beginning.

1.4. Geometrical applications of vectors

(*a*) **Condition that three points are collinear.** Let A, B, C be three collinear points. Then, by virtue of equation (1.21.2), there exist scalars λ, μ such that

$$(\lambda+\mu)\overrightarrow{OC}=\lambda\overrightarrow{OA}+\mu\overrightarrow{OB}.$$

This can be rewritten in the form

$$p\overrightarrow{OA}+q\overrightarrow{OB}+r\overrightarrow{OC}=0 \tag{1.4.1}$$

where

$$p=\frac{\lambda}{\lambda+\mu}, \qquad q=\frac{\mu}{\lambda+\mu}, \qquad r=-1$$

and

$$p+q+r=0. \tag{1.4.2}$$

Thus if A, B, C are collinear, a relation of the form (1.4.1) exists subject to the equation (1.4.2) being satisfied.

Alternatively, if equations (1.4.1) and (1.4.2) are satisfied

$$p\overrightarrow{OA}+q\overrightarrow{OB}-(p+q)\overrightarrow{OC}=0$$

and

$$p\overrightarrow{CA}+q\overrightarrow{CB}=0. \tag{1.4.3}$$

Thus the vectors $\overrightarrow{CA}$ and $\overrightarrow{BC}$ are parallel and so A, B, C are collinear. Thus the conditions (1.4.1) and (1.4.2) are necessary and sufficient for A, B, C to be collinear.

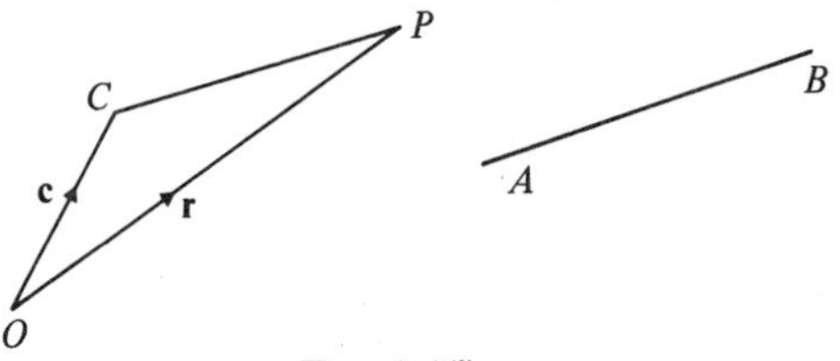

FIG. 1.4(i)

(*b*) **Equation of a straight line.** The equation of the line through C, parallel to AB, is given by

$$\mathbf{r}=\mathbf{c}+\lambda\overrightarrow{AB}. \tag{1.4.4}$$

Alternatively if P is to lie on the line AB, it follows from what has been done previously that

$$\mathbf{r}=(1-\lambda)\overrightarrow{OA}+\lambda\overrightarrow{OB}, \tag{1.4.5}$$

this being a particular case of the formula (1.4.1) when subjected to condition (1.4.2). In each case, as λ varies, the point moves along the line.

(*c*) **Condition that three vectors be coplanar.** If three vectors may be represented by $\overrightarrow{CA}$, $\overrightarrow{CB}$, $\overrightarrow{CP}$, the condition that they are coplanar is that a relation exists of the form

$$\overrightarrow{CP} = \lambda\overrightarrow{CA} + \mu\overrightarrow{CB}. \tag{1.4.6}$$

The plane is that of the parallelogram $CA'PB'$ where

$$\overrightarrow{CA'} = \lambda\overrightarrow{CA}, \quad \overrightarrow{CB'} = \mu\overrightarrow{CB}. \tag{1.4.7}$$

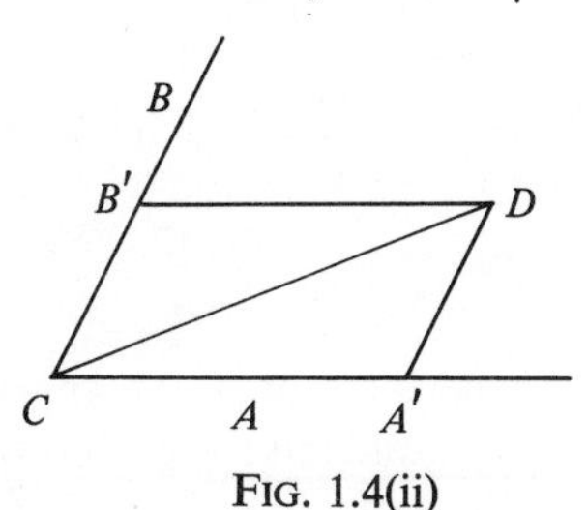

FIG. 1.4(ii)

(*d*) **Condition that four points be coplanar.** If the four points, A, B, C, P are coplanar, then by equation (1.4.6), there exist scalars λ, μ such that

$$\overrightarrow{CP} = \lambda\overrightarrow{CA} + \mu\overrightarrow{CB}.$$

This may be rewritten as

$$\mathbf{r} - \mathbf{c} = \lambda(\mathbf{a} - \mathbf{c}) + \mu(\mathbf{b} - \mathbf{c}) \tag{1.4.8}$$

with an obvious notation, that is

$$\mathbf{r} - \lambda\mathbf{a} - \mu\mathbf{b} - \{1 - (\lambda + \mu)\}\mathbf{c} = 0. \tag{1.4.9}$$

That is, if A, B, C, P are coplanar, there exists a relation of the form

$$\alpha\mathbf{a} + \beta\mathbf{b} + \gamma\mathbf{c} + \delta\mathbf{r} = 0 \tag{1.4.10}$$

where

$$\alpha + \beta + \gamma + \delta = 0. \tag{1.4.11}$$

By arguments similar to those already used, the pair of equations (1.4.10) and (1.4.11) give necessary and sufficient conditions.

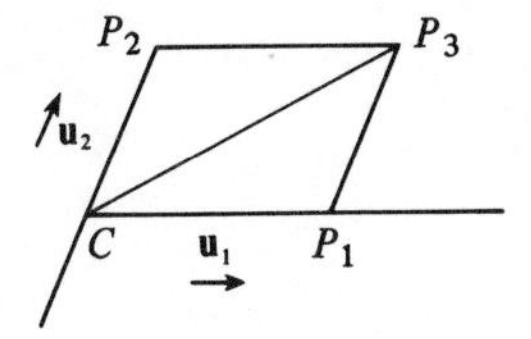

FIG. 1.4(iii)

(*e*) **Bisector of the angle between two straight lines.** Suppose that the two lines are

$$\mathbf{r} = \mathbf{c} + \lambda\mathbf{u}_1 \qquad (1.4.12a)$$
$$\mathbf{r} = \mathbf{c} + \lambda\mathbf{u}_2 \qquad (1.4.12b)$$

$\mathbf{c}$ is the position vector of their point of intersection and $\mathbf{u}_1$ and $\mathbf{u}_2$ are the unit vectors defining their directions.

If

$$\mathbf{r}_1 = \mathbf{c} + \lambda\mathbf{u}_1, \qquad \mathbf{r}_2 = \mathbf{c} + \lambda\mathbf{u}_2, \qquad \mathbf{r}_3 = \mathbf{c} + \lambda(\mathbf{u}_1 + \mathbf{u}_2). \quad (1.4.13)$$

$CP_1P_3P_2$ is a parallelogram as indicated in figure 1.4(iii), and so CP_3 bisects the angle P_1CP_2.

Thus,

$$\mathbf{r} = \mathbf{c} + \lambda(\mathbf{u}_1 + \mathbf{u}_2) \qquad (1.4.14)$$

is the equation of the internal bisector of the angle between the two straight lines (1.4.12). It is not difficult to see that the equation of the external bisector is

$$\mathbf{r} = \mathbf{c} + \lambda(\mathbf{u}_1 - \mathbf{u}_2). \qquad (1.4.15)$$

1.5. The vector as a function

The concept of one variable being a function of another variable is well known and it is not necessary to discuss it here in detail. A function describes the change in one quantity as a consequence of the change of another quantity. If the first quantity never changes, the function is a constant. This extends naturally to the concept of a vector being a function of a scalar, and also to the concept of a scalar, or a vector being a function of a vector. The discussion of functions of a vector will, however, be reserved for chapter 3. It will be convenient to consider the discussion of a vector function of a scalar in terms of a vector function of time. The extension to any other scalar is obvious.

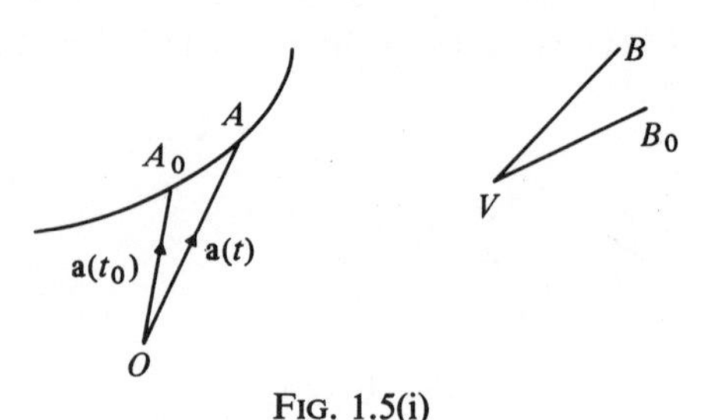

FIG. 1.5(i)

Let $\mathbf{a}(t)$ be a vector function of time. Then a may be represented by a vector $\overrightarrow{OA}$. As the time increases, the position of A may alter. Crudely, $\mathbf{a}$ is continuous if, when t increases a little, $\mathbf{a}$ does not alter very much.

Formally this can be stated as follows. $\mathbf{a}(t)$ is continuous at $t = t_0$ if

$$|\mathbf{a}(t) - \mathbf{a}_0(t)| \to 0 \tag{1.5.1}$$

as $t \to t_0$. That is the length $A_0 A$ tends to zero as $t \to t_0$.

1.51. *Derivative of a vector with respect to time*

The derivative of a vector with respect to time is defined in a manner analogous with that of a scalar with respect to time.

Let $\overrightarrow{VB}$ be the representation of

$$\frac{\mathbf{a}(t) - \mathbf{a}(t_0)}{t - t_0}. \tag{1.51.1}$$

This is the product of a scalar and a vector and is therefore a vector. It should be noted that this quantity is dimensionally different from $\mathbf{a}$ and so the representation should not be in the same diagram as $\mathbf{a}$. If as $t \to t_0$, B tends to a limiting position B_0, $\overrightarrow{VB_0}$ represents the derivative of $\mathbf{a}$ at t_0. The successive positions of B_0 form the hodograph of A.

Equivalent definitions of the derivative are

$$\dot{\mathbf{a}}(t_0) = \lim_{t \to t_0} \frac{\mathbf{a}(t) - \mathbf{a}(t_0)}{t - t_0} \tag{1.51.2}$$

$$= \lim_{\tau \to 0} \frac{\mathbf{a}(t_0 + \tau) - \mathbf{a}(t_0)}{r} \tag{1.51.3}$$

$$= \left[\lim_{\delta t \to 0} \frac{\delta \mathbf{a}}{\delta t}\right]_{t=t_0}. \tag{1.51.4}$$

The derivative of a vector with respect to a scalar is a vector. $\delta\mathbf{a}/\delta t$ is the result of the multiplication of a vector $\delta\mathbf{a}$ by a scalar $1/\delta t$. This is a vector and will be so in the limit. The same argument may be applied to derivatives of any order.

Some simple properties of the derivative are easily established.

(*a*) If $\mathbf{a}$ and $\mathbf{b}$ are any two vectors

$$\frac{d}{dt}(\mathbf{a} + \mathbf{b}) = \frac{d\mathbf{a}}{dt} + \frac{d\mathbf{b}}{dt} \tag{1.51.5}$$

(*b*) If **P**, **Q**, **R** are three *fixed* non-coplanar vectors, then

$$\mathbf{a} = a_P\,\mathbf{P} + a_Q\,\mathbf{Q} + a_R\,\mathbf{R} \tag{1.51.6}$$

and

$$\dot{\mathbf{a}} = \dot{a}_P\,\mathbf{P} + \dot{a}_Q\,\mathbf{Q} + \dot{a}_R\,\mathbf{R}. \tag{1.51.7}$$

In particular for an orthogonal frame

$$\dot{\mathbf{a}} = (\dot{a}_x, \dot{a}_y, \dot{a}_z). \tag{1.51.8}$$

(*c*) If s is a scalar variable

$$\frac{d\mathbf{a}}{ds} = \frac{d\mathbf{a}}{dt}\frac{dt}{ds}. \tag{1.51.9}$$

If $\mathbf{a} = a\hat{\mathbf{a}}$, then

$$\delta\mathbf{a} = \delta a\hat{\mathbf{a}} + a\,\delta\hat{\mathbf{a}} \tag{1.51.10}$$

$\hat{\mathbf{a}}$ is a unit vector. So also is $\hat{\mathbf{a}} + \delta\hat{\mathbf{a}}$. $\delta\hat{\mathbf{a}}$ is thus the difference between two unit vectors. Let $\delta\theta$ be the angle between $\hat{\mathbf{a}}$ and $\hat{\mathbf{a}} + \delta\hat{\mathbf{a}}$. Then

$$UU' = \delta\theta \tag{1.51.11}$$

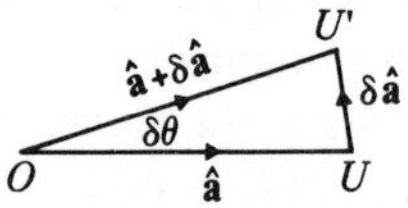

FIG. 1.51(i)

because $OU = OU' = 1$, and so

$$\overrightarrow{UU'} = \delta\theta\mathbf{b} \tag{1.51.12}$$

where **b** is some unit vector which is perpendicular to **a**.

Thus

$$\frac{d\mathbf{a}}{dt} = \frac{da}{dt}\hat{\mathbf{a}} + a\frac{d\hat{\mathbf{a}}}{dt}$$

$$= \frac{da}{dt}\,\hat{\mathbf{a}} + a\dot{\theta}\mathbf{b}. \tag{1.51.13}$$

If the unit vector has a fixed direction, the second term vanishes.

The concept of integration with respect to a scalar is equally applicable.

If

$$\frac{d\mathbf{a}}{dt} = \mathbf{b}(t) \tag{1.51.14}$$

and

$$\mathbf{a}(t_0) = \mathbf{a}_0 \tag{1.51.15}$$

then

$$\mathbf{a}(t) = \mathbf{a}(t_0) + \int_{t_0}^{t} \mathbf{b}(t')\,dt' \qquad (1.51.16)$$

It is not necessary to discuss this in detail.

1.52. *Velocity and acceleration*

The position of a point relative to some base point O in a frame of reference is defined through a position vector $\overrightarrow{OP} = \mathbf{r}(t)$. This will always be a continuous function of t. In general the first and second derivatives $\dot{\mathbf{r}}(t)$ and $\ddot{\mathbf{r}}(t)$ will exist except at certain exceptional values of t.

If

$$\mathbf{v} = \frac{d\mathbf{r}}{dt} \qquad (1.52.1)$$

$$\boldsymbol{\alpha} = \frac{d\mathbf{v}}{dt} = \frac{d^2\mathbf{r}}{dt^2}. \qquad (1.52.2)$$

$\mathbf{v}$ is termed the velocity and $\boldsymbol{\alpha}$ the acceleration, it being understood that this is with respect to the particular frame of reference. $v = |\mathbf{v}|$ is termed the speed. Thus velocity is a vector and speed is a scalar. Even though v is constant, α may be non zero. This is because the direction of $\mathbf{v}$ may vary.

For an orthogonal frame

$$\mathbf{r} = (x, y, z) \qquad (1.52.3)$$

$$\mathbf{v} = (\dot{x}, \dot{y}, \dot{z}) \qquad (1.52.4)$$

$$\boldsymbol{\alpha} = (\ddot{x}, \ddot{y}, \ddot{z}). \qquad (1.52.5)$$

Conversely, integration is possible. If $\mathbf{r}_0$ is the position vector of a point of time t_0 and if $\mathbf{v}_0$ is its velocity at time t_0

$$\mathbf{v}(t) = \mathbf{v}_0 + \int_{t_0}^{t} \boldsymbol{\alpha}(t')\,dt' \qquad (1.52.6)$$

$$\mathbf{r}(t) = \mathbf{r}_0 + \int_{t_0}^{t} \mathbf{v}(t')\,dt' \qquad (1.52.7)$$

1.521. *Relative velocity and acceleration*

More generally, it is possible to extend the concept of velocity and to discuss the velocity of B relative to A, or the relative velocity of

B with respect to A, and similarly the acceleration of B relative to A, or the relative acceleration of B with respect to A.

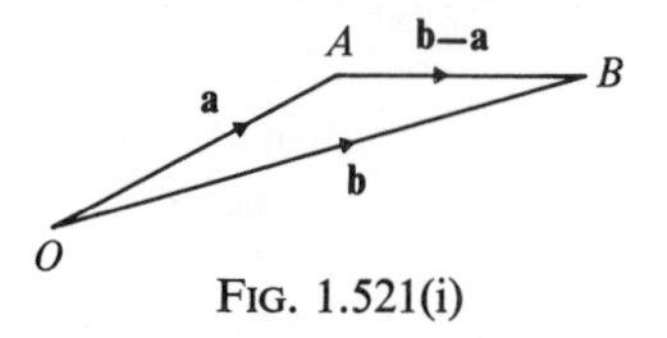

FIG. 1.521(i)

Now

$$\overrightarrow{AB} = \mathbf{b} - \mathbf{a} \tag{1.521.1}$$

$$\frac{d}{dt}\overrightarrow{AB} = \frac{d\mathbf{b}}{dt} - \frac{d\mathbf{a}}{dt}$$

$$= \mathbf{v}_B - \mathbf{v}_A \tag{1.521.2}$$

where $\mathbf{v}_A$, $\mathbf{v}_B$ are the velocities of A and B (with respect to O understood).

$$\frac{d}{dt}\overrightarrow{AB} = \mathbf{v}_{B,\,A}$$

is the velocity of B with respect to A.

Thus

$$\mathbf{v}_{B,\,A} = \mathbf{v}_B - \mathbf{v}_A \tag{1.521.3}$$

is the principle of relative velocities, giving the velocity of B relative to A in terms of the velocity of B and the velocity of A. It may be seen that this principle is independent of the origin O. For, if the origin be O' instead of O

$$\overrightarrow{AB} = \overrightarrow{O'B} - \overrightarrow{O'A} = \overrightarrow{O'O} + \overrightarrow{OB} - \overrightarrow{O'O} - \overrightarrow{OA}$$
$$= \mathbf{b} - \mathbf{a}, \text{ the previous expression.}$$

Similarly, if $\boldsymbol{\alpha}_{B,\,A}$ is the acceleration of B with respect to A, and $\boldsymbol{\alpha}_A$, $\boldsymbol{\alpha}_B$ the accelerations of A and B (with respect to O)

$$\boldsymbol{\alpha}_{B,\,A} = \boldsymbol{\alpha}_B - \boldsymbol{\alpha}_A \tag{1.521.4}$$

which is the principle of relative accelerations. This principle is clearly independent of the origin also.

1.53. *Force*

Attention has been drawn in section 1.0 to the empirical evidence that force is a vector. That this is indeed the case may be seen in another manner. Suppose that a particle is subject to two simultaneous displacements $\overrightarrow{OP_1}$, $\overrightarrow{OP_2}$. The net resultant displacement is $\overrightarrow{OP}$ where OP_1PP_2 is a parallelogram.

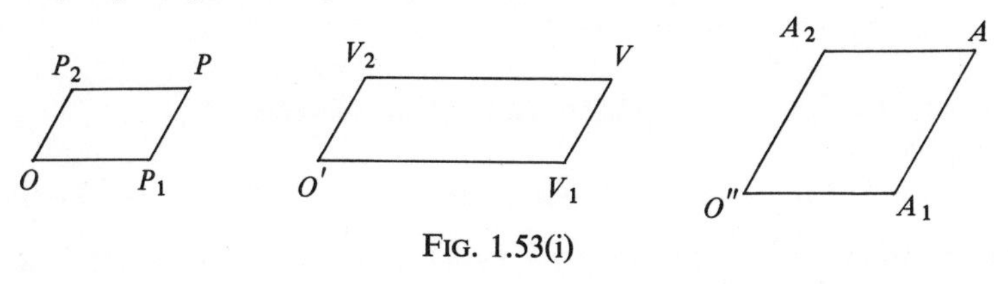

FIG. 1.53(i)

If the particle is subject to two simultaneous velocities represented by $\overrightarrow{O'V_1}$, $\overrightarrow{O'V_2}$ parallel to $\overrightarrow{OP_1}$, $\overrightarrow{OP_2}$ respectively, the resultant velocity is represented by $\overrightarrow{O'V}$ where $O'V_1VV_2$ is a parallelogram.

Now

$$\overrightarrow{OP} = \overrightarrow{OP_1} + \overrightarrow{OP_2} \tag{1.53.1}$$

where

$$\frac{d}{dt}\overrightarrow{OP} = \frac{d}{dt}\overrightarrow{OP_1} + \frac{d}{dt}\overrightarrow{OP_2} \tag{1.53.2}$$

$$= \overrightarrow{O'V_1} + \overrightarrow{O'V_2} = \overrightarrow{O'V}. \tag{1.53.3}$$

Thus

$$\frac{d}{dt}\overrightarrow{OP} = \overrightarrow{O'V} \tag{1.53.4}$$

and so velocities can be compounded vectorially.

Similarly,

$$\frac{d^2}{dt^2}\overrightarrow{OP} = \overrightarrow{O''A} = \overrightarrow{O''A_1} + \overrightarrow{O''A_2} \tag{1.53.5}$$

where $\overrightarrow{O''A_1}$, $\overrightarrow{O''A_2}$ are parallel to $\overrightarrow{OP_1}$, $\overrightarrow{OP_2}$ respectively. Thus accelerations can be compounded vectorially and if $\boldsymbol{\alpha}_1$, $\boldsymbol{\alpha}_2$ are two simultaneous accelerations, the resultant acceleration is

$$\boldsymbol{\alpha} = \boldsymbol{\alpha}_1 + \boldsymbol{\alpha}_2. \tag{1.53.6}$$

Now, when a force $\mathbf{F}$ acts on a particle a relation of the form

$$\mathbf{F} = m\boldsymbol{\alpha} \tag{1.53.7}$$

exists, m being the mass and $\boldsymbol{\alpha}$ the acceleration (see, for example, L. A. Pars, *Introduction to Dynamics*, Cambridge Univ. Press, (1953), p. 36.) and so

$$\mathbf{F} = m\boldsymbol{\alpha} = m\boldsymbol{\alpha}_1 + m\boldsymbol{\alpha}_2 = \mathbf{F}_1 + F_2 \tag{1.53.8}$$

where $\mathbf{F}_1$, $\mathbf{F}_2$ are the forces which would be required to make m move with accelerations $\boldsymbol{\alpha}_1$, $\boldsymbol{\alpha}_2$ respectively. Thus, the acceleration due to the concurrent effect of the three forces $\mathbf{F}_1$, $\mathbf{F}_2$, $\mathbf{F} = -(\mathbf{F}_1 + \mathbf{F}_2)$ will be zero, and forces add as vectors in agreement with the results of the experiment outlined in section 1.0.

1.531. *Relation between force and frame of reference*

The relation

$$\mathbf{F} = m\boldsymbol{\alpha} \tag{1.531.1}$$

exists in any frame of reference, $\mathbf{F}$ being the force, and $\boldsymbol{\alpha}$ the acceleration *in that frame of reference.*

Similarly, in another frame of reference

$$\mathbf{F}' = m\boldsymbol{\alpha}'. \tag{1.531.2}$$

If P is the position of the particle and O, O' are the origins of the two frames of reference

$$\overrightarrow{O'P} = \overrightarrow{OP} - \overrightarrow{OO'} \tag{1.531.3}$$

or

$$\mathbf{r}' = \mathbf{r} - \overrightarrow{OO'}. \tag{1.531.4}$$

It follows on differentiating twice that

$$\boldsymbol{\alpha}' = \boldsymbol{\alpha} - \frac{d^2}{dt^2}\overrightarrow{OO'}. \tag{1.531.5}$$

Multiplying by the mass m

$$m\boldsymbol{\alpha}' = m\boldsymbol{\alpha} - m\frac{d^2}{dt^2}\overrightarrow{OO'}$$

whence

$$\mathbf{F}' = \mathbf{F} + \mathbf{F}^{+} \tag{1.531.6}$$

where

$$\mathbf{F}^{+} = -m\frac{d^2}{dt^2}\overrightarrow{OO'}. \tag{1.531.7}$$

$\mathbf{F}^{+}$ is an apparent force which arises with the change in frames of reference. It is only when the two frames have no acceleration relative to one another that the forces $\mathbf{F}'$, $\mathbf{F}$ as defined by equations (1.53.1) and (1.53.2) are equal. The implications of this are far reaching, but do not concern us here.

1.6. Coordinate systems

So far only the cartesian coordinate system has been utilized. That is, everything has been considered in terms of a resolution along the three directions associated with the edges of a fixed parallelopiped (which is usually a cuboid, so that the directions are mutually perpendicular). That is any vector **P** is written in the form

$$\mathbf{P} = \alpha\overrightarrow{OA} + \beta\overrightarrow{OB} + \gamma\overrightarrow{OC} \tag{1.6.1}$$

where $\overrightarrow{OA}$, $\overrightarrow{OB}$, $\overrightarrow{OC}$ are constant vectors, or a quantity, the magnitude of which varies with position, is written $f(x,y,z)$, x, y, z being rectangular cartesian coordinates. It is often found more convenient, however, to express a quantity, not in terms of constant vectors, but in terms of variable vectors. This does not make any difference to the quantity, it merely expresses it in a form which is, for some purposes, more convenient.

Two particular coordinate systems of this kind will be considered in detail in this book.

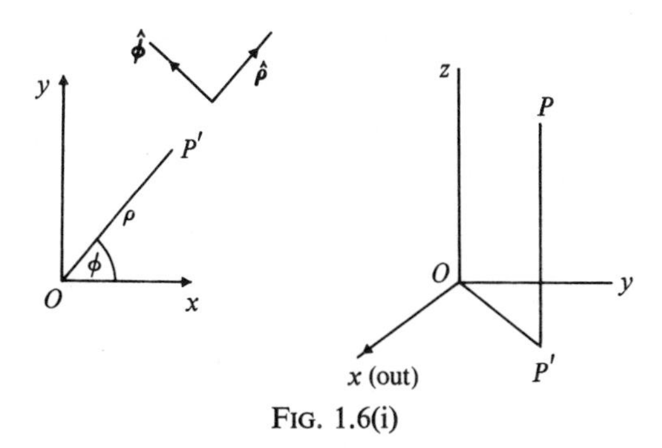

FIG. 1.6(i)

(*a*) **Cylindrical coordinates.** Let P' be the projection of an arbitrary point P upon the x, y plane.

Define quantities ρ, ϕ by

$$\rho = +\sqrt{(x^2+y^2)}, \qquad \phi = \tan^{-1}\frac{y}{x}. \tag{1.6.2}$$

Then the quantities ρ, ϕ, z $(0 \leqslant \rho, -\pi < \phi \leqslant \pi)$ define the position of a point in space uniquely.

Clearly

$$\begin{aligned}
\mathbf{i} &= \hat{\boldsymbol{\rho}}\cos\phi - \hat{\boldsymbol{\phi}}\sin\phi \\
\mathbf{j} &= \hat{\boldsymbol{\rho}}\sin\phi + \hat{\boldsymbol{\phi}}\cos\phi \\
\hat{\boldsymbol{\rho}} &= \mathbf{i}\cos\phi + \mathbf{j}\sin\phi \\
\hat{\boldsymbol{\phi}} &= -\mathbf{i}\sin\phi + \mathbf{j}\cos\phi
\end{aligned} \tag{1.6.3}$$

where $\hat{\boldsymbol{\rho}}$, $\hat{\boldsymbol{\phi}}$ are unit vectors in the direction of ρ, ϕ increasing. $\hat{\boldsymbol{\rho}}$ is parallel to OP', $\hat{\boldsymbol{\phi}}$ is in the xOy plane and perpendicular to OP'. Although unit vectors, they are not constant vectors, but variable vectors as the associated directions are not constant. ρ, ϕ are in fact polar coordinates in the plane.

If $\mathbf{Q}$ is an arbitrary vector

$$\begin{aligned}\mathbf{Q} &= Q_x\mathbf{i} + Q_y\mathbf{j} + Q_z\mathbf{k}\\ &= Q_x(\hat{\boldsymbol{\rho}}\cos\phi - \hat{\boldsymbol{\phi}}\sin\phi) + Q_y(\hat{\boldsymbol{\rho}}\sin\phi + \hat{\boldsymbol{\phi}}\cos\phi) + Q_z\mathbf{k} \quad (1.6.4)\\ &= Q_\rho\hat{\boldsymbol{\rho}} + Q_\phi\hat{\boldsymbol{\phi}} + Q_z\mathbf{k}\end{aligned}$$

where

$$\begin{aligned}Q_\rho &= Q_x\cos\phi + Q_y\sin\phi\\ Q_\phi &= -Q_x\sin\phi + Q_y\cos\phi. \qquad (1.6.5)\end{aligned}$$

Q_ρ, Q_ϕ are the resolutes of $\mathbf{Q}$ in the directions of ρ and ϕ increasing. Similarly,

$$f(x,y,z) = f(\rho\cos\phi, \rho\sin\phi, z) = F(\rho,\phi,z). \qquad (1.6.6)$$

It may be seen that $\hat{\boldsymbol{\rho}}$, $\hat{\boldsymbol{\phi}}$, $\mathbf{k}$ in that order form a right-handed orthogonal set of unit vectors.

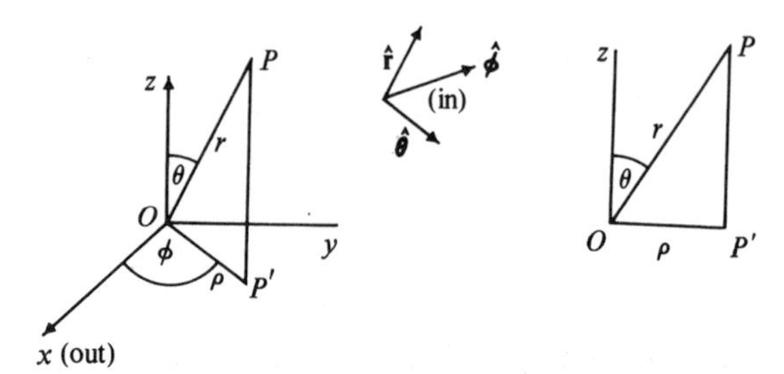

FIG. 1.6(ii)

(*b*) **Spherical coordinates.** Define quantities r, θ, ϕ by

$$r = +\sqrt{(x^2 + y^2 + z^2)} = +\sqrt{(\rho^2 + z^2)}$$

$$\theta = \cos^{-1}\frac{z}{r} = \tan^{-1}\left(\frac{\rho}{z}\right)$$

$$\phi = \tan^{-1}\frac{y}{x}. \qquad (1.6.7)$$

The quantities r, θ, ϕ $(O \leqslant r, 0 \leqslant \theta \leqslant \pi, -\pi < \phi \leqslant \pi)$ define the position of a point in space, uniquely. There are three associated unit vectors $\hat{\mathbf{r}}$, $\hat{\boldsymbol{\theta}}$, $\hat{\boldsymbol{\phi}}$ in the directions of r, θ, ϕ increasing. They are defined as follows. P is the point (r,θ,ϕ) and P' is the projection of P on the xy plane, that is the point $(r,\frac{1}{2}\pi,\phi)$ $\hat{\mathbf{r}}$ is parallel to OP. $\hat{\boldsymbol{\theta}}$ is perpendicular to OP in the plane OPP' and points away from the positive z axis. $\hat{\boldsymbol{\phi}}$ is perpendicular to the plane OPP' and is such that $\hat{\mathbf{r}}$, $\hat{\boldsymbol{\theta}}$, $\hat{\boldsymbol{\phi}}$ in that order form a right-handed set of unit vectors. It is obviously an orthogonal set.

The relations for the unit vectors are as follows

$$\left.\begin{aligned}\mathbf{i} &= \hat{\boldsymbol{\rho}}\cos\phi - \hat{\boldsymbol{\phi}}\sin\phi \\ &= (\hat{\mathbf{r}}\sin\theta + \hat{\boldsymbol{\theta}}\cos\theta)\cos\phi - \hat{\boldsymbol{\phi}}\sin\phi \\ \mathbf{j} &= \hat{\boldsymbol{\rho}}\sin\phi + \hat{\boldsymbol{\phi}}\cos\phi \\ &= (\hat{\mathbf{r}}\sin\theta + \hat{\boldsymbol{\theta}}\cos\theta)\cos\phi + \hat{\boldsymbol{\phi}}\cos\phi \\ \mathbf{k} &= \hat{\mathbf{r}}\cos\theta - \hat{\boldsymbol{\theta}}\sin\theta\end{aligned}\right\} \tag{1.6.8}$$

and

$$\left.\begin{aligned}\hat{\mathbf{r}} &= \mathbf{i}\sin\theta\cos\phi + \mathbf{j}\sin\theta\sin\phi + \mathbf{k}\cos\theta \\ \hat{\boldsymbol{\theta}} &= \mathbf{i}\cos\theta\cos\phi + \mathbf{j}\cos\theta\sin\phi - \mathbf{k}\sin\theta \\ \hat{\boldsymbol{\phi}} &= -\mathbf{i}\sin\phi + \mathbf{j}\cos\phi.\end{aligned}\right\} \tag{1.6.9}$$

It is thus possible to express any vector (Q_x, Q_y, Q_z) in the form

$$Q_r\hat{\mathbf{r}} + Q_\theta\hat{\boldsymbol{\theta}} + Q_\phi\hat{\boldsymbol{\phi}}. \tag{1.6.10}$$

Similarly

$$\begin{aligned}f(x,y,z) &= f(r\sin\theta\cos\phi, r\sin\theta\sin\phi, r\cos\theta) \\ &= F(r,\theta,\phi).\end{aligned} \tag{1.6.11}$$

Worked examples

1. If $ABCD$ is a parallelogram, and E is the middle point of AB, DE trisects and is trisected by AC (E. G. Phillips).

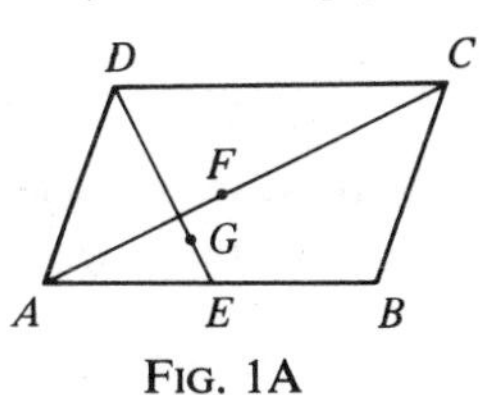

FIG. 1A

Let $\overrightarrow{AF} = \frac{1}{3}\overrightarrow{AC}$, $\overrightarrow{EG} = \frac{1}{3}\overrightarrow{ED}$. Then it is required to prove that F and G are the same point.

By equation (1.21.3)

$$\begin{aligned}\overrightarrow{AG} &= \tfrac{1}{3}(2\overrightarrow{AE} + \overrightarrow{AD}) = \tfrac{1}{3}(\overrightarrow{AB} + \overrightarrow{AD}) \\ &= \tfrac{1}{3}(\overrightarrow{AB} + \overrightarrow{BC}) = \tfrac{1}{3}\overrightarrow{AC} = \overrightarrow{AF}.\end{aligned}$$

Thus F and G are the same point, thereby proving the result.

2. *ABCD* is any quadrilateral: prove that if *H* and *K* are the middle points of *AC* and *BD* respectively, $\overrightarrow{AB}+\overrightarrow{CB}+\overrightarrow{CD}+\overrightarrow{AD}=4\overrightarrow{HK}$ (E. G. Phillips).

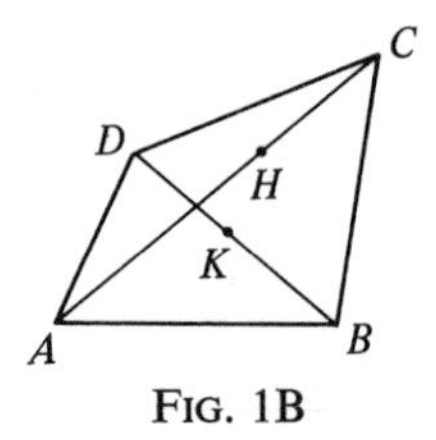

FIG. 1B

Let O be an arbitrary origin and let $\overrightarrow{OA}=\mathbf{a}$, etc.

$$\begin{aligned}\overrightarrow{AB}+\overrightarrow{CB}+\overrightarrow{CD}+\overrightarrow{AD} &= (\mathbf{b}-\mathbf{a})+(\mathbf{b}-\mathbf{c})+(\mathbf{d}-\mathbf{c})+(\mathbf{d}-\mathbf{a})\\ &= 2(\mathbf{b}+\mathbf{d})-2(\mathbf{a}+\mathbf{c})\\ &= 2[2\mathbf{k}-2\mathbf{h}]=4(\mathbf{k}-\mathbf{h})\\ &= 4\overrightarrow{HK}.\end{aligned}$$

3. If $\mathbf{A}=10\mathbf{i}+3\mathbf{j}+5\mathbf{k}$, find A and $\hat{\mathbf{A}}$

$$\hat{\mathbf{A}}=(10^2+3^2+5^2)^{1/2}\left\{\frac{10}{(10^2+3^2+5^2)^{1/2}},\frac{3}{(10^2+3^2+5^2)},\frac{5}{(10^2+3^2+5^2)^{1/2}}\right\}$$

and so $A=\sqrt{134}$, and

$$\hat{\mathbf{A}}=\left\{\frac{10}{\sqrt{134}},\frac{3}{\sqrt{134}},\frac{5}{\sqrt{134}}\right\}.$$

4. If $\mathbf{B}=2\mathbf{j}+3\mathbf{k}$, find the vector $\mathbf{C}$ which is parallel to $\mathbf{B}$ and is of the same modulus as the vector $\mathbf{A}$ defined in example 3.

$$\mathbf{B}=\sqrt{13}\left\{0,\frac{2}{\sqrt{13}},\frac{3}{\sqrt{13}}\right\}$$

and so

$$\begin{aligned}\mathbf{C} &= \sqrt{134}\left\{0,\frac{2}{\sqrt{13}},\frac{3}{\sqrt{13}}\right\}\\ &= 2\sqrt{\left(\frac{134}{13}\right)}\mathbf{j}+3\sqrt{\left(\frac{134}{13}\right)}\mathbf{k}.\end{aligned}$$

5. If $\mathbf{a}$ and $\mathbf{b}$ are given vectors, show that

(*i*) $|\mathbf{a}|-|\mathbf{b}|\leqslant|\mathbf{a}-\mathbf{b}|$,

(*ii*) $|\mathbf{a}+\mathbf{b}|\leqslant|\mathbf{a}|+|\mathbf{b}|$.

Let $\overrightarrow{OA}$, $\overrightarrow{OB}$, $\overrightarrow{OB'}$ represent the vectors, $\mathbf{a}$, $\mathbf{b}$, $-\mathbf{b}$ respectively. Then $OA\leqslant OB+BA$ because the sum of two sides of a triangle is greater than the third side, that is $|\mathbf{a}|\leqslant|\mathbf{b}|+|\mathbf{a}-\mathbf{b}|$ proving (*i*).

Similarly, $B'A\leqslant OA+OB'$

$$|\mathbf{a}-(-\mathbf{b})|\leqslant|\mathbf{a}|+|(-\mathbf{b})|$$

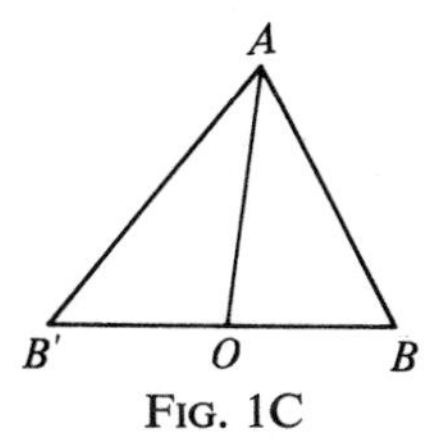

FIG. 1C

which may be rewritten

$$|\mathbf{a}+\mathbf{b}| \leqslant |\mathbf{a}| + |\mathbf{b}|.$$

6. (*i*) What is the equation of the sphere of radius a and centre P_0 (position vector $\mathbf{r}_0$)?

By definition, $P_0P = a$, and so the equation of the sphere is

$$|\mathbf{r}-\mathbf{r}_0| = a.$$

It follows that the position vector of any point on the sphere is of the form

$$\mathbf{r} = \mathbf{r}_0 + a\mathbf{u}$$

where $\mathbf{u}$ is a unit vector.

(*ii*) If $\mathbf{b}$ is the position vector of a point on the sphere, what is the position vector of the point on the opposite end of the diameter?

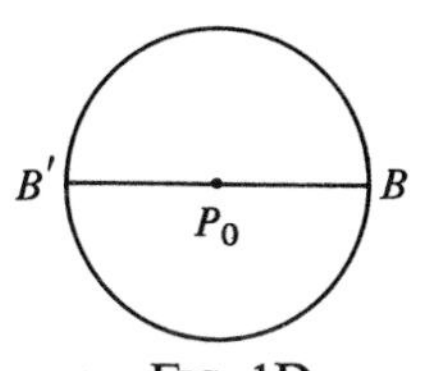

FIG. 1D

If B is the point with position vector b and B' is the required point,

$$\begin{aligned}\overrightarrow{OB'} &= \overrightarrow{P_0B'} + \overrightarrow{OP_0} = \overrightarrow{BP_0} + \overrightarrow{OP_0}\\ &= 2\overrightarrow{OP_0} - \overrightarrow{OB} = 2\vec{r_0} - \vec{b}.\end{aligned}$$

7. Find the equation of the straight line parallel to the line joining the points $(1,-1,3)$ and $(2,1,1)$ which passes through $(1,2,1)$.

The position vector of the point $(2,1,1)$ with respect to the point $(1,-1,3)$ is

$$(2-1, 1-(-1), 1-3) = (1,2,-2)$$

Thus the position vector of a point on the line is

$$\mathbf{r} = (1,2,1) + \lambda(1,2,-2)$$

that is

$$x = 1+\lambda, \qquad y = 2+2\lambda, \qquad z = 1-2\lambda$$

and so the equation of the line is

$$\frac{x-1}{1} = \frac{y-2}{2} = \frac{z-1}{-2}.$$

8. Find the components of velocity and acceleration for a particle moving in a plane expressed

(*a*) in terms of polar coordinates,

(*b*) in terms of components tangential and normal to the path.

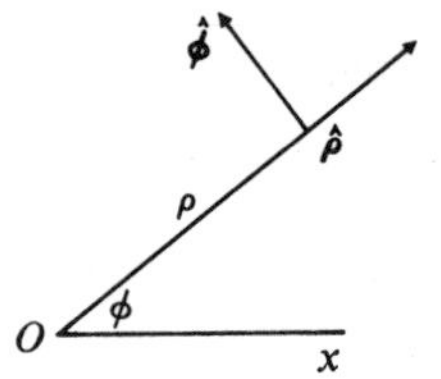

FIG. 1E

(*a*) Let $\hat{\boldsymbol{\rho}}$, $\hat{\boldsymbol{\varphi}}$ be unit vectors along the radius vector, and perpendicular to it in the direction of ϕ increasing.

$$\mathbf{v} = \frac{d\boldsymbol{\rho}}{dt} = \frac{d\rho}{dt}\hat{\boldsymbol{\rho}} + \rho\frac{d\hat{\boldsymbol{\rho}}}{dt}$$

$$= \frac{d\rho}{dt}\hat{\boldsymbol{\rho}} + \rho\dot{\phi}\hat{\boldsymbol{\varphi}}$$

using equation (1.51.13).

$$\boldsymbol{\alpha} = \frac{d^2\mathbf{r}}{dt^2} = \frac{d^2\rho}{dt^2}\hat{\boldsymbol{\rho}} + \frac{d\rho}{dt}\cdot\frac{d\hat{\boldsymbol{\rho}}}{dt} + \frac{d\rho}{dt}\dot{\phi}\hat{\boldsymbol{\varphi}} + \rho\frac{d^2\phi}{dt^2}\hat{\boldsymbol{\varphi}} + \rho\dot{\phi}\frac{d\hat{\boldsymbol{\varphi}}}{dt}$$

$$= \frac{d^2\rho}{dt^2}\hat{\boldsymbol{\rho}} + 2\dot{\rho}\dot{\phi}\hat{\boldsymbol{\varphi}} + \rho\ddot{\phi}\hat{\boldsymbol{\varphi}} + \rho\dot{\phi}(-\dot{\phi}\hat{\boldsymbol{\rho}}),$$

the last expression following from the diagram above,

$$= (\ddot{\rho} - \rho\dot{\phi}^2)\hat{\boldsymbol{\rho}} + \frac{1}{\rho}\frac{d}{dt}(\rho^2\dot{\phi})\,\hat{\boldsymbol{\varphi}}.$$

(*b*) P moves along a curve, and its position at any time may be specified in terms of the distance $s = P_0P$ where P_0 is a fixed point on the curve. Let $\hat{\mathbf{t}}$, $\hat{\mathbf{n}}$ be the tangential and normal unit vectors at time t.

Clearly, $\mathbf{v} = v\hat{\mathbf{t}}$ because the veocity of a particle is tangential to its path

$$\boldsymbol{\alpha} = \frac{d\mathbf{v}}{dt} = \frac{d\mathbf{v}}{dt}\hat{\mathbf{t}} + v\frac{d\hat{\mathbf{t}}}{dt}$$

$$= \frac{dv}{dt}\hat{\mathbf{t}} + v\frac{d\psi}{dt}\cdot\hat{\mathbf{n}}.$$

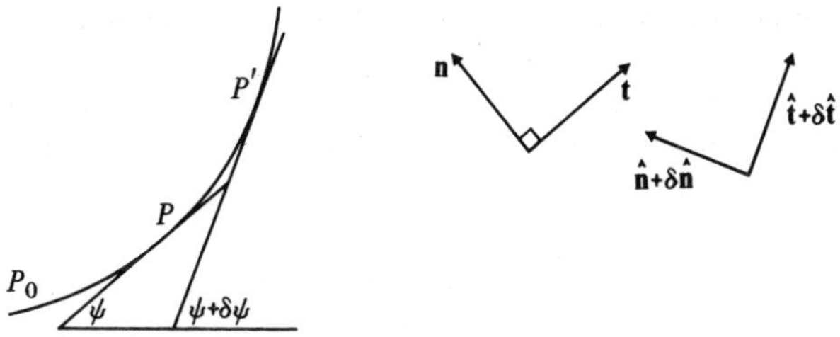

FIG. 1F

Now $v = ds/dt$ and so the second term may be expressed as

$$v^2 \frac{d\psi}{ds} = v^2 \kappa$$

where κ is the curvature.

9. Find the condition (*i*) that the force **F** be coplanar with the forces $(2P, 3P, -P)$ and $(P, -P, P)$, (*ii*) that a particle is in equilibrium when the three forces are applied to it.

(*i*) The condition for coplanarity is that it is possible to express **F** in the form

$$\mathbf{F} = \lambda(2P, 3P, -P) + \mu(P, -P, P) = P(2\lambda + \mu, 3\lambda - \mu, -\lambda + \mu).$$

where λ and μ are scalars.

(*ii*) The equilibrium condition is given by

$$\mathbf{F} + (2P, 3P, -P) + (P, -P, P) = \mathbf{0}$$

that is $\mathbf{F} = (-3P, -2P, O)$.

10. Forces **P**, **Q** act at O and have a resultant **R**. If any transversal cuts their lines of action at A, B, C respectively, show that

$$\frac{P}{OA} + \frac{Q}{OB} = \frac{R}{OC}.$$

Let λ, μ be scalars such that

$$\mathbf{P} = \lambda\overrightarrow{OA}, \qquad \mathbf{Q} = \mu\overrightarrow{OB}.$$

Now a force $\lambda\overrightarrow{OA}$ acting at O is, equivalent to a force $\lambda\overrightarrow{OC} + \lambda\overrightarrow{CA}$ acting at O and a force $\mu\overrightarrow{OB} + \mu\overrightarrow{CB}$ acting at O.

Thus the force system $\mathbf{P} + \mathbf{Q}$ is equivalent to a system

$$(\lambda + \mu)\,\overrightarrow{OC} = \lambda\overrightarrow{CA} + \mu\overrightarrow{CB}.$$

Now the resultant is in the direction of OC, and so $\lambda\overrightarrow{CA} + \mu\overrightarrow{CB} = 0$ and so $\mathbf{P} + \mathbf{Q} = (\lambda + \mu)\overrightarrow{OC} = \mathbf{R}$, whence the result follows.

11. Prove that the centroid of a system symmetrical about a line lies on that line.

A system symmetrical about a line will consist of (1) s pairs of points P_{2k-1}, P_{2k} such that $\overrightarrow{P_{2k-1}P_k^*} + \overrightarrow{P_{2k}P_k^*} = 0$ symmetrical about the line and u points P_l' on this line.

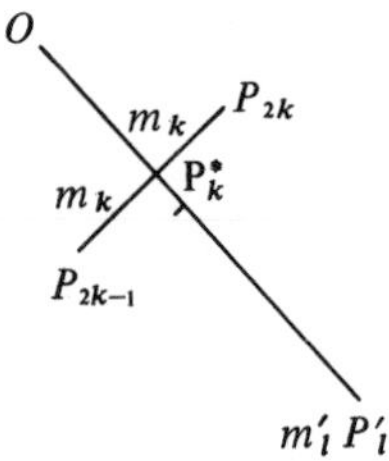

FIG. 1G

Then, if O is some point on the line,

$$\sum_{k=1}^{s} m_k(\overrightarrow{OP}_{2k-1} + \overrightarrow{OP}_{2k}) + \sum_{l=1}^{u} m_l' \overrightarrow{OP}_l'$$

$$= \left(\sum_{k=1}^{s} 2m_k + \sum_{l=1}^{u} m_l' \right) \overrightarrow{OG}$$

$$= \sum_{k=1}^{s} 2m_k \overrightarrow{OP}_k^* + \sum_{l=1}^{u} m_l' \overrightarrow{OP}_l'$$

Now all the $\overrightarrow{OP}_k^*$ and $\overrightarrow{OP}_l'$ are parallel, and so it follows from equation (1.2.11) that $\overrightarrow{OG}$ is parallel to them. That is G lies on the axis of symmetry.

12. Prove that the medians of a triangle concur.

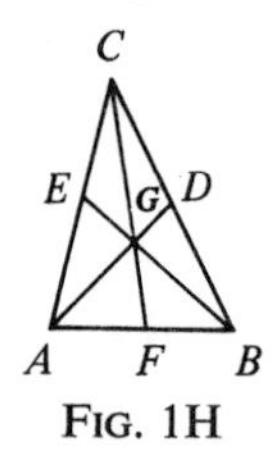

FIG. 1H

Let AD, BE, CF be the medians. Let G be the meet of AD and BE. If O is an arbitrary origin, let $OA = a$, etc.

Then $\mathbf{d} = \frac{1}{2}(\mathbf{b} + \mathbf{c})$, $\mathbf{e} = \frac{1}{2}(\mathbf{a} + \mathbf{c})$, $\mathbf{f} = \frac{1}{2}(\mathbf{a} + \mathbf{b})$.

If G is on AD, and BE, there exist scalars λ, μ such that

$$\overrightarrow{GD} = \lambda \overrightarrow{AD}, \qquad \overrightarrow{GE} = \mu \overrightarrow{BE}$$

that is

$$\tfrac{1}{2}(\mathbf{b} + \mathbf{c}) - \mathbf{g} = \lambda\{\tfrac{1}{2}(\mathbf{b} + \mathbf{c}) - \mathbf{a}\}$$
$$\tfrac{1}{2}(\mathbf{a} + \mathbf{c}) - \mathbf{g} = \mu\{\tfrac{1}{2}(\mathbf{a} + \mathbf{c}) - \mathbf{b}\}.$$

These may be rewritten

$$\mathbf{g} = \lambda\mathbf{a} + \mathbf{b}\{\tfrac{1}{2}(1-\lambda)\} + \frac{\mathbf{c}}{2}(1-\lambda)$$

$$= \mathbf{a}\tfrac{1}{2}(1-\mu) + \mu\mathbf{b} + \frac{\mathbf{c}}{2}(1-\mu).$$

It follows that on equating coefficients

$$\lambda = \tfrac{1}{2}(1-\mu), \qquad \mu = \tfrac{1}{2}(1-\lambda), \qquad \lambda = \mu$$

all of which are satisfied by $\lambda = \mu = \frac{1}{3}$, there being no other solution and so $\mathbf{g} = \frac{1}{3}(\mathbf{a}+\mathbf{b}+\mathbf{c})$. The same result would follow if the meet of BE and CF, or of CF and AD were used, and so the required result is proved.

13. A particle moves so that the acceleration is directed towards a fixed point and is always proportional to the distance from the fixed point. Prove that it moves in a plane.

The acceleration is given by

$$\frac{d^2\mathbf{r}}{dt^2} = \boldsymbol{\alpha} = -\omega^2(\mathbf{r}-\mathbf{c}).$$

The negative sign indicates that the acceleration is *towards* the fixed point $\mathbf{c}$. Clearly, by a change of origin it may be assumed that $\mathbf{c}$ is zero and that the origin is the fixed point and so

$$\frac{d^2\mathbf{r}}{dt^2} = -\omega^2\mathbf{r}.$$

Let $\mathbf{r}_0$, $\mathbf{v}_0$ be the position vector and velocity of the particle at time $t = 0$. Then the solution is

$$\mathbf{r} = \mathbf{r}_0 \cos \omega t + \frac{\mathbf{v}_0}{\omega} \sin \omega t.$$

This may easily be verified

$$\mathbf{v} = \frac{d\mathbf{r}}{dt} = -\mathbf{r}_0\,\omega \sin \omega t + \mathbf{v}_0 \cos \omega t$$

$$\boldsymbol{\alpha} = \frac{d\mathbf{v}}{dt} = \frac{d^2\mathbf{r}}{dt^2} = -\omega^2\left(\mathbf{r}_0 \cos \omega t + \frac{\mathbf{v}_0}{\omega} \sin \omega t\right) = -\omega^2\mathbf{r}.$$

That this is motion in a plane follows from the fact that $\mathbf{r}$ is of the form

$$\mathbf{r} = \lambda\mathbf{a} + \mu\mathbf{b}$$

where $\mathbf{a} = \mathbf{r}_0$, $\mathbf{b} = \mathbf{v}_0/\omega$, $\lambda = \cos \omega t$, $\mu = \sin \omega t$.

Motion of this character is termed elliptic harmonic. For a fuller discussion see L. A. Pars, *Introduction to Dynamics*, p. 181.

15. A particle moves under the influence of a constant gravitational force in a medium which resists it with a force proportional to the velocity. Prove that at great times the velocity becomes constant.

The equation of motion is

$$m\boldsymbol{\alpha} = -m\omega\mathbf{v} + m\mathbf{g}.$$

ω is a constant of proportionality and $\mathbf{g}$ is the constant gravitational acceleration. This may be rewritten as

$$\frac{d\mathbf{v}}{dt} + \omega\mathbf{v} = \mathbf{g}.$$

Multiplying by $e^{\omega t}$

$$\frac{d}{dt}(e^{\omega t}\mathbf{v}) = \mathbf{g}\,e^{\omega t}.$$

If $\mathbf{v} = \mathbf{v}_0$ at time $t = 0$ (i.e. the particle is projected with velocity $\mathbf{v}_0$), this integrates to

$$e^{\omega t}\mathbf{v} - \mathbf{v}_0 = \mathbf{g}(e^{\omega t} - 1)$$

or

$$\mathbf{v} = \left(\mathbf{v}_0 - \frac{\mathbf{g}}{\omega}\right)e^{-\omega t} + \frac{\mathbf{g}}{\omega}$$

and as t becomes great, $\mathbf{v}$ tends to $\mathbf{v}_\infty = \mathbf{g}/\omega$. $\mathbf{g}/\omega$ is the so-called terminal velocity for the system. It follows that

$$\boldsymbol{\alpha} = \mathbf{g} - \omega\mathbf{v} = \mathbf{g} - \omega\left(\mathbf{v}_0 - \frac{\mathbf{g}}{\omega}\right)e^{-\omega t} - \mathbf{g}$$

$$= (\mathbf{g} - \omega\mathbf{v}_0)\,e^{-\omega t}.$$

16. Find the condition that the four points $ABCD$ are the apices of a parallelogram, ABC being given.

The conditions that $ABCD$ form a parallelogram are that

$$\overrightarrow{AD} = \overrightarrow{BC} \qquad \text{and} \qquad \overrightarrow{DC} = \overrightarrow{AB}$$

i.e., $\mathbf{d} - \mathbf{a} = \mathbf{c} - \mathbf{b}$ and $\mathbf{c} - \mathbf{d} = \mathbf{b} - \mathbf{a}$.

Both of these conditions are equivalent to

$$\mathbf{d} = \mathbf{a} - \mathbf{b} + \mathbf{c}.$$

17. Prove that the centroid of four equal masses placed at the apices of a parallelogram is at the meet of the diagonals.

Let $ABCD$ be the parallelogram. Take D as origin, then

$$\begin{aligned} 4m\overrightarrow{DG} &= m\overrightarrow{DA} + m\overrightarrow{DB} + m\overrightarrow{DC} \\ &= 2m\overrightarrow{DH} + m.2\overrightarrow{DH} \end{aligned}$$

where H is the meet of the diagonals. This proves the result.

18. A particle describes a plane curve, the acceleration at any point P of its path being towards a fixed point O in the plane and of magnitude f. OQ is drawn such that $\overrightarrow{OQ} = \tau(d/dt)\overrightarrow{OP}$. If a second particle describes

the curve traced out by Q, being always at Q when the first particle is at P, prove that its acceleration may be represented by an acceleration $f\,(OQ/OP)$

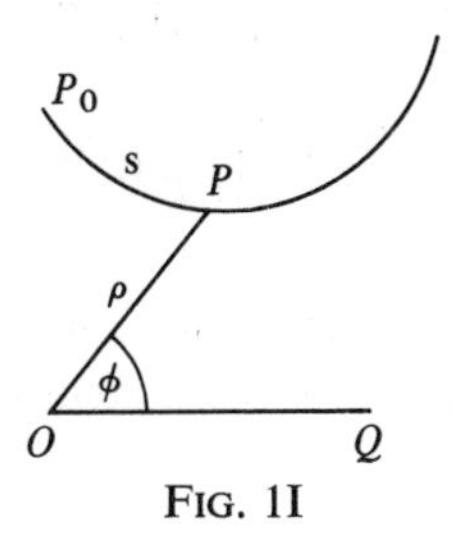

FIG. 1I

towards O, together with an acceleration $OP.OQ(d/ds)(f/OP)$ in the direction PO, s being the arc of the path of P. (S. L. Green.)

$$\frac{d}{dt}\overrightarrow{OP} = \dot{\rho}\hat{\boldsymbol{\rho}} + \rho\dot{\phi}\hat{\boldsymbol{\varphi}}, \qquad \rho = \overrightarrow{OP}$$

and so

$$\overrightarrow{OQ} = \tau(\dot{\rho}\hat{\boldsymbol{\rho}} + \rho\dot{\phi}\hat{\boldsymbol{\varphi}})$$

$$\frac{d^2}{dt^2}\overrightarrow{OP} = -f\hat{\boldsymbol{\rho}} = \frac{1}{\tau}\frac{d}{dt}\overrightarrow{OQ}$$

$$\frac{d^2}{dt^2}\overrightarrow{OQ} = -\tau\frac{d}{dt}(f\hat{\boldsymbol{\rho}})$$

$$= -\tau\frac{df}{dt}\cdot\hat{\boldsymbol{\rho}} - \tau f\frac{d\hat{\boldsymbol{\rho}}}{dt}$$

$$= -\tau\frac{df}{dt}\cdot\hat{\boldsymbol{\rho}} - \tau f\dot{\phi}\hat{\boldsymbol{\varphi}}$$

use being made of the results of worked example 8(a).

It follows that

$$\frac{d^2}{dt^2}\overrightarrow{OQ} = -\tau\frac{df}{dt}\hat{\boldsymbol{\rho}} - f\left[\frac{\overrightarrow{OQ}}{\rho} - \tau\frac{\dot{\rho}}{\rho}\hat{\boldsymbol{\rho}}\right]$$

on eliminating $\hat{\boldsymbol{\varphi}}$

$$= -\tau\left[\frac{df}{dt} - \frac{\dot{\rho}}{\rho}f\right]\hat{\boldsymbol{\rho}} - \frac{f\overrightarrow{OQ}}{\rho}$$

$$= -\tau\rho\frac{d}{dt}\left(\frac{f}{\rho}\right)\hat{\boldsymbol{\rho}} + \frac{f\overrightarrow{QO}}{\rho}.$$

Now

$$\left|\frac{d}{dt}\overrightarrow{OP}\right| = \frac{ds}{dt} \qquad OQ = \tau\frac{ds}{dt}$$

and so

$$\frac{d^2}{dt^2}\overrightarrow{OQ} = -\tau.\rho\frac{ds}{dt}\frac{d}{ds}\left(\frac{f}{\rho}\right)\hat{\mathbf{\rho}} + \frac{f\overrightarrow{QO}}{\rho}$$

$$= OQ\frac{d}{ds}\left(\frac{f}{OP}\right)\overrightarrow{PO} + \frac{f\overrightarrow{QO}}{\rho}$$

which is the required result.

The path of Q is the hodograph of the path of P. Thus the acceleration of the point in the hodograph may be determined when the acceleration of P is central.

19. A ship steams south with speed v, and the apparent wind direction is east. Another ship steams west with speed $2v/\sqrt{3}$ and the apparent wind direction is 30° east of north. Find the true wind velocity.

If $\mathbf{v}_s$ is the velocity of the ship, $\mathbf{v}_w$ the velocity of the wind and $\mathbf{v}_w'$ the apparent velocity of the wind, they are related by

$$\mathbf{v}_w - \mathbf{v}_s = \mathbf{v}_w'$$

that is

$$\mathbf{v}_w = \mathbf{v}_s + \mathbf{v}_w'.$$

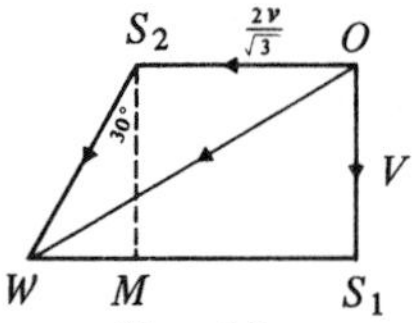

FIG. 1J

The velocity of the first ship is represented by the vector $\overrightarrow{OS_1}$, as indicated. The apparent velocity of the wind is known in direction, but not in magnitude. It is represented by a vector going west from S_1 of unspecified length. The velocity of the second ship is represented by $\overrightarrow{OS_2}$ and the apparent velocity of the wind is again known in direction but not in magnitude. It is represented by a vector going 30° west of south from S_2 of unspecified length. The actual wind velocity will thus be given by $\overrightarrow{OW}$ where W is a point at which the two vectors representing the apparent wind velocities meet for

$$\overrightarrow{OW} = \overrightarrow{OS_1} + \overrightarrow{S_1W} = \overrightarrow{OS_2} + \overrightarrow{S_2W}$$

$\overrightarrow{S_1W}$, $\overrightarrow{S_2W}$ representing the apparent wind velocities.

$$S_2M = vS_2W = v\sec 30° = \frac{2v}{\sqrt{3}} = OS_2.$$

Thus S_2OW is isosceles and

$$S_2WO = S_2OW = 30°$$

and so

$$OW = \frac{2\cdot 2v}{\sqrt{3}}\cos 30^\circ = 2v.$$

Thus the wind velocity is $2v$, 60° east of north.

Two points may be noted in connection with examples involving relative velocities such as this one. Firstly, a wind 60° east of north is *coming from* that direction. It is *going to* 60° west of south. Secondly, many examples of this kind are best solved by accurate drawing – particularly when the figures are awkward. It is for this reason that it is usual for angles to be given in degrees rather than in the usual mathematical radian measure.

If the problem had been specified slightly differently and apparent wind speeds given instead of apparent wind directions, the solution would involve the drawing of circles of appropriate radii. In this case sometimes there would be no solution, sometimes one solution, sometimes several solutions.

20. Express the vector $\mathbf{P} = xyz^2\mathbf{i} + yz\mathbf{j} + xy\mathbf{k}$ in cylindrical polar coordinates.

$$\begin{aligned}\mathbf{P} &= \rho\cos\phi\,.\,\rho\sin\phi z^2(\hat{\boldsymbol{\rho}}\cos\phi - \hat{\boldsymbol{\varphi}}\sin\phi)\\ &\quad + \rho\sin\phi z(\hat{\boldsymbol{\rho}}\sin\phi + \hat{\boldsymbol{\varphi}}\cos\phi)\\ &\quad + \rho\sin\phi\rho\cos\phi\mathbf{k}\\ &= \hat{\boldsymbol{\rho}}(\rho^2 z^2\cos^2\phi\sin\phi + \rho z\sin^2\phi)\\ &\quad + \hat{\boldsymbol{\varphi}}(\rho z\sin\phi\cos\phi - \rho^2 z^2\cos\phi\sin^2\phi)\\ &\quad + \mathbf{k}(\rho^2\sin\phi\cos\phi).\end{aligned}$$

21. Express the vector $\mathbf{Q} = xy\mathbf{i} + 2yz\mathbf{j} + (z^2 - yz)\mathbf{k}$ in spherical polar coordinates.

$$\begin{aligned}\mathbf{Q} &= r\sin\theta\cos\phi r\sin\theta\sin\phi(\hat{\mathbf{r}}\sin\theta\cos\phi + \hat{\boldsymbol{\theta}}\cos\theta\cos\phi - \hat{\boldsymbol{\varphi}}\sin\phi)\\ &\quad + 2r\sin\theta\sin\phi r\cos\theta(\hat{\mathbf{r}}\sin\theta\sin\phi + \hat{\boldsymbol{\theta}}\cos\theta\sin\phi + \hat{\boldsymbol{\varphi}}\cos\phi)\\ &\quad + (r^2\cos^2\theta - r\sin\theta r\cos\theta)(\hat{\mathbf{r}}\cos\theta - \hat{\boldsymbol{\theta}}\sin\theta)\\ &= \hat{\mathbf{r}}(r^2\sin^3\theta\cos^2\phi\sin\phi + 2r^2\sin^2\theta\cos\theta\sin^2\phi\\ &\quad + r^2\cos^3\theta - r^2\cos^2\theta\sin\theta)\\ &\quad + \hat{\boldsymbol{\theta}}(r^2\sin^2\theta\cos\theta\cos^2\phi\sin\phi\\ &\quad + 2r^2\cos^2\theta\sin\theta\sin^2\phi\\ &\quad + r^2\cos\theta\sin^2\theta - r^2\cos^2\theta\sin\theta)\\ &\quad + \hat{\boldsymbol{\varphi}}(2r^2\cos\theta\sin\theta\cos\phi\sin\phi - r^2\sin^2\theta\cos\phi\sin^2\phi).\end{aligned}$$

22. Express the vector

$$\mathbf{F} = r^3\,\hat{\boldsymbol{\varphi}} + 2a^2 r\cos\theta\sin\phi\,\hat{\boldsymbol{\theta}} + 3ar^2\sin\theta\cos\phi\,\hat{\mathbf{r}}$$

in terms of x, y, z, $\mathbf{i}$, $\mathbf{j}$, $\mathbf{k}$.

$$\begin{aligned}
\mathbf{F} &= r^3(-\mathbf{i}\sin\phi + \mathbf{j}\cos\phi) \\
&\quad + 2a^2 r\{-\mathbf{k}\sin\theta + \cos\theta(\mathbf{i}\cos\phi + \mathbf{j}\sin\phi)\}\cos\theta\,\sin\phi \\
&\quad + 3ar^2\sin\theta\cos\phi\{\mathbf{k}\cos\theta + \sin\theta(\mathbf{i}\cos\phi + \mathbf{j}\sin\phi)\} \\
&= \mathbf{i}\{-r^3\sin\phi + 2a^2 r\cos^2\theta\sin\phi\cos\phi + 3ar^2\sin^2\theta\cos^2\phi\} \\
&\quad + \mathbf{j}\{r^3\cos\phi + 2a^2 r\cos^2\theta\sin^2\phi + 3ar^2\sin^2\theta\sin\phi\cos\phi\} \\
&\quad + \mathbf{k}\{-2a^2 r\sin\theta + 3ar^2\sin\theta\,\cos\theta\cos\phi\} \\
&= \mathbf{i}\left\{-(x^2+y^2+z^2)^{3/2}\frac{y}{(x^2+y^2)^{1/2}} + \frac{2a^2z^2}{(x^2+y^2+z^2)^{1/2}}\cdot\frac{xy}{x^2+y^2} + 3ax^2\right\}. \\
&\quad + \mathbf{j}\left\{(x^2+y^2+z^2)^{3/2}\cdot\frac{x}{(x^2+y^2)^{3/2}} + \frac{2a^2z^2}{(x^2+y^2+z^2)^{1/2}}\cdot\frac{y^2}{x^2+y^2}\right. \\
&\quad \left. + 3zxy\right\} + \mathbf{k}\{-2a^2(x^2+y^2)^{1/2} + 3azx\}.
\end{aligned}$$

Exercises

1. Find the sum of the vectors

$$\mathbf{i}+\mathbf{j}+\mathbf{k}; \qquad 2\mathbf{i}-3\mathbf{j}+\mathbf{k}; \qquad 5\mathbf{i}-3\mathbf{k}$$

and calculate the modulus and direction cosines of the resulting vector.

2. If

$$\overrightarrow{OP} = a\mathbf{i} + 2a\mathbf{j} + 4a\mathbf{k}$$

and

$$\overrightarrow{OQ} = 2a\mathbf{i} - 3a\mathbf{j} + 5a\mathbf{k}$$

find the direction cosines of $\overrightarrow{PQ}$.

3. The position vectors of A, B, C, D are respectively

$$\mathbf{a},\ \mathbf{b},\ 3\mathbf{a}+2\mathbf{b},\ \mathbf{a}-3\mathbf{b}$$

Express $\overrightarrow{CA}$, $\overrightarrow{DB}$, $\overrightarrow{AD}$ in terms of $\mathbf{a}$ and $\mathbf{b}$.

4. Show that the lines joining the mid points of opposite edges of a tetrahedron concur and bisect one another.

5. ABC and $A'B'C'$ are two triangles, G, G' are the respective meets of the medians. Prove that

$$\overrightarrow{AA'} + \overrightarrow{BB'} + \overrightarrow{CC'} = 3\overrightarrow{GG'}$$

6. Find the sum of the three vectors determined by the diagonals of three adjacent faces of a parallelopiped passing through the same corner, the vectors being directed to that corner.

7. Find the locus of points P such that $\mathbf{r} - \mathbf{a} = \lambda(\mathbf{r} - \mathbf{b})$ where a, b are constant.

8. What is the vector equation of the straight line through the point $(1, 1, -1)$ parallel to the line joining $(1, -2, 1)$ to $(2, 1, 0)$?

9. Prove that in a skew quadrilateral the joins of the mid points of opposite edges bisect one another, and that if the mid points be joined in order, a parallelogram is generated.

10. An object is acted upon by three coplanar forces as indicated. Determine the force which will exactly counteract them graphically and check this.

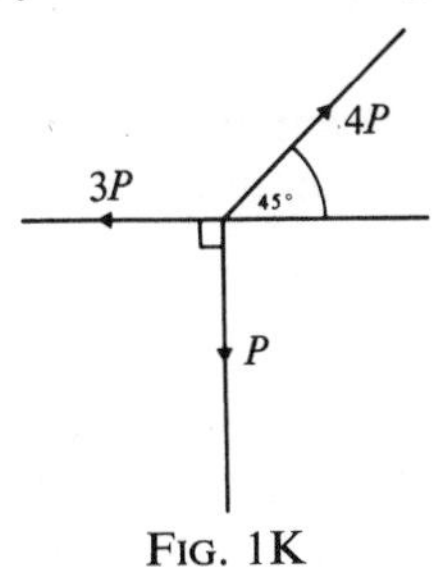

FIG. 1K

11. Find the condition that the points P, A, B, C are coplanar when A, B, C are the points

$$(1,1,1)(2,3,4)(-1,3,0), \text{ respectively.}$$

12. Two towns A and B are situated on opposite sides of a river of width a. B is a distance $2a$ upstream from A. The speed of flow of the river is u. A boat which moves relative to the water with speed $v(>u)$ makes a journey from A to B and back again. Find the time taken.

13. P is a point on the tangent at a variable point Q to a fixed circle of radius a. QP is of length ut and makes an angle θ with a fixed tangent. Find the resolutes of the acceleration of P along and perpendicular to QP.

14. Prove that the vectors $(2, 1, 1)$, $(2, 5, -2)$, $(-4, -6, 1)$ can form the sides of a triangle. If the first vector ends on $(0, 1, 3)$ and the second one begins on $(0, 1, 3)$, find the centroid of the triangle and the length of the median from $(0, 1, 3)$.

15. A particle moves so that

$$\frac{d^2\mathbf{r}}{dt^2} = \omega^2\mathbf{r}$$

If its initial position vector and velocity are $\mathbf{r}_0$ and $\mathbf{v}_0$, find the position vector at any time, and show that if $\mathbf{v}_0 + \omega\mathbf{r}_0 = 0$, the particle moves in a straight line approaching the origin asymptotically for large times.

16. A particle P moves in a plane under the influence of n fixed centres of attraction $A_1, \ldots A_n$ in the plane. The force due to the centre of attraction A_s is $m\omega_s^2\overrightarrow{PA_s}$. Show that the motion is determined by an equation of the form

$$\ddot{\mathbf{r}} = -\omega^2(\mathbf{r} - \mathbf{r}_c)$$

and determine ω^2 and $\mathbf{r}_c$.

17. N points A_r lie at equal intervals on a circle. If $\mathbf{i}$, $\mathbf{j}$ are two mutually perpendicular unit vectors in the plane of the circle, show that the unit vectors in the directions from the centre of the circle to the N points are

$$\mathbf{i}\cos\left(\frac{2\pi r}{N} + \alpha\right) + \mathbf{j}\sin\left(\frac{2\pi r}{N} + \alpha\right) 1 \leqslant r \leqslant N.$$

18. A particle of mass m is subject to a force $m\boldsymbol{\alpha}$. At times t_1, t_2, t it passes through the points P_1, P_2, P respectively. If

$$(t_2 - t)(t - t_1)(t_1 - t_2)\boldsymbol{\alpha} = 2\{t_1\overrightarrow{P_2P} + t_2\overrightarrow{PP_1} + t\overrightarrow{P_1P_2}\}$$

for all P, t, show that $\boldsymbol{\alpha}$ is constant.

19. If $\overrightarrow{OV_1}$, $\overrightarrow{OV_2}$ be vectors representing the velocities at any two points P_1, P_2 of the path of a particle subject to a constant force, and V is the middle point of V_1V_2, prove that $\overrightarrow{OV}$ represents the mean velocity between P_1 and P_2.

20. A swarm of particle expands throughout all space, the velocity at any point any time being $\mathbf{v} = \mathbf{r}f(t)$. Prove that, if the velocity be measured relative to any particle, the same result holds.

2 Products of Vectors

2.0. Introduction

The concept of the multiplication of a vector by a scalar has already been discussed. The multiplication of two vectors is more complicated. There are in fact two different types of product of two vectors which correspond respectively to the two mechanical concepts of work and moment of a force. In one case the result is a scalar, and in the other a vector and the corresponding types of product are referred to as scalar product and vector product respectively.

The definitions of these products were originally given by Hamilton who supposed that any product obeys the relations

$$(x\mathbf{a})(y\mathbf{b}) = xy\mathbf{ab} \tag{2.0.1}$$

(x, y are scalars)

$$\mathbf{a}(\mathbf{b} + \mathbf{c}) = \mathbf{ab} + \mathbf{ac} \tag{2.0.2}$$

(the distributive law).

The scalar product of two vectors **a**, **b** will be written **a**.**b** and the vector product **a** × **b**. It is possible to define these products without reference to any frame of reference other than that provided by the vectors themselves and this approach will be followed.† These definitions will lead to the more usual geometrical definitions. It is then unnecessary to prove that the distributive law is satisfied.

It may be remarked that complex numbers also obey the vectorial product laws (2.0.1) and (2.0.2). These are discussed in section 2.10. A fourth type of product of two vectors is associated with dyads and this is discussed in section 6.1.

2.1. Scalar product

Let

$$\mathbf{b} = \mathbf{b}_1 + \mathbf{b}_2 \tag{2.1.1}$$

† *A Second Report on the Teaching of Mechanics in Schools.* Mathematical Association, G. Bell (1965), p. 81.

where $\mathbf{b}_1$ and $\mathbf{b}_2$ are respectively parallel and perpendicular to $\mathbf{a}$. Consider $\mathbf{a}.\mathbf{b}_2$. Then from equation (2.0.2) it follows that

$$\mathbf{a}.(-\mathbf{b}_2) = -(\mathbf{a}.\mathbf{b}_2). \tag{2.1.2}$$

The vectors $\mathbf{a}$ and $-\mathbf{b}_2$ can be obtained from $\mathbf{a}$ and $\mathbf{b}_2$ (which may be regarded as forming a rigid framework) by rotation about $\mathbf{a}$ as axis.

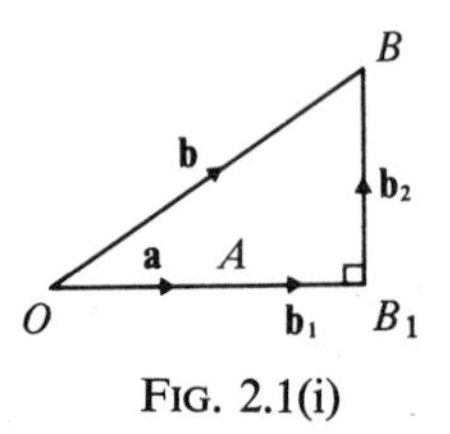

FIG. 2.1(i)

In the absence of any other frame of reference, it is impossible to tell one pair from another and so

$$\begin{aligned}\mathbf{a}.\mathbf{b}_2 &= \mathbf{a}.(-\mathbf{b}_2)\\ &= -(\mathbf{a}.\mathbf{b}_2).\end{aligned} \tag{2.1.3}$$

Thus

$$\mathbf{a}.\mathbf{b}_2 = 0. \tag{2.1.4}$$

Since the common direction of **a** and **b**, cannot be identified without an external frame, it follows that the product must depend on the magnitudes only, and if **a** is the unit vector in the direction of **a** and **b**,

$$\mathbf{a}.\mathbf{b} = \mathbf{a}.\mathbf{b}_1 = |\mathbf{a}||\mathbf{b}_1|(\hat{\mathbf{a}}.\hat{\mathbf{a}}). \tag{2.1.5}$$

The assumption is made that $\mathbf{a}.\mathbf{a}$ has the value unity and so

$$\begin{aligned}\mathbf{a}.\mathbf{b} &= |\mathbf{a}||\mathbf{b}_1|\\ &= |\mathbf{a}||\mathbf{b}|\cos\theta\end{aligned} \tag{2.1.6}$$

where θ is the angle between **a** and **b**, this being the usual definition of scalar product. (It has been assumed that the angle θ is acute. If this is not the case a slight alteration in the proof is necessary.)

It follows immediately from the above definition that the scalar product is commutative, that is

$$\mathbf{a}.\mathbf{b} = \mathbf{b}.\mathbf{a} \tag{2.1.7}$$

and that the scalar product of two vectors which are perpendicular to one another is zero. Thus the statement

$$\mathbf{a}.\mathbf{b} = 0 \tag{2.1.8}$$

implies either that the modulus of **a** or **b** is zero or that **a** and **b** are perpendicular.

If $\mathbf{b} = \mathbf{a}$, the self scalar product of $\mathbf{a}$ is

$$\mathbf{a}.\mathbf{a} = |\mathbf{a}|^2 = a^2 \tag{2.1.9}$$

that is the square of $\mathbf{a}$.

2.11. *The scalar product in rectangular cartesian coordinates*
It is obvious from the definition of scalar product that

$$\mathbf{i}.\mathbf{i} = \mathbf{j}.\mathbf{j} = \mathbf{k}.\mathbf{k} = 1 \tag{2.11.1}$$

and that

$$\mathbf{j}.\mathbf{k} = \mathbf{k}.\mathbf{i} = \mathbf{i}.\mathbf{j} = 0. \tag{2.11.2}$$

It follows that

$$\begin{aligned}\mathbf{a}.\mathbf{b} &= (a_x\mathbf{i} + a_y\mathbf{j} + a_z\mathbf{k}).(b_x\mathbf{i} + b_y\mathbf{j} + b_z\mathbf{k}) \\ &= a_x b_x + a_y b_y + a_z b_z.\end{aligned} \tag{2.11.3}$$

If $\mathbf{u}$ is a unit vector, and α, β, γ are the angles that the direction of $\mathbf{u}$ makes with the three coordinate axes, then

$$\mathbf{u}.\mathbf{i} = \cos\alpha, \qquad \mathbf{u}.\mathbf{j} = \cos\beta, \qquad \mathbf{u}.\mathbf{k} = \cos\gamma, \tag{2.11.4}$$

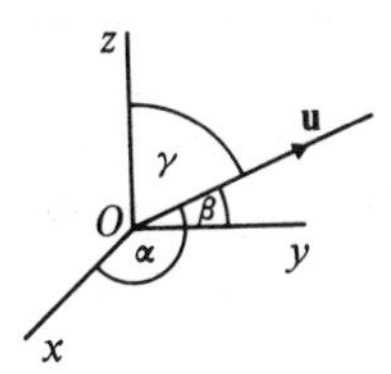

FIG. 2.11(i)

these three quantities being in fact the direction cosines associated with the direction defined by u and

$$1 = u^2 = \mathbf{u}.\mathbf{u} = \cos^2\alpha + \cos^2\beta + \cos^2\gamma \tag{2.11.5}$$

as would be expected.

The resolute of a in the direction of u is given by

$$\mathbf{a}.\mathbf{u} = a_x\cos\alpha + a_y\cos\beta + a_x\cos\gamma \tag{2.11.6}$$

and the angle between two lines whose directions are defined by unit vectors $\mathbf{u}_1$, $\mathbf{u}_2$ is given by

$$\cos\theta = \mathbf{u}_1.\mathbf{u}_2 = \cos\alpha_1\cos\alpha_2 + \cos\beta_1\cos\beta_2 + \cos\gamma_1\cos\gamma_2 \tag{2.11.7}$$

which is a well-known result in solid geometry.

2.12. *The concept of work*
In mechanics, the work done in an elementary displacement is defined as the product of the elementary displacement with the resolute of

the force in the direction of the elementary displacement of the point of application of the force, that is

$$\begin{aligned}\delta W &= |\delta \mathbf{r}|\, F\cos\theta \\ &= \mathbf{F}.\delta\mathbf{r}.\end{aligned} \tag{2.12.1}$$

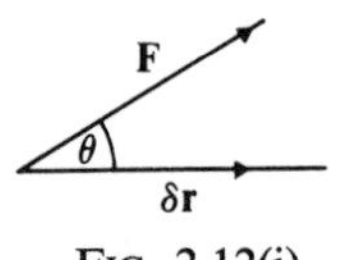

FIG. 2.12(i)

The activity of the force is

$$\frac{dW}{dt} = \mathbf{F}.\mathbf{v} \tag{2.12.2}$$

where $\mathbf{v}$ is the velocity of the point of application of the force.

2.13. *Geometrical applications of the scalar product*

(*a*) If $\mathbf{r}$ is the position vector of a point on a plane, $\mathbf{n}$ is the unit vector normal to the plane and p the length of the perpendicular from the origin to the plane, then the equation of the plane is

$$\mathbf{r}.\mathbf{n} = p. \tag{2.13.1}$$

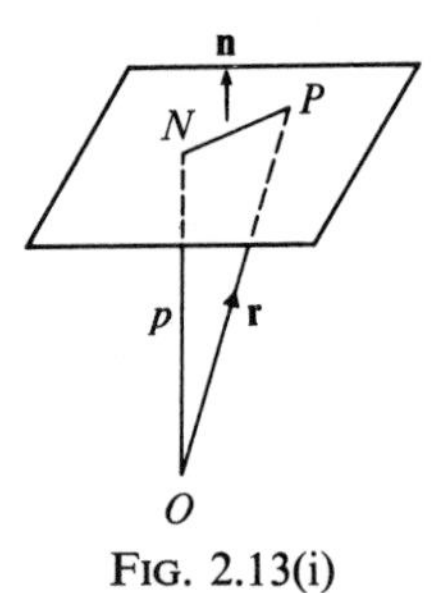

FIG. 2.13(i)

The proof is simple:

$$\begin{aligned}0 &= \overrightarrow{NP}.\mathbf{n} \\ &= (\overrightarrow{OP} - \overrightarrow{ON}).\mathbf{n} \\ &= (\mathbf{r} - p\mathbf{n}).\mathbf{n} \\ &= \mathbf{r}.\mathbf{n} - p.\end{aligned} \tag{2.13.2}$$

(*b*) The equation

$$(\mathbf{r} - \mathbf{a}).(\mathbf{r} - \mathbf{b}) = 0 \tag{2.13.3}$$

is the equation of the sphere with the ends of a diameter having position vectors **a**, **b**. The result is obvious from the diagram (figure 2.13(ii)).

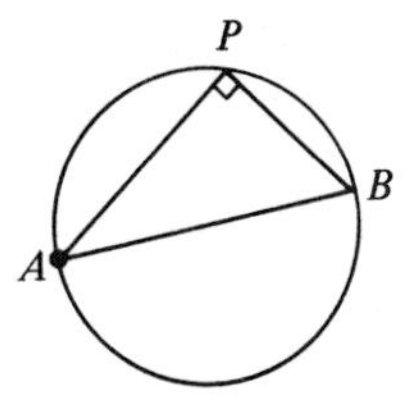

FIG. 2.13(ii)

(*c*) The statement

$$\begin{aligned} BA^2 = |\mathbf{b}-\mathbf{a}|^2 &= (\mathbf{b}-\mathbf{a}).(\mathbf{b}-\mathbf{a}) \\ &= b^2 - 2\mathbf{b}.\mathbf{a} + a^2 \\ &= (b+a)^2 - 2ab(1+\cos\theta) \\ &< (b+a)^2 = (OA+OB)^2 \end{aligned} \tag{2.13.4}$$

is equivalent to the statement that the length of any side of a triangle is less than the sum of the lengths of the other two.

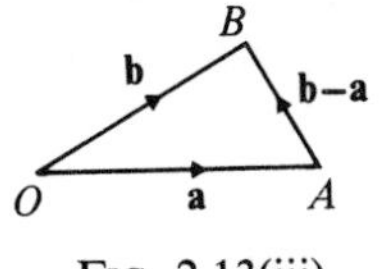

FIG. 2.13(iii)

2.14. *Differentiation of the scalar product*

If **A**, **B** are functions of some parameter t, the derivative of $\mathbf{A}.\mathbf{B}$ with respect to t may be obtained as follows:

If $\phi = \mathbf{A}.\mathbf{B}$

$$\phi + \delta\phi = (\mathbf{A}+\delta\mathbf{A}).(\mathbf{B}+\delta\mathbf{B}) \tag{2.14.1}$$

and if $\delta\mathbf{A}$, $\delta\mathbf{B}$ are infinitesimally small

$$\delta\phi = \mathbf{A}.\delta\mathbf{B} + \mathbf{B}.\delta\mathbf{A} \tag{2.14.2}$$

whence

$$\frac{d\phi}{dt} = \mathbf{A}.\frac{d\mathbf{B}}{dt} + \mathbf{B}.\frac{d\mathbf{A}}{dt}. \tag{2.14.3}$$

In particular

$$\frac{d}{dt}(A^2) = \frac{d}{dt}(\mathbf{A}.\mathbf{A}) = 2\mathbf{A}.\frac{d\mathbf{A}}{dt} \tag{2.14.4}$$

and if $\mathbf{A}$ is constant, it follows that

$$\mathbf{A}.\frac{d\mathbf{A}}{dt} = 0. \tag{2.14.5}$$

Thus, if a particle moves in a circle, the velocity is always perpendicular to the radius vector from the centre for

$$a^2 = r^2 = \mathbf{r}.\mathbf{r} \tag{2.14.6}$$

and so

$$0 = \mathbf{r}.\frac{d\mathbf{r}}{dt} = \mathbf{r}.\mathbf{v}. \tag{2.14.7}$$

This, of course, would be expected from other considerations. Similarly, if a particle moves with constant speed v_0, the acceleration is perpendicular to the velocity for

$$v_0^2 = v^2 = \mathbf{v}.\mathbf{v} \tag{2.14.8}$$

whence

$$0 = \mathbf{v}.\frac{d\mathbf{v}}{dt} = \mathbf{v}.\boldsymbol{\alpha}. \tag{2.14.9}$$

2.2. Vector product

The vector product of two vectors $\mathbf{a}$ and $\mathbf{b}$ is to be a vector $\mathbf{a} \times \mathbf{b}$.

Resolving $\mathbf{b}$ again into two vectors $\mathbf{b}_1$ and $\mathbf{b}_2$ respectively parallel and perpendicular to $\mathbf{a}$, it follows from the fact that the distributive law is to be obeyed that

$$\mathbf{a} \times \mathbf{b} = \mathbf{a} \times \mathbf{b}_1 + \mathbf{a} \times \mathbf{b}_2 \tag{2.2.1}$$

(It will be assumed that the angle θ is acute.) Now $\mathbf{a} \times \mathbf{b}_1$, if it is a vector, can only be in the common direction of $\mathbf{a}$ and $\mathbf{b}_1$ because there

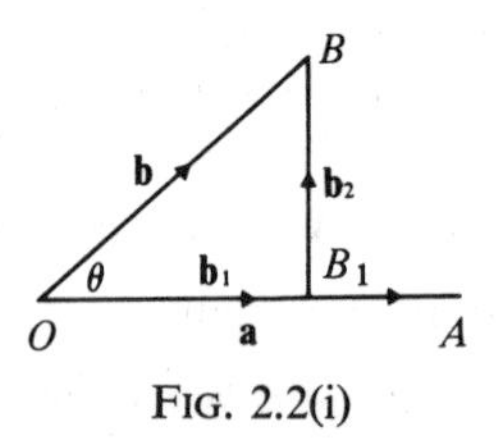

FIG. 2.2(i)

is no other direction involved. There cannot be a component perpendicular as, if there were, it could be in any direction perpendicular to $\mathbf{a}$ and $\mathbf{b}$, and so the product would not be defined uniquely. Similarly, the direction of $-\mathbf{a} \times -\mathbf{b}_1$ must be in the common direction of $-\mathbf{a}$ and $-\mathbf{b}_1$. But from equation (2.0.1)

$$-\mathbf{a} \times -\mathbf{b}_1 = \mathbf{a} \times \mathbf{b}_1. \tag{2.2.2}$$

Thus $\mathbf{a} \times \mathbf{b}_1$ has the directions of both $\mathbf{a}$ and $\mathbf{b}_1$, and of $-\mathbf{a}$ and $-\mathbf{b}_1$, and so

$$\mathbf{a} \times \mathbf{b}_1 = \mathbf{0}. \tag{2.2.3}$$

It follows from this that the vector product of two parallel vectors (and in particular of a vector with itself) vanishes. Thus

$$\mathbf{a} \times \mathbf{b} = \mathbf{a} \times \mathbf{b}_2. \tag{2.2.4}$$

Now $\mathbf{a} \times \mathbf{b}_2$ depends upon the magnitudes of $\mathbf{a}$, $\mathbf{b}_2$ and their directions

$$\mathbf{a} \times \mathbf{b}_2 = ab_2 \hat{\mathbf{a}} \times \hat{\mathbf{b}}_2. \tag{2.2.5}$$

It now remains to define $\hat{\mathbf{a}} \times \hat{\mathbf{b}}_2$ and this is taken to be the unit vector $\hat{\mathbf{c}}$ which is perpendicular to both $\mathbf{a}$ and $\mathbf{b}_2$ in the sense shown in the diagram (figure 2.2(ii)). $\mathbf{a}$ and $\mathbf{b}_2$ are in the plane of the paper and $\hat{\mathbf{c}}$ sticks out of the plane of the paper.

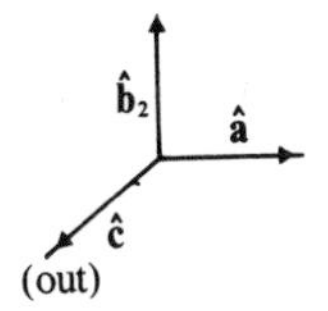

FIG. 2.2(ii)

Now $b_2 = b \sin \theta$, and it follows that

$$\mathbf{a} \times \mathbf{b} = ab \sin \theta \hat{\mathbf{c}} \tag{2.2.6}$$

which is the usual definition of vector product.

It follows that

$$\begin{aligned} \mathbf{b} \times \mathbf{a} &= ba \sin (2\pi - \theta) \hat{\mathbf{c}} \\ &= -\mathbf{a} \times \mathbf{b} \end{aligned} \tag{2.2.7}$$

and the vector product is not commutative. It is clear from the foregoing that the statement

$$\mathbf{a} \times \mathbf{b} = \mathbf{0} \tag{2.2.8}$$

implies either that the modulus of $\mathbf{a}$ or $\mathbf{b}$ is zero, *or* that $\mathbf{a}$ and $\mathbf{b}$ are parallel.

If $\mathbf{u}$ is a unit vector, then

$$|\mathbf{a} \times \mathbf{u}| = a \sin \theta \tag{2.2.9}$$

where θ is the acute angle between the directions of $\mathbf{a}$ and $\mathbf{u}$.

Following the methods of section 2.14, it is not difficult to see that, if

$$\mathbf{c} = \mathbf{a} \times \mathbf{b} \tag{2.2.10}$$

$$\frac{d\mathbf{c}}{dt} = \frac{d\mathbf{a}}{dt} \times \mathbf{b} + \mathbf{a} \times \frac{d\mathbf{b}}{dt}. \tag{2.2.11}$$

2.21. *The vector product in rectangular cartesian coordinates*

It is obvious from the definition of vector product that

$$\left.\begin{aligned} \mathbf{j} \times \mathbf{k} = -\mathbf{k} \times \mathbf{j} = \mathbf{i} \\ \mathbf{k} \times \mathbf{i} = -\mathbf{i} \times \mathbf{k} = \mathbf{j} \\ \mathbf{i} \times \mathbf{j} = -\mathbf{j} \times \mathbf{i} = \mathbf{k} \end{aligned}\right\} \tag{2.21.1}$$

and that

$$\mathbf{i} \times \mathbf{i} = \mathbf{j} \times \mathbf{j} = \mathbf{k} \times \mathbf{k} = \mathbf{0}. \tag{2.21.2}$$

It follows that

$$\begin{aligned} \mathbf{a} \times \mathbf{b} &= \mathbf{i}(a_y b_z - a_z b_y) \\ &\quad + \mathbf{j}(a_z b_x - a_x b_z) \\ &\quad + \mathbf{k}(a_x b_y - a_y b_x) \\ &= \begin{vmatrix} \mathbf{i} & \mathbf{j} & \mathbf{k} \\ a_x & a_y & a_z \\ b_x & b_y & b_z \end{vmatrix} \end{aligned} \tag{2.21.3}$$

If **a** and **b** lie in the xy plane, their vector product is in the z direction.

2.22. *Mechanical applications*

(*a*) **Moment of a force.** Suppose that **F** is a force, **r** is the position vector of some point on its line of action, and **c** is the position vector of some point C. Then $(\mathbf{r} - \mathbf{c}) \times \mathbf{F}$ is the moment of **F** about C. If $\mathbf{F} = (X, Y, Z)$, then the z component of $\mathbf{r} \times \mathbf{F}$ is $xY - yX$, which is the usual expression for the moment of a force in two-dimensional mechanics.

(*b*) **Moment of momentum.** Suppose that a particle of mass m, instantaneously at a point with position vector **r**, is moving with velocity **v**. The moment of momentum, or angular momentum, **H** of the particle about some other point **C** with position vector **c** is

$$\mathbf{H} = (\mathbf{r} - \mathbf{c}) \times m\mathbf{v} \tag{2.22.1}$$

$$\frac{d\mathbf{H}}{dt} = \left(\frac{d\mathbf{r}}{dt} - \frac{d\mathbf{c}}{dt}\right) \times m\mathbf{v} + (\mathbf{r} - \mathbf{c}) \times m\frac{d\mathbf{v}}{dt} \tag{2.22.2}$$

$$= (\mathbf{v} - \mathbf{v}_c) \times m\mathbf{v} + (\mathbf{r} - \mathbf{c}) \times m\boldsymbol{\alpha}$$

$$= (\mathbf{r} - \mathbf{c}) \times \mathbf{F} - m\mathbf{v}_C \times \mathbf{v} \tag{2.22.3}$$

where $\boldsymbol{\alpha}$ is the acceleration of the particle, $\mathbf{F}$ the force acting on it and $\mathbf{v}_C$ the velocity of C. Thus the rate of change of angular momentum of a particle about a point is equal to the moment of the force applied to the particle, provided that the quantity $\mathbf{v}_C \times \mathbf{v}$ vanishes. This is so if C is a fixed point, or if C and the particle are moving parallel to one another.

2.23. *Vector product as an area*

It follows from the definition (2.2.6) together with Fig. 2.23(i) that

$$|\mathbf{a} \times \mathbf{b}| = ab \sin\theta$$
$$= \text{area of parallelogram } OAFB. \tag{2.23.1}$$

Also the direction of $\mathbf{a} \times \mathbf{b}$ is normal to the plane area $OAFB$ and consequently it is possible to regard the quantity $\mathbf{a} \times \mathbf{b}$ as representing

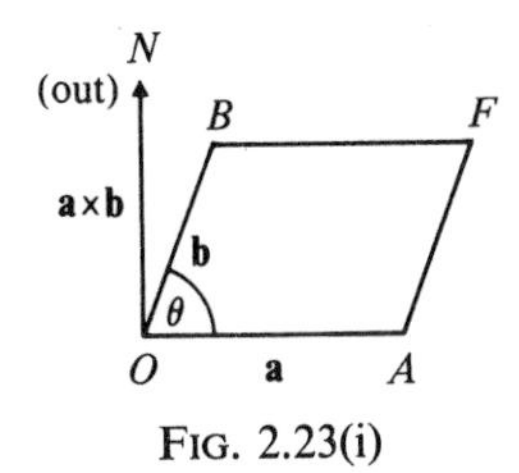

FIG. 2.23(i)

the vector area of the surface $OAFB$. Now any smooth surface may, by the methods of infinitesimal calculus, be represented as the limit of plane rectangles and consequently any infinitesimal surface may be written in the form

$$\delta\boldsymbol{\mathscr{A}} = \mathbf{n}\,\delta\mathscr{A} \tag{2.23.2}$$

where $\delta\boldsymbol{\mathscr{A}}$ is the elementary vector area, $\mathbf{n}$ is the unit normal and $\delta\mathscr{A}$ the surface area. For a closed surface, it is usual to assume that the normal is the outward direction.

2.3. **Scalar triple product**

The scalar triple product of three vectors $\mathbf{a}$, $\mathbf{b}$, $\mathbf{c}$ is defined as $(\mathbf{a} \times \mathbf{b}).\mathbf{c}$. Let $\mathbf{a}$, $\mathbf{b}$, $\mathbf{c}$ be represented by vectors $\overrightarrow{OA}$, $\overrightarrow{OB}$, $\overrightarrow{OC}$ and complete the parallelopiped. The magnitude of $\mathbf{a} \times \mathbf{b}$ is given by the area of the parallelogram $OAFB$ and its direction is along ON which is perpendicular to the plane of the parallelogram.

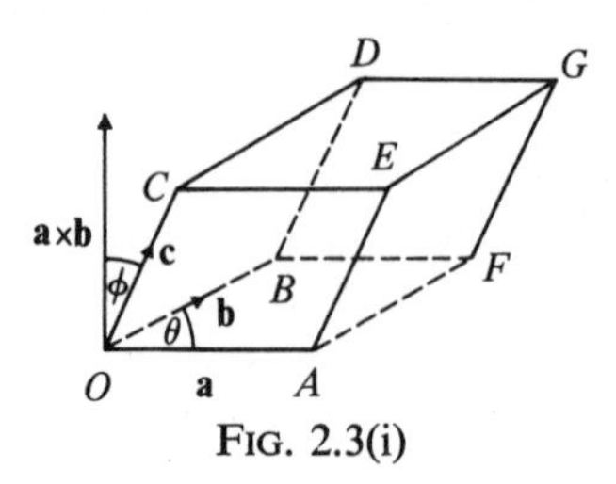

FIG. 2.3(i)

Let OC make an angle ϕ with ON. Then

$$(\mathbf{a} \times \mathbf{b}).\mathbf{c} = (\text{area of } OAFB)\,\mathbf{n}.\mathbf{c} \tag{2.3.1}$$
$$= (\text{area of } OAFB)\,c\cos\phi$$
$$= (\text{area of base of parallelopiped}) \times (\text{height of parallelopiped})$$
$$= \text{volume of parallelopiped.} \tag{2.3.2}$$

It is clear that the same result would follow for $(\mathbf{b} \times \mathbf{c}).\mathbf{a}$ and $(\mathbf{c} \times \mathbf{a}).\mathbf{b}$ when the cyclic order is maintained and also for $\mathbf{a}.(\mathbf{b} \times \mathbf{c})$, $\mathbf{b}.(\mathbf{c} \times \mathbf{a})$ and $\mathbf{c}.(\mathbf{a} \times \mathbf{b})$. Also, because $\mathbf{a} \times \mathbf{b} = -\mathbf{b} \times \mathbf{a}$, it follows that

$$(\mathbf{b} \times \mathbf{a}).\mathbf{c} = (\mathbf{a} \times \mathbf{c}).\mathbf{b} = (\mathbf{c} \times \mathbf{b}).\mathbf{a} = -(\mathbf{a} \times \mathbf{b}).\mathbf{c}. \tag{2.3.4}$$

The loss of cyclic order in the symbols thus changes the sign of the result. On the other hand, the order of the two different kinds of multiplication does not matter by virtue of the equality of, e.g., $(\mathbf{a} \times \mathbf{b}).\mathbf{c}$ and $\mathbf{a}.(\mathbf{b} \times \mathbf{c})$.

It is clear that if $\mathbf{c}$ is coplanar with $\mathbf{a}$ and $\mathbf{b}$, the scalar triple product vanishes because the parallelopiped then collapses and its volume is zero. A particular case of this is when two of the vectors are parallel.

In rectangular cartesian coordinates

$$(\mathbf{a} \times \mathbf{b}).\mathbf{c} = \{(a_y b_z - a_z b_y)\mathbf{i} + (a_z b_x - a_x b_z)\mathbf{j} + (a_x b_y - a_y b_x)\mathbf{k}\}.(c_x\mathbf{i} + c_y\mathbf{j} + c_z\mathbf{k})$$
$$= c_x(a_y b_z - a_z b_y) + c_y(a_z b_x - a_x b_z) + c_z(a_x b_y - a_y b_x)$$
$$= \begin{vmatrix} a_x & a_y & a_z \\ b_x & b_y & b_z \\ c_x & c_y & c_z \end{vmatrix} \tag{2.3.5}$$

This is sometimes written $[\mathbf{abc}]$.

The invariance property of this determinant for cyclic row interchange is equivalent to the invariance property discussed above.

If $\mathbf{n}$ is a unit vector, $\mathbf{n}.(\mathbf{r} \times \mathbf{F})$ represents the component of the moment of $\mathbf{F}$ about O in the direction of $\mathbf{n}$.

2.31. *Reciprocal vectors*

Suppose that $\mathbf{a}_1$, $\mathbf{a}_2$, $\mathbf{a}_3$ are three non-coplanar vectors. Then $(\mathbf{a}_1 \mathbf{a}_2 \mathbf{a}_3) \neq 0$. Then if

$$\begin{aligned} \mathbf{b}_1 &= \frac{\mathbf{a}_2 \times \mathbf{a}_3}{[\mathbf{a}_1 \mathbf{a}_2 \mathbf{a}_3]} \\ \mathbf{b}_2 &= \frac{\mathbf{a}_3 \times \mathbf{a}_1}{[\mathbf{a}_1 \mathbf{a}_2 \mathbf{a}_3]} \\ \mathbf{b}_3 &= \frac{\mathbf{a}_1 \times \mathbf{a}_2}{[\mathbf{a}_1 \mathbf{a}_2 \mathbf{a}_3]} \end{aligned} \tag{2.31.1}$$

and

$$\begin{aligned} \delta_{rs} &= 1 \qquad (r = s) \\ &= 0 \qquad (r = s) \end{aligned} \tag{2.32.2}$$

$$\mathbf{a}_r . \mathbf{b}_s = \delta_{rs}. \tag{2.32.3}$$

Such sets of vectors are termed reciprocal. It is not difficult to see that $[\mathbf{b}_1 \mathbf{b}_2 \mathbf{b}_3] \neq 0$ also, and that

$$\frac{\mathbf{b}_2 \times \mathbf{b}_3}{[\mathbf{b}_1 \mathbf{b}_2 \mathbf{b}_3]} = \mathbf{a}_1, \text{ etc.} \tag{2.32.4}$$

Suppose that $\mathbf{F}$ is an arbitrary vector. Then $\mathbf{F} = F_1 \mathbf{a}_1 + F_2 \mathbf{a}_2 + F_3 \mathbf{a}_3$. This expansion is possible because $\mathbf{a}_1$, $\mathbf{a}_2$, $\mathbf{a}_3$ are non-coplanar.

The value of F_1, F_2, F_3 may be found as follows

$$\begin{aligned} \mathbf{F} . \mathbf{b}_1 &= (F_1 \mathbf{a}_1 + F_2 \mathbf{a}_2 + F_3 \mathbf{a}_3) . \mathbf{b}_1 \\ &= F_1. \end{aligned} \tag{2.32.5}$$

On using (2.32.3) similar results follow for the quantities F_2 and F_3. Thus

$$\mathbf{F} = (\mathbf{F}_1 . \mathbf{b}_1) \mathbf{a}_1 + (\mathbf{F} . \mathbf{b}_2) \mathbf{a}_2 + (\mathbf{F} . \mathbf{b}_3) \mathbf{a}_3. \tag{2.32.6}$$

An alternative proof of this formula will be given in section 2.5. It is not difficult to see that the right-handed orthogonal set of unit vectors $\mathbf{c}_1$, $\mathbf{c}_2$, $\mathbf{c}_3$ is self-reciprocal, that is, it is its own reciprocal.

2.4. The vector triple product

The vector triple product will be defined as $\mathbf{a} \times (\mathbf{b} \times \mathbf{c})$ and it will be shown that

$$\mathbf{a} \times (\mathbf{b} \times \mathbf{c}) = (\mathbf{a}.\mathbf{c})\mathbf{b} - (\mathbf{a}.\mathbf{b})\mathbf{c} \tag{2.4.1}$$

Clearly the position of the brackets is important and $(\mathbf{a} \times \mathbf{b}) \times \mathbf{c}$ will be quite different.

The simplest method of proving the relation (2.4.1) is by resolving into rectangular cartesian coordinates.

$$\begin{aligned}(\mathbf{a} \times (\mathbf{b} \times \mathbf{c}))_x &= a_y(\mathbf{b} \times \mathbf{c})_z - a_z(\mathbf{b} \times \mathbf{c})_y \\ &= a_y(b_x c_y - b_y c_x) - a_z(b_z c_x - b_x c_z) \\ &= (a_x c_x + a_y c_y + a_z c_z) b_x - (a_x b_x + a_y b_y + a_z b_z) c_x \\ &= (\mathbf{a}.\mathbf{c}) b_x - (\mathbf{a}.\mathbf{b}) c_x. \end{aligned} \tag{2.4.2}$$

Similar relations follow for the other two components proving the result (2.4.1). It will be realized that this proof is somewhat unsatisfactory. The relation is one between vectors, and it should not be necessary to resolve these vectors into their rectangular cartesian components in order to prove general relations such as (2.4.1) which are independent of any coordinate system.

A proof which does not have this disadvantage is the following. $\mathbf{a} \times (\mathbf{b} \times \mathbf{c})$ is, by definition, perpendicular to $\mathbf{a}$, and to $\mathbf{b} \times \mathbf{c}$. $\mathbf{b} \times \mathbf{c}$ is by definition perpendicular to both $\mathbf{b}$ and $\mathbf{c}$. It follows that $\mathbf{a} \times (\mathbf{b} \times \mathbf{c})$ is coplanar with $\mathbf{b}$ and $\mathbf{c}$, and so

$$\mathbf{a} \times (\mathbf{b} \times \mathbf{c}) = \beta\mathbf{b} + \gamma\mathbf{c}. \tag{2.4.3}$$

Multiplying scalarly by $\mathbf{a}$

$$0 = \mathbf{a}.[\mathbf{a} \times (\mathbf{b} \times \mathbf{c})] = \beta(\mathbf{b}.\mathbf{a}) + \gamma(\mathbf{c}.\mathbf{a})$$

which determines the ratio of β to γ, and so it follows that

$$\mathbf{a} \times (\mathbf{b} \times \mathbf{c}) = \lambda\{\mathbf{b}(\mathbf{c}.\mathbf{a}) - \mathbf{c}(\mathbf{a}.\mathbf{b})\} \tag{2.4.4}$$

where λ is some scalar to be determined. Now $\mathbf{a}$ can always be assumed to be coplanar with $\mathbf{b}$ and $\mathbf{c}$ since the part normal to the plane defined by $\mathbf{b}$ and $\mathbf{c}$ has the direction of $(\mathbf{b} \times \mathbf{c})$ and $(\mathbf{b} \times \mathbf{c}) \times (\mathbf{b} \times \mathbf{c}) = \mathbf{0}$, $\mathbf{b}.(\mathbf{b} \times \mathbf{c}) = \mathbf{c}.(\mathbf{b} \times \mathbf{c}) = 0$.

Equation (2.4.4) is homogeneous in $\mathbf{a}$, $\mathbf{b}$, $\mathbf{c}$ and dividing by abc, it follows that

$$\hat{\mathbf{a}} \times (\hat{\mathbf{b}} \times \hat{\mathbf{c}}) = \lambda\{\hat{\mathbf{b}}(\hat{\mathbf{c}}.\hat{\mathbf{a}}) - \hat{\mathbf{c}}(\hat{\mathbf{a}}.\hat{\mathbf{b}})\}. \tag{2.4.5}$$

Multiplying scalarly by $\hat{\mathbf{b}}$ and using a property of the scalar triple product as applied to $\hat{\mathbf{b}}$, $\hat{\mathbf{a}}$, $\hat{\mathbf{b}} \times \hat{\mathbf{c}}$

$$(\hat{\mathbf{b}} \times \hat{\mathbf{a}}).(\hat{\mathbf{b}} \times \hat{\mathbf{c}}) = \lambda[\hat{\mathbf{c}}.\hat{\mathbf{a}} - (\hat{\mathbf{c}}.\hat{\mathbf{b}})(\hat{\mathbf{a}}.\hat{\mathbf{b}})] \tag{2.4.6}$$

Because $\hat{\mathbf{a}}$, $\hat{\mathbf{b}}$, $\hat{\mathbf{c}}$ are coplanar the directions of $\hat{\mathbf{b}} \times \hat{\mathbf{a}}$, $\hat{\mathbf{b}} \times \hat{\mathbf{c}}$ are parallel, both being normal to the plane of the three vectors and (if the sense of $\hat{\mathbf{a}}$, $\hat{\mathbf{b}}$, $\hat{\mathbf{c}}$ is as in figure 2.4(i)) in the same sense. Thus

$$(\hat{\mathbf{b}} \times \hat{\mathbf{a}}).(\hat{\mathbf{b}} \times \hat{\mathbf{c}}) = |\hat{\mathbf{b}} \times \hat{\mathbf{a}}||\hat{\mathbf{b}} \times \hat{\mathbf{c}}|$$

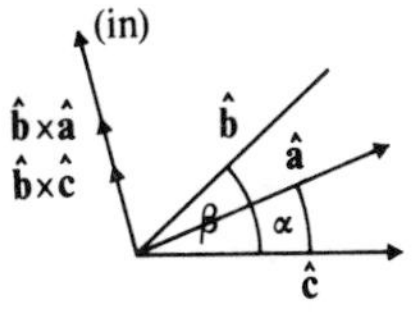

FIG. 2.4(i)

and equation (2.4.6) becomes

$$\begin{aligned}\sin(\beta - \alpha)\sin\beta &= \lambda[\cos\alpha - \cos(\beta - \alpha)\cos\beta] \\ &= \lambda \sin(\beta - \alpha)\sin\beta \qquad (2.4.7)\end{aligned}$$

whence $\lambda = 1$, and so

$$\mathbf{a} \times (\mathbf{b} \times \mathbf{c}) = \mathbf{b}.(\mathbf{c}.\mathbf{a}) - \mathbf{c}(\mathbf{a}.\mathbf{b}).$$

A particular case of interest is when $\mathbf{a} = \mathbf{b}$.

$$\mathbf{a} \times (\mathbf{a} \times \mathbf{c}) = \mathbf{a}(\mathbf{c}.\mathbf{a}) - a^2\mathbf{c} \qquad (2.4.8)$$

and if $\mathbf{a}$ is a unit vector equation (2.4.8) can be written as

$$\mathbf{c} = (\mathbf{a}.\mathbf{c})\mathbf{a} + \mathbf{a} \times (\mathbf{c} \times \mathbf{a}) \qquad (2.4.9)$$

thereby resolving $\mathbf{c}$ into two components, one parallel to the unit vector $\mathbf{a}((\mathbf{a}.\mathbf{c}).\mathbf{a})$, the other perpendicular to it $(\mathbf{a} \times (\mathbf{c} \times \mathbf{a}))$.

2.5. **Quadruple products of vectors**

Quadruple products of vectors may be calculated using the triple products:

$$\begin{aligned}(\mathbf{a} \times \mathbf{b}).(\mathbf{c} \times \mathbf{d}) &= \{(\mathbf{a} \times \mathbf{b}) \times \mathbf{c}\}.\mathbf{d} \\ &= \{(\mathbf{a}.\mathbf{c})\mathbf{b} - (\mathbf{b}.\mathbf{c})\mathbf{a}\}\mathbf{d} \\ &= \{(\mathbf{a}.\mathbf{c})(\mathbf{b}.\mathbf{d}) - (\mathbf{b}.\mathbf{c})(\mathbf{a}.\mathbf{d})\}. \qquad (2.5.1)\end{aligned}$$

This is known as Lagrange's Identity.

If $\mathbf{c} = \mathbf{a}$, $\mathbf{d} = \mathbf{b}$, Lagrange's Identity assumes the form

$$(\mathbf{a} \times \mathbf{b}).(\mathbf{a} \times \mathbf{b}) = a^2 b^2 - (\mathbf{a}.\mathbf{b})(\mathbf{a}.\mathbf{b}) \qquad (2.5.2)$$

which corresponds to the trigonometrical result $\sin^2\theta = 1 - \cos^2\theta$.

Another useful relationship is derived as follows

$$\begin{aligned}(\mathbf{a} \times \mathbf{b}) \times (\mathbf{c} \times \mathbf{d}) &= -\mathbf{a}\{\mathbf{b}.(\mathbf{c} \times \mathbf{d})\} + \mathbf{b}\{\mathbf{a}.(\mathbf{c} \times \mathbf{d})\} \qquad (2.5.3a) \\ &= \mathbf{c}\{\mathbf{d}.(\mathbf{a} \times \mathbf{b})\} - \mathbf{d}\{\mathbf{c}.(\mathbf{a} \times \mathbf{b})\} \qquad (2.5.3b)\end{aligned}$$

by use of the triple vector product formulae as applied to $\mathbf{a}$, $\mathbf{b}$, $\mathbf{c} \times \mathbf{d}$ and $\mathbf{a} \times \mathbf{b}$, $\mathbf{c}$, $\mathbf{d}$ respectively. Thus $(\mathbf{a} \times \mathbf{b}) \times (\mathbf{c} \times \mathbf{d})$ lies in the plane of $\mathbf{a}$ and $\mathbf{b}$, and also in the plane of $\mathbf{c}$ and $\mathbf{d}$. Thus if non-zero, it lies in the direction of the line of intersection of these two planes. Furthermore the expansions on the right-hand sides of equations (2.5.3) are equal and so

$$\mathbf{d}\{(\mathbf{a} \times \mathbf{b}).\mathbf{c}\} = \mathbf{a}\{(\mathbf{b} \times \mathbf{c}).\mathbf{d}\} + \mathbf{b}\{(\mathbf{c} \times \mathbf{a}).\mathbf{d}\} + \mathbf{c}\{(\mathbf{a} \times \mathbf{b}).\mathbf{d}\}. \quad (2.5.4)$$

Equation (2.5.4) gives, if $(\mathbf{a} \times \mathbf{b}).\mathbf{c}$ is non-zero the expansion of $\mathbf{d}$ in terms of three arbitrary vectors $\mathbf{a}$, $\mathbf{b}$, $\mathbf{c}$. The condition of $(\mathbf{a} \times \mathbf{b}).\mathbf{c}$ not vanishing is satisfied if $\mathbf{a}$, $\mathbf{b}$, $\mathbf{c}$ are not coplanar. This representation is clearly unique. Equation (2.5.4) could also have been proved by writing

$$\mathbf{d} = \alpha\mathbf{a} + \beta\mathbf{b} + \gamma\mathbf{c}$$

and making successive scalar multiplications by $\mathbf{b} \times \mathbf{c}$, $\mathbf{c} \times \mathbf{a}$, $\mathbf{a} \times \mathbf{b}$, the process being similar to the proof of (2.3.7).

2.6. **Solution of vector equations**

(*a*) Let $\mathbf{x}$ be an unknown vector which obeys the relation

$$\mathbf{x}.\mathbf{c} = p \quad (2.6.1)$$

$\mathbf{c}$ and p being known.

If $\mathbf{x}_1$, $\mathbf{x}_2$ are any two solutions, then

$$(\mathbf{x}_1 - \mathbf{x}_2).\mathbf{c} = 0. \quad (2.6.2)$$

In other words, the solution is indefinite by a vector which is perpendicular to $\mathbf{c}$. Such a vector is $\mathbf{c} \times \mathbf{f}$ where f is arbitrary. A particular solution is clearly

$$\mathbf{x} = \frac{p\mathbf{c}}{c^2} \quad (2.6.3)$$

and the general solution is

$$\mathbf{x} = \frac{p\mathbf{c}}{c^2} + \mathbf{c} \times \mathbf{f}. \quad (2.6.4)$$

If $\mathbf{c}$ is a unit vector, an interpretation is given by section 2.13(a).

(*b*) Let $\mathbf{x}$ be an unknown vector which obeys the relation

$$\mathbf{x} \times \mathbf{a} = \mathbf{b}. \quad (2.6.5)$$

Because of the properties of the vector product, $\mathbf{b}$ must be perpendicular to $\mathbf{a}$. If it is not, the scalar equations which are resolutes along the coordinate axes of equation (2.6.5) are inconsistent, and so for a solution to be possible

$$\mathbf{a}.\mathbf{b} = 0.$$

If $\mathbf{x}_1$, $\mathbf{x}_2$ are any two solutions

$$(\mathbf{x}_1 - \mathbf{x}_2) \times \mathbf{a} = \mathbf{0} \tag{2.6.6}$$

and so the solution of (2.6.5) is indefinite by a vector which is parallel to $\mathbf{a}$. Such a vector is $\lambda\mathbf{a}$ where λ is a scalar. A particular solution may be found as follows. Multiply equation (2.6.5) vectorially by $\mathbf{a}$ and it follows that

$$\mathbf{a} \times \mathbf{b} = \mathbf{a} \times (\mathbf{x} \times \mathbf{a}) = \mathbf{x}a^2 - \mathbf{a}(\mathbf{x}.\mathbf{a}). \tag{2.6.7}$$

Assume that $\mathbf{x}$ is such that $\mathbf{x}.\mathbf{a}$ vanishes, then equation (2.6.7) reduces to $\mathbf{x}a^2 = (\mathbf{a} \times \mathbf{b})$ and for this value of $\mathbf{x}$, $\mathbf{x}.\mathbf{a}$ does vanish. Thus $(\mathbf{a} \times \mathbf{b})/a^2$ is in fact a solution and the general solution is

$$\mathbf{x} = \frac{\mathbf{a} \times \mathbf{b}}{a^2} + \lambda\mathbf{a}. \tag{2.6.8}$$

Suppose now that $\mathbf{x}$ obeys both equation (2.6.1) and equation (2.6.5). (This is in fact equivalent to four equations for the three unknown components of $\mathbf{x}$ and the solution will only be possible under certain special conditions.)

From equations (2.6.4) and (2.6.8)

$$\frac{p\mathbf{c}}{c^2} + \mathbf{c} \times \mathbf{f} = \frac{\mathbf{a} \times \mathbf{b}}{a^2} + \lambda\mathbf{a} \tag{2.6.9}$$

and either $\mathbf{f}$ or λ must be determined. It is easier to determine λ, this being a scalar. Multiply scalarly by $\mathbf{c}$, and it follows that

$$p = \frac{(\mathbf{a} \times \mathbf{b}).\mathbf{c}}{a^2} + \lambda\mathbf{a}.\mathbf{c}$$

whence

$$\lambda = \frac{1}{\mathbf{a}.\mathbf{c}}\left[p - \frac{(\mathbf{a} \times \mathbf{b}).\mathbf{c}}{a^2}\right] \tag{2.6.10}$$

provided that $\mathbf{a}.\mathbf{c}$ is non-zero. If $\mathbf{a}.\mathbf{c}$ is zero, no solution is possible except in the special case of $pa^2 = (\mathbf{a} \times \mathbf{b}).\mathbf{c}$ when λ is indeterminate.

2.7. **Representation of a rotation of a rigid body by a vector**

It is intuitively obvious that any displacement of a rigid body may be regarded as a pure translation by which some arbitrary point fixed in the body moves from its initial to its final position and a rotation about an axis through this point, the order in which the operations are performed being immaterial (H. Lamb, *Higher Mechanics*, Cambridge Univ. Press (1929), p. 8). Thus in obtaining formulae concerning the motion of a rigid body, it is possible to discuss the translation and rotation separately. The vectorial representation of the translation is trivial, the position vector of all points merely altering by the vector **a** defining the translation. A relation in vector form may also be found to represent the rotation of a rigid body. As only displacements are of interest, it is only necessary to consider the initial and final positions of the particles of the rigid body. The intermediate positions and the velocities are at present irrelevant.

Consider a rigid body which has rotated in the angle θ in the positive sense about an axis passing through some fixed point O, the unit vector along the axis being **u**. It is assumed that $0 < \theta < \pi$, the extension to other angles being straightforward. The meaning of the term positive is as follows. If the vector **u** points vertically upwards, a positive rotation would move the extremity of a line which lies initially in an eastwards direction from O towards the north.

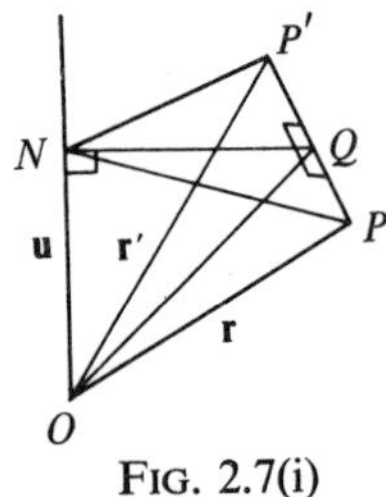

FIG. 2.7(i)

Let P, P' be the initial and final positions of some point fixed in the rigid body and let $\mathbf{r}$, $\mathbf{r}'$ be the position vectors of P, P' with respect to O. Let Q be the mid point of PP'. Let N be the foot of the perpendicular from Q onto the axis. The angles ONQ, PQN, $P'QN$ are all right angles and the angles PNQ, $P'NQ$ are $\theta/2$. The rotation through an angle θ moves the particles in the plane ONP to corresponding positions in the plane ONP'. The three unit vectors $\overrightarrow{NQ}$, $\overrightarrow{QP}'$, **u** in that order, form a right-handed orthogonal triad and so

$$\mathbf{u} \times \hat{\overrightarrow{NQ}} = \hat{\overrightarrow{QP}}'. \tag{2.7.1}$$

This may be rewritten as

$$2\frac{QP'}{NQ}\mathbf{u} \times \overrightarrow{NQ} = 2\overrightarrow{QP}' = \overrightarrow{PR}' \tag{2.7.2}$$

whence

$$\begin{aligned}\overrightarrow{PP}' &= = \tan\frac{\theta}{2}\mathbf{u} \times (\overrightarrow{OQ} - \overrightarrow{ON}) \\ &= 2\tan\frac{\theta}{2}\mathbf{u} \times \overrightarrow{OQ}.\end{aligned}$$

This follows because $\overrightarrow{ON}$ is parallel to $\mathbf{u}$. Thus

$$r' - r = \mathbf{\Omega} \times (\mathbf{r} + \mathbf{r}') \tag{2.7.3}$$

where

$$\mathbf{\Omega} = \tan\frac{\theta}{2}\mathbf{u}. \tag{2.7.4}$$

The quantity $\mathbf{\Omega}$ may be said in a sense to represent the rotation.

Equation (2.7.3) is an equation for $\mathbf{r}'$ in terms of $\mathbf{r}$, and solving by the processes of worked example 11, it follows that

$$\mathbf{r}' = \frac{(1 - \Omega^2)\mathbf{r} + 2(\mathbf{\Omega}.\mathbf{r})\mathbf{\Omega} + 2\mathbf{\Omega} \times \mathbf{r}}{1 + \Omega^2}. \tag{2.7.5}$$

This procedure, however, is inapplicable when θ is an odd multiple of π. In this case the formula (2.7.5) becomes, on taking the limit as θ tends to infinity,

$$\mathbf{r}' = 2(\mathbf{u}.\mathbf{r})\mathbf{u} - \mathbf{r} \tag{2.7.6}$$

and it can easily be seen that this follows from figure 2.7(i) when $\theta = \pi$ when N and Q coincide.

Suppose now that a further rotation represented by $\mathbf{\Omega}'$ be applied to the body. Let the particle at P' move to P'' which has position vector $\mathbf{r}''$ with respect to O.

Then

$$\mathbf{r}' - \mathbf{r} = \mathbf{\Omega} \times (\mathbf{r}' + \mathbf{r}) \tag{2.7.7a}$$

$$\mathbf{r}'' - \mathbf{r}' = \mathbf{\Omega}' \times (\mathbf{r}'' + \mathbf{r}') \tag{2.7.7b}$$

whence

$$\mathbf{\Omega}.\mathbf{r}' = \mathbf{\Omega}.\mathbf{r} \tag{2.7.8a}$$

and

$$\mathbf{\Omega}'.\mathbf{r}'' = \mathbf{\Omega}.\mathbf{r}'. \tag{2.7.8b}$$

It follows from equations (2.7.7) that

$$\begin{aligned}\mathbf{r}'' - \mathbf{r} &= \mathbf{\Omega} \times (\mathbf{r}' + \mathbf{r}) + \mathbf{\Omega}' \times (\mathbf{r}'' + \mathbf{r}') \\ &= (\mathbf{\Omega} + \mathbf{\Omega}') \times (\mathbf{r}'' + \mathbf{r}) + \mathbf{\Omega} \times (\mathbf{r}' - \mathbf{r}'') + \mathbf{\Omega}' \times (\mathbf{r}' - \mathbf{r}). \\ &= (\mathbf{\Omega} + \mathbf{\Omega}') \times (\mathbf{r}'' + \mathbf{r}) - \mathbf{\Omega} \times \{\mathbf{\Omega}' \times (\mathbf{r}'' + \mathbf{r})\} \\ &\quad + \mathbf{\Omega}' \times \{\mathbf{\Omega} \times (\mathbf{r}' + \mathbf{r})\}.\end{aligned} \tag{2.7.9}$$

This last line may be expanded giving

$$\begin{aligned}\mathbf{r}'' - \mathbf{r} &= (\mathbf{\Omega} + \mathbf{\Omega}') \times (\mathbf{r}'' + \mathbf{r}) + (\mathbf{\Omega}'.\mathbf{\Omega})(\mathbf{r}'' - \mathbf{r}) + \mathbf{\Omega}\{\mathbf{\Omega}'.(\mathbf{r}' + \mathbf{r})\} \\ &\quad - \mathbf{\Omega}'\{\mathbf{\Omega}.(\mathbf{r}'' + \mathbf{r}')\} \\ &= (\mathbf{\Omega} + \mathbf{\Omega}') \times (\mathbf{r}'' + \mathbf{r}) + (\mathbf{\Omega}'.\mathbf{\Omega})(\mathbf{r}'' - \mathbf{r}) + \mathbf{\Omega}\{\mathbf{\Omega}'.(\mathbf{r}'' + \mathbf{r})\} \\ &\quad - \mathbf{\Omega}'\{\mathbf{\Omega}.(\mathbf{r}'' + \mathbf{r})\}.\end{aligned} \tag{2.7.10}$$

This last stage follows by virtue of equation (2.7.8).

Thus $(\mathbf{r}'' - \mathbf{r})(1 - \boldsymbol{\Omega}.\boldsymbol{\Omega}') = (\boldsymbol{\Omega} + \boldsymbol{\Omega}' + \boldsymbol{\Omega}' \times \boldsymbol{\Omega}) \times (\mathbf{r}'' + \mathbf{r})$ and it follows that, if

$$\boldsymbol{\Omega}^* = \frac{\boldsymbol{\Omega} + \boldsymbol{\Omega}' + \boldsymbol{\Omega}' \times \boldsymbol{\Omega}}{1 - \boldsymbol{\Omega}.\boldsymbol{\Omega}'} \tag{2.7.11}$$

$$\mathbf{r}'' - \mathbf{r} = \boldsymbol{\Omega}^* \times (\mathbf{r}'' + \mathbf{r}). \tag{2.7.12}$$

Thus the resultant of two rotations represented by $\boldsymbol{\Omega}$, $\boldsymbol{\Omega}'$ in that order is a rotation represented by $\boldsymbol{\Omega}^*$. It is immediately obvious that the expression for $\boldsymbol{\Omega}^*$ does not contain $\boldsymbol{\Omega}$, $\boldsymbol{\Omega}'$ symmetrically and so the effect of a rotation represented by $\boldsymbol{\Omega}$ followed by a rotation represented by $\boldsymbol{\Omega}'$ is different to that in which the rotations take place in a different order. Thus although a finite rotation may, in a sense, be represented by a vector quantity $\boldsymbol{\Omega}$, the resultant of two finite rotations represented by $\boldsymbol{\Omega}$, $\boldsymbol{\Omega}'$ is not $\boldsymbol{\Omega} + \boldsymbol{\Omega}'$.

2.71. *Elementary rotations*

Suppose that a body is subjected to an elementary rotation of angle $\delta\theta$ about an axis associated with unit vector $\mathbf{u}$, and that as a result the particle which was originally at the point with position vector $\mathbf{r}$ moves to the point with position vector $\mathbf{r} + \delta\mathbf{r}$, then, from equation (2.7.3)

$$\delta\mathbf{r} = 2\tan\frac{\delta\theta}{2}\,\mathbf{u} \times \left\{\mathbf{r} + \frac{\delta\mathbf{r}}{2}\right\}. \tag{2.71.1}$$

Taking infinitesimals of the first order only, this becomes

$$\delta\mathbf{r} = \mathbf{u}\,\delta\theta \times \mathbf{r}. \tag{2.71.2}$$

Suppose now that there are two elementary rotations, the first being represented by $\mathbf{u}\,\delta\theta$ and the second by $\mathbf{u}'\,\delta\theta'$. The resultant rotation is represented by $\boldsymbol{\Omega}^*$, where

$$\boldsymbol{\Omega}^* = \frac{\mathbf{u}\,\delta\theta + \mathbf{u}'\,\delta\theta' + (\mathbf{u}'\,\delta\theta') \times (\mathbf{u}\,\delta\theta)}{1 - \mathbf{u}.\mathbf{u}'\,\delta\theta\,\delta\theta'} \tag{2.71.3}$$

$$= \mathbf{u}\,\delta\theta + \mathbf{u}'\,\delta\theta' \tag{2.71.4}$$

to the first order, and this is independent of the order in which the elementary rotations take place. Thus elementary rotations may be added as vectors, the resultant of two elementary rotations represented by $\mathbf{u}\,\delta\theta$, $\mathbf{u}'\,\delta\theta'$ being an elementary rotation represented by $\mathbf{u}''\,\delta\theta''$, where

$$\mathbf{u}''\,\delta\theta'' = \mathbf{u}\,\delta\theta + \mathbf{u}'\,\delta\theta'. \tag{2.71.5}$$

2.72. *Angular velocity*

Suppose that a rotation represented by $\mathbf{u}\,\delta\theta$ takes place in time δt. The angular velocity of the body is defined to be

$$\boldsymbol{\omega} = \lim_{\delta t \to 0} \mathbf{u}\frac{\delta\theta}{\delta t} = \mathbf{u}\dot{\theta}. \tag{2.72.1}$$

If a body is subject to two angular velocities $\boldsymbol{\omega}$, $\boldsymbol{\omega}'$ simultaneously, it follows, on dividing equation (2.71.4) by δt and taking the limit, that the resultant angular velocity $\boldsymbol{\omega}''$ is given by

$$\boldsymbol{\omega}'' = \boldsymbol{\omega} + \boldsymbol{\omega}'. \tag{2.72.2}$$

Dividing equation (2.71.2) by δt and taking the limit as δt tends to zero

$$\frac{\delta \mathbf{r}}{\delta t} = \mathbf{u}\frac{\delta\theta}{\delta t} \times \mathbf{r}$$

and

$$\mathbf{v} = \boldsymbol{\omega} \times \mathbf{r} \tag{2.72.3}$$

$\mathbf{v}$ being the velocity of the particle with position vector $\mathbf{r}$.

2.73. *Rotating frame of reference*

It is clear from what has gone before that, if some vector $\mathbf{A}$ is fixed in the rotating body, the arguments which have been applied to a position vector $\mathbf{r}$, apply equally well to $\mathbf{A}$ and the change in $\mathbf{A}$, when there is an elementary rotation is given by

$$\mathbf{u}\,\delta\theta \times \mathbf{A} = \boldsymbol{\omega}\,\delta t \times \mathbf{A}. \tag{2.73.1}$$

If now, the vector $\mathbf{A}$ is measured in a frame (which is of course equivalent to a rigid body) rotating with angular velocity $\boldsymbol{\omega}$, and varies with time also, the actual change in $\mathbf{A}$ in a time δt is composed of two elements. Firstly, the change due to the time at which $\mathbf{A}$ is considered is

$$\frac{\partial \mathbf{A}}{\partial t}\,\delta t \tag{2.73.2}$$

where $\partial \mathbf{A}/\partial t$ is the partial derivative of $\mathbf{A}$ with respect to time. Secondly the change due to the rotation of the frame of reference is, as indicated above

$$\boldsymbol{\omega}\,\delta t \times \mathbf{A} \tag{2.73.3}$$

and so the total change in $\mathbf{A}$ in the fixed frame

$$\delta \mathbf{A} = \frac{\partial \mathbf{A}}{\partial t}\,\delta t + \boldsymbol{\omega}\,\delta t \times \mathbf{A}$$

and

$$\frac{d\mathbf{A}}{dt} = \frac{\partial \mathbf{A}}{dt} + \boldsymbol{\omega} \times \mathbf{A}. \tag{2.73.4}$$

Clearly

$$\frac{d\boldsymbol{\omega}}{dt} = \frac{\partial \boldsymbol{\omega}}{\partial t} \tag{2.73.5}$$

and the rate of change of $\boldsymbol{\omega}$ is the same in the fixed and rotating frames.

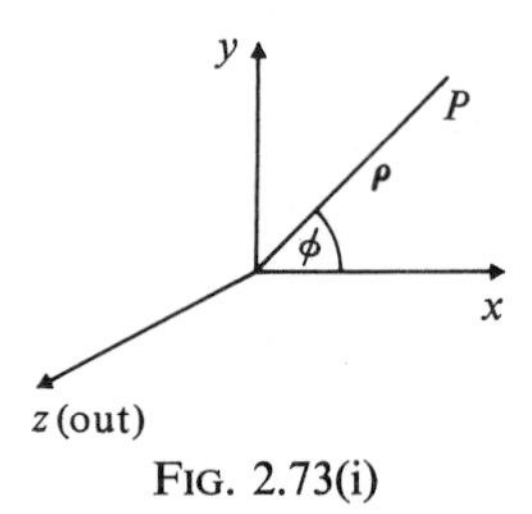

FIG. 2.73(i)

An example of the application of this formula is the calculation of velocity and acceleration formulae in polar coordinates.

Here the position vector $\boldsymbol{\rho}$ may be regarded as being fixed in a frame which rotates with angular velocity $\dot{\phi}\mathbf{k}$

$$\mathbf{v} = \frac{d\boldsymbol{\rho}}{dt} = \frac{\partial \boldsymbol{\rho}}{\partial t} + \dot{\phi}\mathbf{k} \times \boldsymbol{\rho}. \tag{2.73.6}$$

Now $\boldsymbol{\rho} = \rho\hat{\boldsymbol{\rho}}$ and $\hat{\boldsymbol{\rho}}$ does not depend explicitly upon time *in the rotating frame*, and so

$$\mathbf{v} = \frac{\partial \rho}{\partial t}\hat{\boldsymbol{\rho}} + \dot{\phi}\mathbf{k} \times \rho\hat{\boldsymbol{\rho}} \tag{2.73.7}$$

$$= \dot{\rho}\hat{\boldsymbol{\rho}} + \rho\dot{\phi}\hat{\boldsymbol{\varphi}} \tag{2.73.8}$$

which is a result that has been obtained previously. The acceleration can be obtained in a similar manner. See worked example 8 in chapter 1.

2.8. **Axial and polar vectors**

It is not difficult to see that there are in fact two different types of vector which may be characterized by, for example, displacement on the one hand, and angular velocity on the other. Displacement can, in effect, be represented by a directed segment with an arrowhead.

One moves from A to B and the representation is the vector starting at A and finishing at B. On the other hand, angular velocity about an axis is represented by a vector, the direction of which is conventional. There is no reason whatsoever for a rotation about the axis in a certain sense, to be represented by a vector parallel to the axis in a specific sense. It could equally well be represented by a vector, parallel to the axis, but in the opposite sense, the sense being purely a matter of convention. Similarly velocity is a quantity, the vector of which has an intrinsic sense, but the vector associated with the area of an element of surface has only a conventional sense. Vectors of the first kind are termed polar vectors (sometimes true vectors) and of the second kind axial vectors (sometimes pseudo-vectors).

Briefly, the distinction is as follows. Polar vectors can be represented by directed lines, the sense of which is quite definite. Consequently, under an inversion of the coordinate system through the origin when $-x$, $-y$, $-z$ replace, x, y, z respectively the components of polar vectors change sign. On the other hand, axial vectors are represented by an axis together with a magnitude and a sense of rotation about the axis. The sense of rotation is clear. It is the sense of the vector parallel to the axis which represents the rotation about it which is purely a matter of convention. Conventially an anticlockwise rotation of the hand of a clock is represented by a vector normal to the clock face, in the outward sense – it would be equally easy to take a convention that it be in an inward sense. It is the anticlockwise rotation that has physical significance, not the conventional vector representing it. Some convention is implicit such as that of the right-handed screw. In an inversion of the coordinate system through the origin, the rectangular components of axial vectors do not change sign. What happens is that a left-handed screw or frame of reference replaces a right-handed screw or frame of reference. This is because an inversion through the origin creates a left-handed system from a right-handed system.

Consider two polar vectors $\mathbf{A}$, $\mathbf{B}$. In the inverse coordinate system, quantities in which will be primed

$$A'_x = -A_x, \text{ etc.} \tag{2.8.1}$$

Let

$$\mathbf{C} = \mathbf{A} \times \mathbf{B}. \tag{2.8.2}$$

$$\begin{aligned} C_z = (\mathbf{A} \times \mathbf{B})_z &= A_x B_y - A_y B_x \\ &= A'_x B'_y - A'_y B'_x = (\mathbf{A}' \times \mathbf{B}')_z. \end{aligned} \tag{2.8.3}$$

Thus the components of the vector representing a vector product of two polar vectors does not change sign under an inversion of coordinates through the origin and is thus an axial vector. Similarly, it may be shown that the vector product of two axial vectors is an axial vector and that the vector product of an axial vector and a polar vector is a polar vector (e.g., $\mathbf{v} = \boldsymbol{\omega} \times \mathbf{r}$).

2.81. *Scalars and pseudo-scalars*

The scalar product of two axial vectors, or of two polar vectors, is invariant under the inversion of coordinates through the origin. On the other hand, the scalar product of an axial vector and a polar vector, changes sign and the resultant is therefore termed a pseudo-scalar. An example of such a quantity is the scalar triple product of the three polar vectors **A**, **B**, **C** as $(\mathbf{A} \times \mathbf{B}).\mathbf{C}$ is the scalar product of the pseudo-vector $\mathbf{A} \times \mathbf{B}$ and the vector **C**. A pseudo-scalar is in fact a scalar whose sign is determined by convention.

2.9. **Orthogonal curvilinear coordinates**

Suppose that the rectangular coordinates of any point P with position vector $\mathbf{r}$ can be expressed in terms of three quantities u_1, u_2, u_3 so that

$$\left.\begin{aligned} x &= x(u_1, u_2, u_3) \\ y &= y(u_1, u_2, u_3) \\ z &= z(u_1, u_2, u_3) \end{aligned}\right\} \qquad (2.9.1)$$

that is, $\mathbf{r} = \mathbf{r}(u_1, u_2, u_3)$.

Then equations (2.9.1) may be solved so that

$$\begin{aligned} u_m &= u_m(x, y, z) \\ &= u_m(\mathbf{r}) \qquad (m = 1, 2, 3). \end{aligned} \qquad (2.9.2)$$

That is to each point there corresponds a set of values u_m and the point is the point of intersection of the three surfaces $u_m(\mathbf{r}) = \text{constant}$.

The values of the u_m determine $\mathbf{r}$ and so may be regarded as its coordinates in some coordinate system. The three surfaces form coordinate surfaces, their three curves of intersection are the coordinate lines, and the tangents to these curves at the point P form the axes of coordinates. The directions of these axes vary from point to point and so the unit vectors associated with them are not constant.

It is most convenient to consider in practice only systems of coordinates such that, at every point, the coordinate axes are mutually perpendicular. That is, if P, P_1, P_2, P_3, Q are points with position vectors given respectively by

$$\left.\begin{aligned}\mathbf{r} &= \mathbf{r}(u_1, u_2, u_3)\\ \mathbf{r} + \delta\mathbf{r}_1 &= \mathbf{r}(u_1 + \delta u_1, u_2, u_3)\\ \mathbf{r} + \delta\mathbf{r}_2 &= \mathbf{r}(u_1, u_2 + \delta u_2, u_3)\\ \mathbf{r} + \delta\mathbf{r}_3 &= \mathbf{r}(u_1, u_2, u_3 + \delta u_3)\\ \mathbf{r} + \delta\mathbf{r} &= \mathbf{r}(u_1 + \delta u_1, u_2 + \delta u_2, u_3 + \delta u_3)\end{aligned}\right\} \tag{2.9.3}$$

$\delta\mathbf{r}_1$, $\delta\mathbf{r}_2$, $\delta\mathbf{r}_3$ are mutually orthogonal.

Now

$$\begin{aligned}\delta\mathbf{r}_m &= \frac{\partial r}{\partial u_m}\,\delta u_m\\ &= |\delta\mathbf{r}_m|\,\mathbf{i}_m\end{aligned} \tag{2.9.4}$$

where $\mathbf{i}_m$ is the associated unit vector. Because of the perpendicular property of the $\delta\mathbf{r}_m$, the $\mathbf{i}_m$ are mutually perpendicular and they will be ordered so that $\mathbf{i}_1$, $\mathbf{i}_2$, $\mathbf{i}_3$ in that order form a right-handed set.

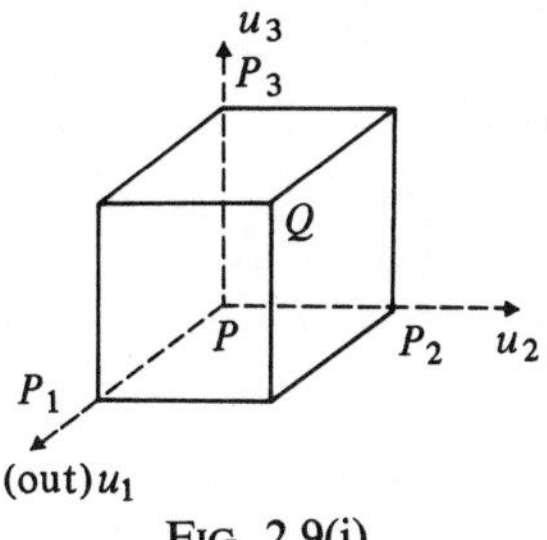

FIG. 2.9(i)

It follows that

$$\mathbf{i}_1 \times \mathbf{i}_2 = \mathbf{i}_3, \mathbf{i}_1 . \mathbf{i}_2 = 0, \text{ etc.} \tag{2.9.5}$$

It is convenient to write

$$|\delta\mathbf{r}_m| = \delta s_m = h_m\,\delta u_m \tag{2.9.6}$$

where

$$h_m = \left|\frac{\partial\mathbf{r}}{\partial u_m}\right| \tag{2.9.7}$$

Thus

$$\begin{aligned}\delta\mathbf{r} &= \delta\mathbf{r}_1 + \delta\mathbf{r}_2 + \delta\mathbf{r}_3\\ &= \mathbf{i}_1 h_1\,\delta u_1 + \mathbf{i}_2 h_2\,\delta u_2 + \mathbf{i}_3 h_3\,\delta u_3\end{aligned} \tag{2.9.8}$$

and

$$\begin{aligned}\delta s^2 &= \delta s_1^2 + \delta s_2^2 + \delta s_3^2\\ &= h_1^2\,\delta u_1^2 + h_2^2\,\delta u_2^2 + h_3^2\,\delta u_3^2.\end{aligned} \tag{2.9.9}$$

It is this last equation which gives the test for whether a system of coordinates is orthogonal. If the system is orthogonal, then the quadratic form for an element of distance is given by an expression of the form (2.9.9), cross-product terms such as those which involve $\delta u_1\,\delta u_2$ vanishing. It is to be noted that, in general the h_m coefficients are functions of u_1, u_2, u_3. It is easy to see that the volume of the elementary cuboid in figure 2.9(i) is

$$(\delta\mathbf{r}_1 \times \delta\mathbf{r}_2).\delta\mathbf{r}_3 = h_1\,h_2\,h_3\,\delta u_1\,\delta u_2\,\delta u_3. \tag{2.9.10}$$

It may be seen that for the cylindrical coordinate system

$$\delta s^2 = \delta\rho^2 + \rho^2\,\delta\phi^2 + \delta z^2 \tag{2.9.11}$$

and that for the spherical coordinate system

$$\delta s^2 = \delta r^2 + r^2\,\delta\theta^2 + r^2\sin^2\theta\,\delta\phi^2. \tag{2.9.12}$$

The corresponding elementary volumes are $\rho\,\delta\rho\,\delta\phi\,\delta z$ and $r^2\sin\theta\,\delta r\,\delta\theta\,\delta\phi$, respectively.

2.10. **Complex numbers**

Complex numbers of the form $z = x + iy$ where $i^2 = -1$ have a certain affinity with two-dimensional vectors.

If $z = z_1 + z_2$, then with an obvious notation $x = x_1 + x_2$, $y = y_1 + y_2$.

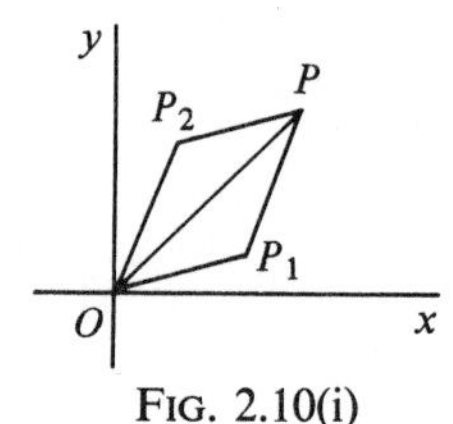

FIG. 2.10(i)

This is strictly analogous with the vector formula

$$\overrightarrow{OP} = \overrightarrow{OP}_1 + \overrightarrow{OP}_2 \tag{2.10.1}$$

where

$$\overrightarrow{OP} = x\mathbf{i} + y\mathbf{j}, \text{ etc.} \tag{2.10.2}$$

An alternative expression for a complex number is

$$z = \rho\cos\phi + i\rho\sin\phi, \text{ etc} \tag{2.10.3}$$

$$\rho = OP, \qquad \phi = \angle xOP. \tag{2.10.4}$$

It is well known that

$$z_1\,z_2 = \rho_1\,\rho_2\cos(\phi_1 + \phi_2) + i\rho_1\,\rho_2\sin(\phi_1 + \phi_2) \tag{2.10.5a}$$

and

$$\frac{z_1}{z_2} = \frac{\rho_1}{\rho_2}\cos(\phi_1 - \phi_2) + i\frac{\rho_1}{\rho_2}\sin(\phi_1 - \phi_2) \qquad (2.10.5b)$$

Clearly it is only necessary to discuss multiplication.

Consider now the quantity

$$(\cos\theta + \sin\theta\mathbf{k}\times)(x\mathbf{i} + y\mathbf{j})$$
$$(\cos\theta + \sin\theta\mathbf{k}\times)(\mathbf{i}\cos\phi + \mathbf{j}\sin\phi)\rho$$
$$= \mathbf{i}(\cos\phi\cos\theta - \sin\phi\sin\theta)\rho + \mathbf{j}(\cos\phi\sin\theta + \sin\phi\sin\theta)\rho$$
$$= \mathbf{i}\rho\cos(\phi + \theta) + \mathbf{j}\rho\sin(\phi + \theta). \qquad (2.10.6)$$

Pre-multiplication of $x\mathbf{i} + y\mathbf{j}$ by $\cos\theta + \sin\theta\mathbf{k}\times$ is equivalent to turning the vector through an angle θ. This corresponds to the statement

$$e^{i\phi}\rho\, e^{i\theta} = \rho\, e^{i(\phi+\theta)}. \qquad (2.10.7)$$

It follows from this that the relation

$$z_1 z_2 = \rho_1\rho_2 e^{i(\phi_1+\phi_2)}$$
$$= \rho_1\rho_2\{\cos(\phi_1 + \phi_2) + i\sin(\phi_1 + \phi_2)\}. \qquad (2.10.8)$$

This corresponds to

$$(\rho_1\cos\phi_1 + \rho_1\sin\phi_1\mathbf{k}\times)(\rho_2\cos\phi_2\mathbf{i} + \rho_2\sin\phi_2\mathbf{j})$$
$$= (\rho_1\cos\phi_1 + \rho_1\sin\phi_1\mathbf{k}\times)(\rho_2\cos\phi_2 + \rho_2\sin\phi_2\mathbf{k}\times)\mathbf{i}.$$

Also, if $z = z_1 + z_2$

$$(x + y\mathbf{k}\times)\mathbf{i} = \{(x_1 + x_2) + (y_1 + y_2)\mathbf{k}\times\}\mathbf{i}.$$

Thus, in a sense it is possible to represent the complex number $x + iy$ by an operator $x + y\mathbf{k}\times$, the operator always acting upon $\mathbf{i}$.

Let α, β be positive numbers

$$(\alpha z_1)(\beta z_2) = \alpha\rho_1 e^{i\phi_1}\beta\rho_2 e^{i\phi_2}$$
$$= \alpha\beta\rho_1 e^{i\phi_1}\rho_2 e^{i\phi_2}$$
$$= \alpha\beta z_1 z_2. \qquad (2.10.9)$$

If α or β is negative or zero an appropriate modification may be made to the proof, but the result still holds. Also

$$z_1(z_2 + z_3) = (x + iy_1)(x_2 + iy_2 + x_3 + iy_3)$$
$$= x_1x_2 - y_1y_2 + x_1x_3 - y_1y_3$$
$$+ i(y_1x_2 + x_1y_2 + y_1x_3 + x_1y_3)$$
$$= z_1z_2 + z_1z_3. \qquad (2.10.10)$$

Thus complex numbers obey the vectorial product laws (2.0.1) and (2.0.2). The product is also commutative. This concept cannot, however, be extended to more than two dimensions as the product is then non-commutative. (See e.g., C. C. Macduffee, *An Introduction to Abstract Algebra*, Wiley (1940), p. 251.)

Worked examples

1. If $\mathbf{A} = 2\mathbf{i} + 3\mathbf{j} + 4\mathbf{k}$, $\mathbf{B} = 4\mathbf{i} + \mathbf{j} + 3\mathbf{k}$, calculate $\mathbf{A}.\mathbf{B}$, $\mathbf{A} \times \mathbf{B}$ and verify that

$$(\mathbf{A}.\mathbf{B})^2 + (\mathbf{A} \times \mathbf{B})^2 = A^2 B^2$$

$$\mathbf{A}.\mathbf{B} = 2.4 + 3.1 + 4.3 = 23$$

$$\mathbf{A} \times \mathbf{B} = \begin{vmatrix} \mathbf{i} & \mathbf{j} & \mathbf{k} \\ 2 & 3 & 4 \\ 4 & 1 & 3 \end{vmatrix}$$

$$= \mathbf{i}(5) + \mathbf{j}(10) + \mathbf{k}(-10)$$

$$A^2 = 2^2 + 3^2 + 4^2 = 29$$

$$B^2 = 4^2 + 1^2 + 9^2 = 26$$

$$(\mathbf{A} \times \mathbf{B})^2 = 5^2 + 10^2 + (10)^2 = 225$$

$$(\mathbf{A}.\mathbf{B})^2 = 23^2 = 529$$

and

$$529 + 225 = 754 = 26.29.$$

2. Find the angle between the two vectors $\mathbf{A}$ and $\mathbf{B}$ where $\mathbf{A} = \mathbf{i} + \mathbf{j} + \mathbf{k}$, $\mathbf{B} = 2\mathbf{i} - 3\mathbf{j} - \mathbf{k}$.

If θ is the angle between the vectors

$$AB \cos\theta = \mathbf{A}.\mathbf{B}$$

$$A = (1^2 + 1^2 + 1^2)^{1/2} = \sqrt{3}$$

$$B = (2^2 + (-3)^2 + (-1)^2)^{1/2} = \sqrt{14}$$

$$\mathbf{A}.\mathbf{B} = 1.2 + 1(-3) + 1(-1) = -2$$

Thus

$$\cos\theta = \frac{-2}{\sqrt{42}} = -\sqrt{\left(\frac{2}{21}\right)}$$

and

$$\theta = \pi - \cos^{-1}\sqrt{\left(\frac{2}{21}\right)}.$$

3. The sides of a triangle ABC satisfy the relation $c^2 = 3(b^2 - a^2)$ and P is any point in its plane. Prove that, if K be the middle point of the perpendicular from C upon AB,

$$\overrightarrow{PA} + 2\overrightarrow{PB} + 3\overrightarrow{PC} = 6\overrightarrow{PK}$$

(E. G. Phillips).

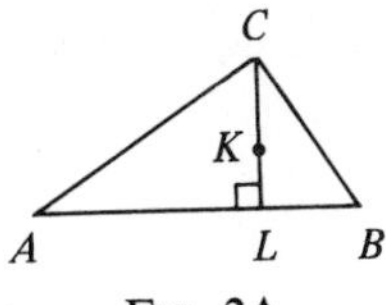

FIG. 2A

The expression $c^2 = 3(b^2 - a^2)$ may be rewritten in the form

$$\overrightarrow{AB}.\overrightarrow{AB} = 3(\overrightarrow{CA}.\overrightarrow{CA} - \overrightarrow{CB}.\overrightarrow{CB})$$

or

$$\begin{aligned}\overrightarrow{AB}.\overrightarrow{AB} &= 3(\overrightarrow{CA} + \overrightarrow{CB}).(\overrightarrow{CA} - \overrightarrow{CB})\\ &= 3(\overrightarrow{CA} + \overrightarrow{CB}).\overrightarrow{BA}\end{aligned}$$

whence

$$\overrightarrow{BA}.\{3\overrightarrow{CA} + 3\overrightarrow{CB} + \overrightarrow{AB}\} = 0$$

and

$$\begin{aligned}\overrightarrow{BA}.\{3\overrightarrow{CA} + 3\overrightarrow{CB} + \overrightarrow{CB} - \overrightarrow{CA}\} &= 0\\ 2\overrightarrow{BA}.\{\overrightarrow{CA} + 2\overrightarrow{CB}\} &= 0.\end{aligned}$$

Thus if L divides AB in the ratio 2:1

$$\overrightarrow{CA} + 2\overrightarrow{CB} = 3\overrightarrow{CL}$$

and so

$$\overrightarrow{BA}.\overrightarrow{CL} = 0$$

that is, $\overrightarrow{CL}$ is the perpendicular from C to AB, and so

$$\overrightarrow{CA} + 2\overrightarrow{CB} = 6\overrightarrow{CK}$$

as K is half way between C and L.

Now

$$\begin{aligned}6\overrightarrow{PK} &= 6\overrightarrow{PC} + 6\overrightarrow{CK}\\ &= 6\overrightarrow{PC} + \overrightarrow{CA} + 2\overrightarrow{CB}\\ &= \overrightarrow{PA} + 2\overrightarrow{PB} + 3\overrightarrow{PC}.\end{aligned}$$

4. What is the unit vector which is perpendicular to the two vectors

$$\mathbf{A} = 2\mathbf{i} + \mathbf{j} + \mathbf{k}, \qquad \mathbf{B} = 3\mathbf{i} - \mathbf{j} - 2\mathbf{k}?$$

Let (λ, μ, ν) be some vector which is perpendicular to **A** and **B**. Then

$$2\lambda + \mu + \nu = 0, \qquad 3\lambda - \mu - 2\nu = 0.$$

The ratios $\lambda:\mu:\nu$ are given by

$$\frac{\lambda}{\begin{vmatrix} 1 & 1 \\ -1 & -2 \end{vmatrix}} = \frac{\mu}{\begin{vmatrix} 1 & 2 \\ -2 & 3 \end{vmatrix}} = \frac{\nu}{\begin{vmatrix} 2 & 1 \\ 3 & -1 \end{vmatrix}}$$

i.e.

$$\frac{\lambda}{-1} = \frac{\mu}{7} = \frac{\nu}{-5} = m \text{ (say)}.$$

Now if (λ, μ, ν) is a vector, the associated unit vector is

$$\frac{1}{(\lambda^2 + \mu^2 + \nu^2)^{1/2}} \qquad (\lambda, \mu, \nu)$$

and so it follows that

$$(\lambda^2 + \mu^2 + \nu^2)^{1/2} = |(-1)^2 + (7)^2 + (5)^2|^{1/2} m = \sqrt{(75)}\, m$$

and the unit vector is

$$\frac{1}{\sqrt{(75)}\, m}\{-m, 7m, -5m\} = \frac{1}{\sqrt{(75)}}(-1, 7, -5).$$

5. Prove that the vector area of the triangle whose vertices are the points with position vectors **a**, **b**, **c** is

$$\tfrac{1}{2}(\mathbf{b} \times \mathbf{c} + \mathbf{c} \times \mathbf{a} + \mathbf{a} \times \mathbf{b}).$$

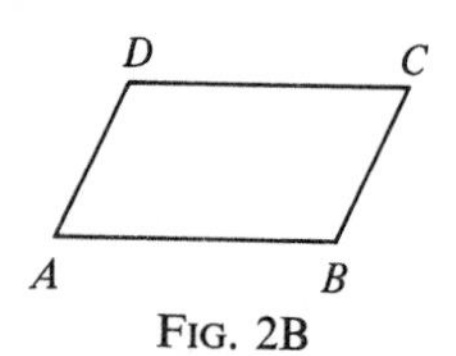

FIG. 2B

The vector area of the triangle ABC is half the vector area of the parallelogram $ABCD$. Thus, required vector area

$$\begin{aligned}
&= \tfrac{1}{2}\overrightarrow{AB} \times \overrightarrow{AD} = \tfrac{1}{2}\overrightarrow{AB} \times \overrightarrow{BC} \\
&= \tfrac{1}{2}(\overrightarrow{OB} - \overrightarrow{OA}) \times (\overrightarrow{OC} - \overrightarrow{OB}) \\
&= \tfrac{1}{2}(\mathbf{b} - \mathbf{a}) \times (\mathbf{c} - \mathbf{b}) = \tfrac{1}{2}(\mathbf{b} \times \mathbf{c} - \mathbf{a} \times \mathbf{c} + \mathbf{a} \times \mathbf{b} - \mathbf{b} \times \mathbf{b}) \\
&= \tfrac{1}{2}(\mathbf{b} \times \mathbf{c} + \mathbf{c} \times \mathbf{a} + \mathbf{a} \times \mathbf{b}).
\end{aligned}$$

6. If $ABCD$ is any quadrilateral and H is the midpoint of the diagonal AC and K is the midpoint of the diagonal BD prove that

$$AB^2 + BC^2 + CD^2 + DA^2 = AC^2 + BD^2 + 4HK^2.$$

Let **a** be the position vector of the point A, etc. Then

$$\begin{aligned}
AB^2 &= \overrightarrow{AB}.\overrightarrow{AB} = (\mathbf{b} - \mathbf{a}).(\mathbf{b} - \mathbf{a}) \\
&= b^2 + 2\mathbf{a}.\mathbf{b} + a^2, \text{ etc.}
\end{aligned}$$

Also

$$\begin{aligned}
\mathbf{h} &= \tfrac{1}{2}(\mathbf{a} + \mathbf{c})\ \mathbf{k} = \tfrac{1}{2}(\mathbf{b} + \mathbf{d}) \\
4HK^2 &= 2\overrightarrow{HK}.2\overrightarrow{HK} \\
&= (\mathbf{b} + \mathbf{d} - \mathbf{a} - \mathbf{c})^2.
\end{aligned}$$

Then

$$\begin{aligned}
&b^2 - 2\mathbf{a}.\mathbf{b} + a^2 + c^2 - 2\mathbf{c}.\mathbf{b} + b^2 + d^2 - 2\mathbf{d}.\mathbf{c} + c^2 + a^2 - 2\mathbf{a}.\mathbf{d} + d^2 \\
&= c^2 - 2\mathbf{a}.\mathbf{c} + a^2 + b^2 - 2\mathbf{b}.\mathbf{d} + d^2 + b^2 + d^2 + a^2 + c^2 + 2\mathbf{b}.\mathbf{d} \\
&\quad + 2\mathbf{a}.\mathbf{c} - 2\mathbf{b}.\mathbf{a} - 2\mathbf{b}.\mathbf{c} - 2\mathbf{a}.\mathbf{d} - 2\mathbf{d}.\mathbf{c}
\end{aligned}$$

which proves the result.

7. A system of forces $\mathbf{F}_i$ act at points with position vector $\mathbf{r}_i$. Find the locus of points such that the moment about them is parallel to the resultant force.

The resultant force is

$$\mathbf{F} = \sum_i \mathbf{F}_i.$$

The moment of the force system about the point with position vector $\mathbf{r}$ is

$$\sum_i (\mathbf{r}_i - \mathbf{r}) \times \mathbf{F}_i = \mathbf{G} - \mathbf{r} \times \mathbf{F}$$

where

$$\mathbf{G} = \sum_i \mathbf{r}_i \times \mathbf{F}_i.$$

It is required to find $\mathbf{r}$, if possible, such that

$$\mathbf{G} - \mathbf{r} \times \mathbf{F} = \lambda \mathbf{F}$$

where λ is a scalar of dimensions length.

Multiplying vectorially by $\mathbf{F}$

$$\mathbf{F} \times \mathbf{G} - \mathbf{F} \times (\mathbf{r} \times \mathbf{F}) = 0$$
$$\mathbf{F} \times \mathbf{G} - \mathbf{r}F^2 + (\mathbf{r}.\mathbf{F})\mathbf{F} = 0.$$

Now clearly, if $\mathbf{r}$ is a solution, so also is $\mathbf{r} + \mu\hat{\mathbf{F}}$, where μ is some scalar of dimension length.

If a particular solution is possible such that $\mathbf{r}$ is perpendicular to $\mathbf{F}$, $\mathbf{r}.\mathbf{F} = 0$ and $\mathbf{F} \times \mathbf{G} - \mathbf{r}F^2 = 0$,

and so $$\mathbf{r} = \frac{\mathbf{F} \times \mathbf{G}}{F^2}.$$

It is now necessary to verify that this solution is perpendicular to $\mathbf{F}$. This is so because $\mathbf{F} \times \mathbf{G}$ is perpendicular to both of its factors. Thus a possible solution is

$$\mathbf{r} = \frac{\mathbf{F} \times \mathbf{G}}{F^2}$$

and by what has been said it follows that the general solution is

$$\mathbf{r} = \frac{\mathbf{F} \times \mathbf{G}}{F^2} + \mu\hat{\mathbf{F}}.$$

The value of λ may easily be determined.

Multiplying scalarly by F

$$\mathbf{F}.\mathbf{G} = \lambda F^2, \qquad \lambda = \frac{\mathbf{F}.\mathbf{G}}{F^2}$$

we note that λ is an invariant, for were the moment taken about some point with position vector $\mathbf{c}$ instead of the origin,

$$\mathbf{F}.\mathbf{G}_C = \mathbf{F}.(\mathbf{G} - \mathbf{c} \times \mathbf{F}) = \mathbf{F}.\mathbf{G}.$$

λ is referred to as the pitch. Thus any system of forces may be reduced to a force acting along an axis, together with a couple about that axis and parallel to the force. Such a system is referred to as a wrench. It is a meaningful concept unless $\mathbf{F} = 0$. In this case the system reduces to a couple $\mathbf{G}$ which has the same moment about all points.

8. Find the values of x, y, z from the simultaneous equations

$$a_i x + b_i y + c_i z = d_i \qquad (i = 1, 2, 3).$$

If $\mathbf{P}_i$ is the vector (a_i, b_i, c_i), these three equations can be written in the form

$$\mathbf{P}_i . \mathbf{r} = d_i \qquad [\mathbf{r} = (x, y, z)].$$

Clearly, the $\mathbf{P}_i$ may be taken as unit vectors.

Thus the solution of the set of equations is equivalent to finding a point common to three planes.

There are three cases to consider

(*i*) There exist scalars α_1, α_2, α_3 such that

$$\alpha_1 \mathbf{P}_1 + \alpha_2 \mathbf{P}_2 + \alpha_3 \mathbf{P}_3 = 0.$$

In this case $\mathbf{P}_1$, $\mathbf{P}_2$, $\mathbf{P}_3$ are coplanar. If $\alpha_1 d_1 + \alpha_2 d_2 + \alpha_3 d_3 = 0$ also, then

$$\alpha_1(\mathbf{P}_1 . \mathbf{r} - d_1) + \alpha_2(\mathbf{P}_2 . \mathbf{r} - d_2) + \alpha_3(\mathbf{P}_3 . \mathbf{r} - d_3) = 0$$

which is the condition that the three planes pass through a common line. In this case the solution is indeterminate because, in effect, there are two equations for three unknowns, the third equation being the sum of multiples of the first two equations.

(*ii*) There exist scalars α_1, α_2, α_3 such that

$$\alpha_1 \mathbf{P}_1 + \alpha_2 \mathbf{P}_2 + \alpha_3 \mathbf{P}_3 = 0$$

and

$$\alpha_1 d_1 + \alpha_2 d_2 + \alpha_3 d_3 \neq 0.$$

In this case the equations are inconsistent and there cannot be any solution; two or more of the planes are parallel or the three planes form a prism.

(*iii*) If the three vectors $\mathbf{P}_1$, $\mathbf{P}_2$, $\mathbf{P}_3$ are non-coplanar, then their triple scalar product, being the volume of the parallelopiped with these vectors representing three concurrent sides, is non-zero. Also, any vector may be written in the form

$$\mathbf{r} = \lambda_1 \mathbf{P}_2 \times \mathbf{P}_3 + \lambda_2 \mathbf{P}_3 \times \mathbf{P}_1 + \lambda_3 \mathbf{P}_1 \times \mathbf{P}_2.$$

Substituting this expression for $\mathbf{r}$ into the three equations $\mathbf{P}_i . \mathbf{r} = d_i$ it follows that, for example

$$\mathbf{P}_1 . \mathbf{r} = \lambda_1 \mathbf{P}_1 . (\mathbf{P}_2 \times \mathbf{P}_3) + \lambda_2 \mathbf{P}_1 . (\mathbf{P}_3 \times \mathbf{P}_1) + \lambda_3 \mathbf{P}_1 . (\mathbf{P}_1 \times \mathbf{P}_2).$$

The last two products vanish and so

$$\lambda_1 = \frac{\mathbf{P}_1 . \mathbf{r}}{\mathbf{P}_1 . (\mathbf{P}_2 \times \mathbf{P}_3)} = \frac{d_1}{[\mathbf{P}_1 \mathbf{P}_2 \mathbf{P}_3]}$$

Similar results will hold for λ_2 and λ_3. Thus

$$\mathbf{r} = \frac{d_1 \mathbf{P}_2 \times \mathbf{P}_3 + d_2 \mathbf{P}_3 \times \mathbf{P}_1 + d_3 \mathbf{P}_1 \times \mathbf{P}_2}{[\mathbf{P}_1 \mathbf{P}_2 \mathbf{P}_3]}$$

9. Prove that

$$[\mathbf{PQR}]\mathbf{A} = (\mathbf{Q} \times \mathbf{R})(\mathbf{P}.\mathbf{A}) + (\mathbf{R} \times \mathbf{P})(\mathbf{Q}.\mathbf{A}) + (\mathbf{P} \times \mathbf{Q})(\mathbf{R}.\mathbf{A})$$

(E. A. Milne)

$$\begin{aligned}[(\mathbf{P} \times \mathbf{Q}) \times \mathbf{A}] \times \mathbf{R} &= [-\mathbf{P}(\mathbf{Q}.\mathbf{A}) + \mathbf{Q}(\mathbf{P}.\mathbf{A})] \times \mathbf{R} \\ &= -(\mathbf{P} \times \mathbf{Q})(\mathbf{A}.\mathbf{R}) + \mathbf{A}[(\mathbf{P} \times \mathbf{Q}).\mathbf{R}]\end{aligned}$$

where the result follows.

10. Resolve the vector **P** into components parallel and perpendicular to the vector **Q**.

$$\mathbf{Q} \times (\mathbf{Q} \times \mathbf{P}) = (\mathbf{Q}.\mathbf{P})\mathbf{Q} - \mathbf{P}Q^2$$

whence

$$Q^2\mathbf{P} = (\mathbf{Q}.\mathbf{P})\mathbf{Q} - \mathbf{Q} \times (\mathbf{Q} \times \mathbf{P}).$$

Dividing by Q^2

$$\mathbf{P} = (\hat{\mathbf{Q}}.\mathbf{P})\hat{\mathbf{Q}} - \hat{\mathbf{Q}} \times (\hat{\mathbf{Q}} \times \mathbf{P}).$$

The first term is parallel to **Q**. The second term is perpendicular to **Q**.

11. Solve for **X** the vector equation

$$\alpha\mathbf{X} + \mathbf{X} \times \mathbf{A} = \mathbf{B} \qquad (\alpha \neq 0)$$

(E. A. Milne)

Before commencing to solve this equation, it is worth while considering what can be done to this equation. It is no use multiplying the equation by **B** either vectorially or scalarly because this will not produce any simplification of the left-hand side which is where the unknown is. Similarly, multiplication by a scalar is pointless. Thus the only possibilities are to multiply scalarly and vectorially by **A**.

Multiplying scalarly,

$$\alpha\mathbf{X}.\mathbf{A} = \mathbf{A}.\mathbf{B}.$$

Multiplying vectorially by **A**

$$\begin{aligned}\alpha\mathbf{X} \times \mathbf{A} + (\mathbf{X} \times \mathbf{A}) \times \mathbf{A} &= \mathbf{B} \times \mathbf{A} \\ \alpha\mathbf{X} \times \mathbf{A} - A^2\mathbf{X} + \mathbf{A}(\mathbf{A}.\mathbf{X}) &= \mathbf{B} \times \mathbf{A}.\end{aligned}$$

Substituting for $\mathbf{X} \times \mathbf{A}$ and $\mathbf{A}.\mathbf{X}$

$$\alpha(\mathbf{B} - \alpha\mathbf{X}) - A^2\mathbf{X} + \frac{\mathbf{A}(\mathbf{A}.\mathbf{B})}{\alpha} = \mathbf{B} \times \mathbf{A}$$

whence

$$\mathbf{X} = \frac{\mathbf{A}(\mathbf{A}.\mathbf{B}) + \alpha^2\mathbf{B} + \alpha\mathbf{A} \times \mathbf{B}}{\alpha(\alpha^2 + A^2)}.$$

It may be verified that this is in fact a solution.

An alternative method of solution would be to write

$$\mathbf{X} = p\mathbf{A} + q\mathbf{B} + r\mathbf{A} \times \mathbf{B}$$

as the three vectors form a non-coplanar set unless **B** is parallel to **A**. If **B** is parallel to **A**, the solution is obviously $\mathbf{X} = \mathbf{B}/\alpha$, since $\mathbf{B} \times \mathbf{A} = 0$.

12. The axis of a wrench $(\mathbf{P}, p\mathbf{P})$ passes through the point with position vector **a** and the axis of a wrench $(\mathbf{Q}, q\mathbf{Q})$, passes through the point with position vector **b**. Find the resultant wrench.

The force associated with the resultant wrench is $\mathbf{P} + \mathbf{Q}$. The moment about the origin of the resultant wrench is

$$\mathbf{G}_0 = p\mathbf{P} + q\mathbf{Q} + \mathbf{a} \times \mathbf{P} + \mathbf{b} \times \mathbf{Q}.$$

The pitch of the resultant wrench is

$$\frac{(\mathbf{P}+\mathbf{Q}).\mathbf{G}_0}{F^2} = \frac{pP^2 + qQ^2 + (p+q)\mathbf{P}.\mathbf{Q} + (\mathbf{a}-\mathbf{b}).(\mathbf{P}\times\mathbf{Q})}{P^2 + 2\mathbf{P}.\mathbf{Q} + Q^2}.$$

A point on the axis of the wrench is

$$\mathbf{c} = \frac{(\mathbf{P}+\mathbf{Q}) \times \mathbf{G}_0}{F^2}$$

$$= \frac{(q-p)\mathbf{P}\times\mathbf{Q} + \mathbf{P}\times(\mathbf{a}\times\mathbf{P}) + \mathbf{P}\times(\mathbf{b}\times\mathbf{Q}) + \mathbf{Q}\times(\mathbf{a}\times\mathbf{P}) + \mathbf{Q}\times(\mathbf{b}\times\mathbf{Q})}{P^2 + 2\mathbf{P}.\mathbf{Q} + Q^2}.$$

The triple products may be expanded if desired.

The equation of the axis of the resultant will be $\mathbf{r} = \mathbf{c} + \lambda(\mathbf{P} + \mathbf{Q})$, where λ is arbitrary.

13. Solve the equations

$$\alpha\mathbf{X} + \beta\mathbf{Y} = \mathbf{A}$$
$$\mathbf{X} \times \mathbf{Y} = \mathbf{B}$$

(E. A. Milne)

Multiplying vectorially by **X** and **Y** respectively

$$\alpha\mathbf{X}\times\mathbf{X} + \beta\mathbf{X}\times\mathbf{Y} = \mathbf{X}\times\mathbf{A}$$
$$\alpha\mathbf{X}\times\mathbf{Y} + \beta\mathbf{Y}\times\mathbf{Y} = \mathbf{A}\times\mathbf{Y}$$

whence

$$\mathbf{X}\times\mathbf{A} = \beta\mathbf{B}$$

and

$$\mathbf{Y}\times\mathbf{A} = -\alpha\mathbf{B}.$$

Thus solutions will only be possible if $\mathbf{A}.\mathbf{B} = 0$.

This follows also because

$$\mathbf{A}.\mathbf{B} = (\alpha\mathbf{X} + \beta\mathbf{Y}).(\mathbf{X}\times\mathbf{Y}) = 0.$$

It follows that

$$\mathbf{X} = \frac{\beta\mathbf{A}\times\mathbf{B}}{A^2} + \lambda\mathbf{A}$$

$$\mathbf{Y} = \frac{\alpha\mathbf{B}\times\mathbf{A}}{A^2} + \mu\mathbf{A}.$$

However, $\alpha\mathbf{X} + \beta\mathbf{Y} = \mathbf{A}$, and so

$$\frac{\alpha\beta\mathbf{A} \times \mathbf{B} + \beta\alpha\mathbf{B} \times \mathbf{A}}{A^2} + (\lambda + \mu)\mathbf{A} = \mathbf{A}$$

whence $\lambda + \mu = 1$. Also,

$$\mathbf{B} = \frac{(\beta\mathbf{A} \times \mathbf{B} + \lambda\mathbf{A})}{A^2} \times \frac{(\alpha\mathbf{B} \times \mathbf{A} + \mu\mathbf{A})}{A^2}$$

$$= \frac{\beta\mu + \alpha\lambda}{A^2}\{(\mathbf{A} \times \mathbf{B}) \times \mathbf{A}\}$$

$$= \frac{\beta\mu + \alpha\lambda}{A^2}\{A^2\mathbf{B} - (\mathbf{A}.\mathbf{B})\mathbf{A}\}$$

whence $\alpha\lambda + \beta\mu = 1$.

Thus if $\alpha \neq \beta$, $(\alpha - \beta)\lambda = 1 - \beta$, $(\beta - \alpha)\mu = (1 - \alpha)$. If $\alpha = \beta = 1$, all that can be said is that $\lambda + \mu = 1$ and there is a one parameter solution. If $\alpha = \beta \neq 1$, write $\mathbf{A}' = \mathbf{A}/\alpha$ and the problem reduces to the case $\alpha = \beta = 1$.

14. The position vector of a particle at any time is $(ut, vt, \frac{1}{2}ft^2)$, u, v, f constant.

Prove that the force is parallel to the z axis and find when the particle is moving perpendicular to the plane $\alpha x + \beta y + \gamma z = p$.

$$\mathbf{r} = ut\mathbf{i} + vt\mathbf{j} + \tfrac{1}{2}ft^2\mathbf{k}$$
$$\dot{\mathbf{r}} = u\mathbf{i} + v\mathbf{j} + ft\mathbf{k}$$
$$\ddot{\mathbf{r}} = f\mathbf{k}.$$

The force acting is $\mathbf{F} = m\ddot{\mathbf{r}} = mf\mathbf{k}$ which is parallel to the z axis. If the particle is moving perpendicular to the plane $\mathbf{n}.\mathbf{r} = p$

$$\dot{\mathbf{r}}.\mathbf{n} = 0$$

that is, $u\alpha + v\beta + ft\gamma = 0$. This happens at time

$$t = -\frac{u\alpha + v\beta}{\gamma f}.$$

15. Find the projection of the vector $\mathbf{A} = (2, 1, 1)$ upon the vector $\mathbf{B} = (1, 3, 5)$.

The projection of $\mathbf{A}$ upon $\mathbf{B}$ is given by $\mathbf{A}.\hat{\mathbf{B}}$

$$\hat{\mathbf{B}} = \frac{(1, 3, 5)}{\{1^2 + 3^2 + 5^2\}} = \frac{(1, 3, 5)}{\sqrt{35}}$$

$$(\mathbf{A}.\hat{\mathbf{B}}) = \frac{2.1 + 1.3 + 1.5}{\sqrt{35}} = \frac{10}{\sqrt{35}}.$$

16. A rigid body which is fixed at a point O moves such that the velocities of two points with position vectors $\mathbf{r}_1, \mathbf{r}_2$ are $\mathbf{v}_1, \mathbf{v}_2$. Find the angular velocity

$$\mathbf{v}_1 = \boldsymbol{\omega} \times \mathbf{r}_1$$
$$\mathbf{v}_2 = \boldsymbol{\omega} \times \mathbf{r}_2$$
$$\mathbf{v}_1 \times \mathbf{v}_2 = \mathbf{v}_1 \times (\boldsymbol{\omega} \times \mathbf{r}_2) = (\mathbf{v}_1 \,.\, \mathbf{r}_2)\,\boldsymbol{\omega} - (\mathbf{v}_1 \,.\, \boldsymbol{\omega})\,\mathbf{r}_2.$$

Now $\mathbf{v}_1$ is perpendicular to $\boldsymbol{\omega}$ and so

$$\mathbf{v}_1 \times \mathbf{v}_2 = (\mathbf{v}_1 \,.\, \mathbf{r}_2)\,\boldsymbol{\omega}$$

whence

$$\boldsymbol{\omega} = \frac{\mathbf{v}_1 \times \mathbf{v}_2}{\mathbf{v}_1 \,.\, \mathbf{r}_2} = \frac{\mathbf{v}_2 \times \mathbf{v}_1}{\mathbf{v}_2 \,.\, \mathbf{r}_1}$$

by symmetry, and so

$$\mathbf{v}_1 \,.\, \mathbf{r}_2 + \mathbf{v}_2 \,.\, \mathbf{r}_1 = 0.$$

17. Two frames of reference S_0 and S have a common origin O and S rotates with angular velocity $\boldsymbol{\omega}$ with respect to S_0. If $\mathbf{v}$ and $\boldsymbol{\alpha}$ denote the velocity and acceleration respectively of a particle in S, find the velocity and acceleration in S_0.

If $\mathbf{r}$ is the position vector of the particle in both frames, then

$$\frac{d\mathbf{r}}{dt} = \frac{\partial \mathbf{r}}{\partial t} + \boldsymbol{\omega} \times \mathbf{r}$$

d/dt denoting a time variation in S_0 and $\partial/\partial t$ denoting a time variation in S. Whence

$$\frac{d\mathbf{r}}{dt} = \mathbf{v} + \boldsymbol{\omega} \times \mathbf{r}.$$

The acceleration with respect to S_0 is given by

$$\begin{aligned}
\frac{d}{dt}\frac{d\mathbf{r}}{dt} &= \frac{d}{dt}\{\mathbf{v} + \boldsymbol{\omega} \times \mathbf{r}\} \\
&= \left\{\frac{\partial}{\partial t} + \boldsymbol{\omega} \times\right\}\{\mathbf{v} + \boldsymbol{\omega} \times \mathbf{r}\} \\
&= \frac{\partial \mathbf{v}}{\partial t} + \frac{\partial \boldsymbol{\omega}}{\partial t} \times \mathbf{r} + \boldsymbol{\omega} \times \frac{\partial \mathbf{r}}{\partial t} \\
&\qquad + \boldsymbol{\omega} \times \mathbf{v} + \boldsymbol{\omega} \times (\boldsymbol{\omega} \times \mathbf{r}) \\
&= \boldsymbol{\alpha} + \dot{\boldsymbol{\omega}} \times \mathbf{r} + 2\boldsymbol{\omega} \times \mathbf{v} + \boldsymbol{\omega} \times (\boldsymbol{\omega} \times \mathbf{r}).
\end{aligned}$$

The last term corresponds to the so-called centrifugal force and the next to the last term to the so-called Coriolis force.

18. A uniform sphere rolls upon a plane which is rotating with uniform angular velocity $\boldsymbol{\Omega}$ about an axis perpendicular to itself. Show that, if the

plane is not horizontal, the motion of the point of contact of the sphere and the plane may be thought of as a motion of uniform rotation about a point moving with uniform velocity in a horizontal direction.

Take a right-handed system of axes as follows. Ox is up the plane, Oz is the axis of rotation and is on the upper side of the plane, Oy completes the triad.

$\mathbf{\Omega} = \Omega\mathbf{k}$ is the angular velocity of the plane.

Let $\boldsymbol{\omega}$ be the angular velocity of the sphere and $\mathbf{r}$ the position vector of the point of contact of sphere and plane.

Let α be the inclination of the plane to the horizontal and $\mathbf{n}$ the upward vertical unit vector $\mathbf{n} = \mathbf{k}\cos\alpha + \mathbf{i}\sin\alpha$.

Let $\mathbf{R}$ be the reaction of the plane on the sphere, A be the moment of inertia of the sphere about any diameter, M its mass and a its radius.

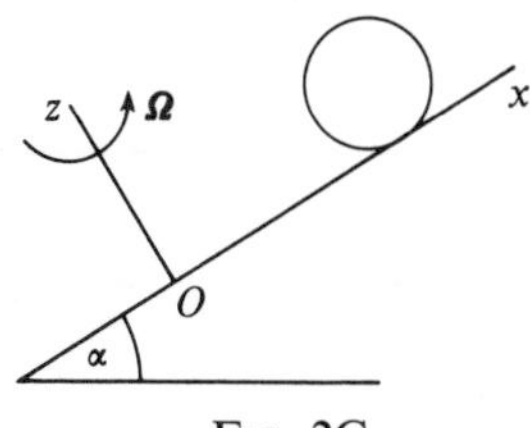

FIG. 2C

The equation of motion is

$$M\frac{d^2\mathbf{r}}{dt^2} = \mathbf{R} - Mg\mathbf{n}$$

the position vector of the centre of gravity of the sphere being $\mathbf{r} + a\mathbf{k}$.

The equation of angular momentum is

$$A\frac{d\boldsymbol{\omega}}{dt} = (-a\mathbf{k}) \times \mathbf{R}.$$

The rolling equation which expresses the fact that the velocity of the point of contact is the same for sphere and plane is

$$\mathbf{\Omega} \times \mathbf{r} = \frac{d\mathbf{r}}{dt} - a\boldsymbol{\omega} \times \mathbf{k}.$$

These three vector equations are sufficient to determine the three unknown vector quantities $\mathbf{r}$, $\mathbf{R}$ and $\boldsymbol{\omega}$. Eliminating $\mathbf{R}$

$$\frac{d\boldsymbol{\omega}}{dt} = \frac{aM}{A}\left(\frac{d^2\mathbf{r} + g\mathbf{n}}{dt^2}\right) \times \mathbf{k}$$

whence

$$\mathbf{k}.\frac{d\boldsymbol{\omega}}{dt} = 0.$$

This is equivalent to the statement that $\boldsymbol{\omega}.\mathbf{k} = \text{constant}$. (It can easily be seen that the same result would hold for a sphere in contact with any surface.)

Integrating

$$\boldsymbol{\omega} - \boldsymbol{\omega}_0 = \frac{Ma}{A}\left(\frac{d\mathbf{r}}{dt} + gt\mathbf{n}\right) \times \mathbf{k}$$

if $\boldsymbol{\omega}_0$ is the value of $\boldsymbol{\omega}$ at $t = 0$ and the point of contact of the sphere is initially at rest.

It follows that

$$\boldsymbol{\Omega} \times \mathbf{r} = \frac{d\mathbf{r}}{dt} + a\mathbf{k} \times \left\{\boldsymbol{\omega}_0 + \frac{Ma}{A}\left(\frac{d\mathbf{r}}{dt} \times \mathbf{k}\right) + \frac{Ma}{A} gt\mathbf{n} \times \mathbf{k}\right\}.$$

Using the fact that $\mathbf{k}.(d\mathbf{r}/dt) = 0$, and that $\mathbf{n} = \mathbf{k}\cos\alpha + \mathbf{i}\sin\alpha$, this simplifies to

$$\frac{d\mathbf{r}}{dt}\left(1 + \frac{Ma^2}{A}\right) = \boldsymbol{\Omega} \times \mathbf{r} - \frac{Ma^2}{A} gt\mathbf{i}\sin\alpha - a\mathbf{k} \times \boldsymbol{\omega}_0$$

$$= \boldsymbol{\Omega} \times \left(\mathbf{r} + \frac{Ma^2}{A\Omega} g\sin\alpha t\mathbf{j} - \frac{a\boldsymbol{\omega}_0}{\Omega}\right).$$

This is a motion of uniform rotation about a centre moving along a horizontal line with uniform velocity $(Ma^2/A\Omega)g\sin\alpha\mathbf{j}$.

19. Find the angle between the two vectors

$$\mathbf{a} = 6\mathbf{i} + 5\mathbf{j} + 4\mathbf{k} \qquad \text{and} \qquad \mathbf{b} = \mathbf{i} + 2\mathbf{j} + 3\mathbf{k}.$$

$$\mathbf{a}.\mathbf{b} = 6.1 + 5.2 + 4.3 = 28$$

$$a = \{36 + 25 + 16\}^{1/2} = (77)^{1/2}$$

$$b = \{1 + 4 + 9\}^{1/2} = (14)^{1/2}.$$

The angle between the two vectors is given by θ, where

$$\cos\theta = \frac{\mathbf{a}.\mathbf{b}}{ab} = \frac{28}{(77.14)^{1/2}} = \left(\frac{8}{11}\right)^{1/2}.$$

20. Show that the altitudes of a triangle concur.

Let D and E be the feet of the altitudes at A and B respectively. Let AD and BE cut at H and produce CH to F. Then it is required to prove that CF is perpendicular to AB.

$$\overrightarrow{AD}.\overrightarrow{BC} = 0, \qquad \overrightarrow{BE}.\overrightarrow{AC} = 0.$$

Hence

$$\overrightarrow{AH}.\overrightarrow{BC} = 0, \qquad \overrightarrow{BH}.\overrightarrow{AC} = 0.$$

These two equations may be written as

$$\overrightarrow{CH}.\overrightarrow{BC} = \overrightarrow{CA}.\overrightarrow{BC}$$

$$\overrightarrow{CH}.\overrightarrow{AC} = \overrightarrow{CB}.\overrightarrow{AC}.$$

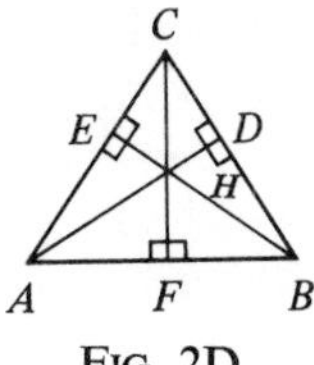

FIG. 2D

Subtracting

$$\overrightarrow{CH}.\overrightarrow{AB} = 0$$

whence

$$\overrightarrow{CF}.\overrightarrow{AB} = 0$$

giving the required result.

21. Find the unit vector which is perpendicular to $\mathbf{i}+\mathbf{j}+\mathbf{k}$ and $2\mathbf{i}-\mathbf{j}+3\mathbf{k}$.

A vector which is perpendicular to both of these vectors is their vector product.

$$\begin{vmatrix} \mathbf{i} & \mathbf{j} & \mathbf{k} \\ 1 & 1 & 1 \\ 2 & -1 & 3 \end{vmatrix} = 4\mathbf{i} - \mathbf{j} - 3\mathbf{k}$$

and the associated unit vector is

$$\frac{(4, -1, -3)}{(16+1+9)^{1/2}} = \frac{1}{\sqrt{26}}(4, -1, -3)$$

which is the required result.

22. Prove that the sum of the vector areas of the faces of the tetrahedron $ABCD$ is zero.

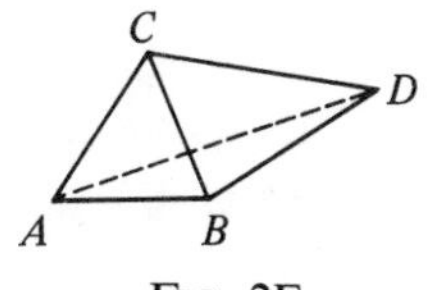

FIG. 2E

By example (5), the vector areas of the various faces are as follows. (The normals are all assumed to point out.)

Triangle BDC: $\frac{1}{2}(\mathbf{b}\times\mathbf{d}+\mathbf{d}\times\mathbf{c}+\mathbf{c}\times\mathbf{b})$
Triangle DAC: $\frac{1}{2}(\mathbf{a}\times\mathbf{a}+\mathbf{a}\times\mathbf{c}+\mathbf{c}\times\mathbf{d})$
Triangle BAD: $\frac{1}{2}(\mathbf{b}\times\mathbf{a}+\mathbf{a}\times\mathbf{d}+\mathbf{d}\times\mathbf{b})$
Triangle ABC: $\frac{1}{2}(\mathbf{a}\times\mathbf{b}+\mathbf{b}\times\mathbf{c}+\mathbf{c}\times\mathbf{a}$.

Adding all these, the sum is zero, which is the required result.

23. Find the volume of the parallelopiped, three concurrent sides of which can be represented by vectors

$$\mathbf{a} = \mathbf{i} + \mathbf{j} + \mathbf{k}$$
$$\mathbf{b} = 2\mathbf{i} - \mathbf{j} + 3\mathbf{k}$$
$$\mathbf{c} = \mathbf{i} - \mathbf{j}.$$

The volume is given by

$$(\mathbf{a} \times \mathbf{b}).\mathbf{c}.$$

Now in section (2.24), it was shown that

$$\mathbf{a} \times \mathbf{b} = 4\mathbf{i} - \mathbf{j} - 3\mathbf{k}$$

and it follows that the volume is

$$(4\mathbf{i} - \mathbf{j} - 3\mathbf{k}).(\mathbf{i} - \mathbf{j}) = 4 + 1 = 5.$$

24. Express the vector $\mathbf{d} = 6\mathbf{i} + 2\mathbf{j} + 3\mathbf{k}$ in terms of $\mathbf{b} \times \mathbf{c}$, $\mathbf{c} \times \mathbf{a}$, $\mathbf{a} \times \mathbf{b}$ where **a, b, c** are defined in example 23.

It may easily be seen that

$$\mathbf{b} \times \mathbf{c} = 3\mathbf{i} + 3\mathbf{j} - \mathbf{k}$$
$$\mathbf{c} \times \mathbf{a} = -\mathbf{i} - \mathbf{j} + 2\mathbf{k}$$
$$\mathbf{a} \times \mathbf{b} = 4\mathbf{i} - \mathbf{j} - 3\mathbf{k}$$
$$\mathbf{a}.\mathbf{d} = 6 + 2 + 3 = 11$$
$$\mathbf{b}.\mathbf{d} = 12 - 2 + 9 = 19$$
$$\mathbf{c}.\mathbf{d} = 6 - 2 + 0 = 4$$

and using (2.3.7)

$$\mathbf{d} = \tfrac{1}{5}[11(3\mathbf{i} + 3\mathbf{j} - \mathbf{k}) + 19(-\mathbf{i} - \mathbf{j} + 2\mathbf{k}) + 4(4\mathbf{i} - \mathbf{j} - 3\mathbf{k})].$$

25. Three light rods OA, OB, OC, smoothly jointed together at O have their other ends smoothly jointed together to a horizontal plane so as to form a tripod. A force F of given magnitude is applied at O. Show that for maximum stress in one of the rods, F is perpendicular to the plane of the other two. (Leicester University.)

Let

$$\hat{\mathbf{a}} = \widehat{\overrightarrow{OA}}, \qquad \hat{\mathbf{b}} = \widehat{\overrightarrow{OB}}, \qquad \hat{\mathbf{c}} = \widehat{\overrightarrow{OC}}.$$

Then,

$$\mathbf{F} = P\hat{\mathbf{a}} + Q\hat{\mathbf{b}} + R\hat{\mathbf{c}}$$

where P, Q, R are the components of F parallel to $\overrightarrow{OA}$, $\overrightarrow{OB}$, $\overrightarrow{OC}$ respectively. These will, in fact, be the stresses in the respective rods.

Now

$$\mathbf{F}.(\hat{\mathbf{a}} \times \hat{\mathbf{b}}) = R(\hat{\mathbf{a}} \times \hat{\mathbf{b}}).\hat{\mathbf{c}}$$

because

$$\hat{\mathbf{a}}.(\hat{\mathbf{a}} \times \hat{\mathbf{b}}) = \hat{\mathbf{b}}.(\hat{\mathbf{a}} \times \hat{\mathbf{b}}) = 0$$

and so

$$\mathbf{R} = \frac{\mathbf{F}.(\hat{\mathbf{a}} \times \hat{\mathbf{b}})}{(\hat{\mathbf{a}} \times \hat{\mathbf{b}}).\hat{\mathbf{c}}}.$$

Now $\mathbf{F} = F\hat{\mathbf{F}}$, where F is constant and $\hat{\mathbf{F}}$ is variable, and so

$$R = F\frac{\hat{\mathbf{F}}.(\hat{\mathbf{a}} \times \hat{\mathbf{b}})}{(\hat{\mathbf{a}} \times \hat{\mathbf{b}}).\hat{\mathbf{c}}}.$$

$(\hat{\mathbf{a}} \times \hat{\mathbf{b}}).\hat{\mathbf{c}}$ is constant and so R will be a maximum when $\hat{\mathbf{F}}.(\hat{\mathbf{a}} \times \hat{\mathbf{b}})$ is a maximum. This occurs when $\hat{\mathbf{F}}$ is parallel to $\hat{\mathbf{a}} \times \hat{\mathbf{b}}$, that is when $\hat{\mathbf{F}}$ is perpendicular to $\hat{\mathbf{a}}$ and $\hat{\mathbf{b}}$, that is when $\mathbf{F}$ is perpendicular to the plane of $\overrightarrow{OA}$ and $\overrightarrow{OB}$.

26. **A**, **B**, **C** are three arbitrary non-coplanar vectors such that not one is perpendicular to any other one. Construct, without the use of the vector product, two vectors which are mutually perpendicular and are perpendicular to **A**.

Let $\mathbf{B}' = \mathbf{B} + \lambda\mathbf{A}$

$$\mathbf{B}'.\mathbf{A} = \mathbf{B}.\mathbf{A} + \lambda A^2 = 0 \qquad \text{if } \lambda = -\frac{\mathbf{A}.\mathbf{B}}{A^2}$$

$$\mathbf{B}' = \mathbf{B} - \frac{\lambda(\mathbf{A}.\mathbf{B})\,\mathbf{A}}{A^2}.$$

Let $\mathbf{C}' = \mathbf{C} + \mu\mathbf{A} + \nu\mathbf{B}'$

$$\mathbf{C}'.\mathbf{A} = \mathbf{C}.\mathbf{A} + \mu A^2 = 0 \qquad \text{if } \mu = \frac{-\mathbf{A}.\mathbf{C}}{A^2}$$

$$\mathbf{C}'.\mathbf{B}' = \mathbf{C}.\mathbf{B}' + \nu B'^2 = 0 \qquad \text{if } \nu = \frac{-\mathbf{B}'.\mathbf{C}}{B'^2}$$

$$\mathbf{C}' = \mathbf{C} - \frac{(\mathbf{A}.\mathbf{C})\,\mathbf{A}}{A^2} - \frac{(\mathbf{B}'.\mathbf{C})\,\mathbf{B}'}{B'^2}$$

and **A**, **B**′, **C**′ are mutually perpendicular. This is known as the Gram-Schmidt orthogonolization process.

Exercises

1. If $\mathbf{A} = \mathbf{i} + \mathbf{j} + \mathbf{k}$, $\mathbf{B} = 2\mathbf{i} - \mathbf{j} + 3\mathbf{k}$, find $\mathbf{A}.\mathbf{B}$, $\mathbf{A} \times \mathbf{B}$ and the angle between **A** and **B**.

2. Construct a vector which is perpendicular to the two vectors $\mathbf{P} = \mathbf{i} - \mathbf{j} + 3\mathbf{k}$ and $\mathbf{Q} = \mathbf{i} + \mathbf{j} + 2\mathbf{k}$, and is of length 3.

3. If $\mathbf{A} = 2\mathbf{i} + 3\mathbf{j} + 4\mathbf{k}$, $\mathbf{B} = \mathbf{i} + \mathbf{j} - \mathbf{k}$, $\mathbf{C} = \mathbf{i} - \mathbf{j} + \mathbf{k}$, find $\mathbf{A} \times (\mathbf{B} \times \mathbf{C})$ and $(\mathbf{A} \times \mathbf{B}).\mathbf{C}$.

   Verify that $\mathbf{A} \times (\mathbf{B} \times \mathbf{C})$ is perpendicular to **A** and $\mathbf{B} \times \mathbf{C}$.

4. Find the equation of the plane which passes through the point (2, 6, 1) and which is perpendicular to the direction (2, 3, 1). Find also the length of the perpendicular to the plane from the point (1, 1, 1).

5. $\mathbf{F} = \alpha\mathbf{i} + \beta\mathbf{j} + \gamma\mathbf{k}$, $\mathbf{R}_1 = 6\mathbf{i}$, $\mathbf{R}_2 = 3\mathbf{i} + 4\mathbf{j}$, $\mathbf{R}_3 = 2\mathbf{i} - 7\mathbf{j} + 5\mathbf{k}$.
Show that if $\mathbf{F}.\mathbf{R}_1 = \mathbf{F}.\mathbf{R}_2 = \mathbf{F}.\mathbf{R}_3 = 0$, then $\mathbf{F} = 0$.

6. Find the equation of the circle whose centre is $(a,a,2a)$ and which passes through the point $(2a,2a,-3a)$.

7. Prove that the perpendicular bisectors of the sides of a triangle concur.

8. Show that a body cannot be in equilibrium under the action of six forces acting along the edges of a tetrahedron.

9. S_r is a face of an arbitrary closed polyhedron. $\mathbf{P}_r$ is a vector proportional to the area of the face S_r and in the direction of the outward normal. Prove that $\sum_r \mathbf{P}_r = \mathbf{0}$ where the summation extends over all faces.

10. Find the general solution of the equations

$$\alpha\mathbf{X} + \beta\mathbf{Y} = \mathbf{A}$$
$$\mathbf{X}.\mathbf{Y} = \gamma.$$

(E. A. Milne)

11. The position of a particle at any time t is given by $(\alpha t^3, \beta t^2, \gamma t)$. Find when the particle is moving perpendicular to the plane $2x + 3y + z = 1$ and when the acceleration is parallel to the plane $x = y$.

12. Find the volume of the parallelopiped whose edges are represented by the vectors $(-3,4,2)(1,2,1)(2,1,3)$.

13. The motion of a rigid body may be regarded as a rotation with angular velocity $(1,2,-3)$ about the axis $x + y = 2a$, $2x + z = a$. Find the velocity of the point $(a,a,2a)$.

14. If a vector $\mathbf{A}$ has constant modulus, prove that, $\mathbf{A}$ and $d\mathbf{A}/dt$ are perpendicular unless

$$\frac{d\mathbf{A}}{dt} = \mathbf{0}.$$

Show also that

$$\mathbf{A}.\frac{d\mathbf{A}}{dt} = A\frac{dA}{dt}.$$

15. Show that

$$[\mathbf{ABC}](\mathbf{P} \times \mathbf{Q}) = \begin{vmatrix} \mathbf{P}.\mathbf{A} & \mathbf{P}.\mathbf{B} & \mathbf{P}.\mathbf{C} \\ \mathbf{Q}.\mathbf{A} & \mathbf{Q}.\mathbf{B} & \mathbf{Q}.\mathbf{C} \\ \mathbf{A} & \mathbf{B} & \mathbf{C} \end{vmatrix}.$$

(London University)

16. A system of forces reduces to a force $\mathbf{F}_1$ acting through a point O together with a couple $\mathbf{G}_1$. A second system reduces to a force $\mathbf{F}_2$ acting

through O together with a couple $\mathbf{G}_2$. If the axes of the two wrenches intersect, prove that

$$\left(\frac{\mathbf{F}_1.\mathbf{G}_1}{F_1^2}+\frac{\mathbf{F}_2.\mathbf{G}_2}{F_2^2}\right)(\mathbf{F}_1.\mathbf{F}_2)=\mathbf{F}_1.\mathbf{G}_2+\mathbf{F}_2.\mathbf{G}_1.$$

17. Prove that

$$2a^2=|\mathbf{a}\times\mathbf{i}|^2+|\mathbf{a}\times\mathbf{j}|^2+|\mathbf{a}\times\mathbf{k}|^2.$$

18. A sphere is placed between two horizontal planes which rotate with angular velocities $\boldsymbol{\omega}_1$, $\boldsymbol{\omega}_2$ about two different vertical axes. If there is no slipping, prove that the centre of the sphere moves in a horizontal circle whose centre lies in the plane of the two axes.

19. A plane $\mathbf{n}.\mathbf{r}=p$ is fixed in space, and the frame of reference rotates with angular velocity $\boldsymbol{\omega}$ about O. Prove that $d\mathbf{n}/dt=\mathbf{n}\times\boldsymbol{\omega}$.

20. A heavy sphere of radius a moves on the inside of a rough vertical fixed circular cylinder of radius b. If the sphere is uniform, prove that it rises and falls in the cylinder with a vertical simple harmonic motion of period $\sqrt{(14)}\pi/n$ where n is the angular velocity of the centre of the sphere about the axis of the cylinder. If the cylinder rotates with angular velocity ω about its axis, prove that the period is $\sqrt{(14)}\pi/(n+\omega)$.

21. The force on a charged particle is given by

$$q(\mathbf{E}+\mathbf{v}\times\mathbf{B})$$

where q is the charge,
$\mathbf{E}$ is the electric intensity,
$\mathbf{v}$ is the velocity of the particle,
$\mathbf{B}$ is the magnetic induction.
Assuming that q is a scalar, prove that $\mathbf{E}$ is a polar vector and that $\mathbf{B}$ is an axial vector.

22. Prove that

$$\mathbf{f}\times\mathbf{g}=\{(\mathbf{i}\times\mathbf{f}).\mathbf{g}\}\mathbf{i}+\{(\mathbf{j}\times\mathbf{f}).\mathbf{g}\}\mathbf{j}+\{(\mathbf{k}\times\mathbf{f}).\mathbf{g}\}\mathbf{k}.$$

23. Two non-collinear vectors $\mathbf{P}$, $\mathbf{Q}$ are given. Prove that

$$\mathbf{u}_1=\hat{\mathbf{P}},\qquad \mathbf{u}_2=\frac{\mathbf{Q}\times\mathbf{P}}{|\mathbf{Q}\times\mathbf{P}|},\qquad \mathbf{u}_3=\mathbf{u}_1\times\mathbf{u}_2$$

form a right-handed orthogonal set of unit vectors.

24. Prove that the system of coordinates defined by

$$x=a\xi\eta,\qquad y=\frac{a}{2}(\xi^2-\eta^2),\qquad z=z$$

gives rise to the following quadratic form for an element of distance and hence that the system is orthogonal:

$$\delta s^2 = a^2(\xi^2 + \eta^2)(\delta\xi^2 + \delta\eta^2) + \delta z^2$$

and find $\mathbf{i}_\xi$, $\mathbf{i}_\eta$ in terms of $\mathbf{i}$ and $\mathbf{j}$. (This system of coordinates is referred to as the parabolic cylindrical system.)

25. Prove that the field represented by

$$f\left(t - \frac{\mathbf{n}.\mathbf{r}}{c}\right)$$

propagates unchanged in the direction $\mathbf{n}$ with speed c.

3 The Vector Operator ∇

3.0. **Point functions**

The values of many physical quantities vary with the positions at which they are measured. The temperature of a room is not the same everywhere in it. The velocity of the wind varies from place to place and so on. Thus the expression for temperature T may be expressed in the form $T(x,y,z)$ and the wind velocity $\mathbf{v}$ in the form $\mathbf{v}(x,y,z)$, x, y, z being the coordinates of the point P at which the entity is measured with respect to some frame of reference. It is often convenient to consider the position vector $\mathbf{r}$ instead of the coordinates and the statements

$$T = T(P) = T(\mathbf{r}) \tag{3.0.1}$$

are clearly alternative forms of writing the statement

$$T = T(x,y,z). \tag{3.0.2}$$

Other independent variables such as, for example, time, can of course exist. These will not in general be of interest in this book and will consequently be omitted unless explicitly of interest. Entities which vary in this manner are termed point functions. A scalar whose value depends upon the point in space with which it is associated is termed a scalar point function. Similarly, a vector depending upon the point in space with which it is associated is termed a vector point function, and has associated with it three scalar point functions, namely the three components. Point functions are also termed fields.

3.01. *Differentiation of point functions*

Suppose that f is some scalar point function, and it is desired to relate its value at the point whose coordinates are $(x+\delta x, y+\delta y, z+\delta z)$ to its value at the point whose coordinates are (x,y,z).

In other words the relation is required between $f(P)$ or $f(\mathbf{r})$ and $f(P')$ or $f(\mathbf{r}+\delta\mathbf{r})$.

Let $\overrightarrow{PP'} = \delta s(\lambda, \mu, \nu) = \hat{\mathbf{s}}\,\delta s \qquad (3.01.1)$

where λ, μ, ν are direction cosines and

$$\left.\begin{aligned} \delta x &= \lambda \delta s \\ \delta y &= \mu\, \delta s \\ \delta z &= \nu\, \delta s \end{aligned}\right\} \tag{3.01.2}$$

To the first order

$$f(P') = f(P) + \delta s \frac{\partial f}{\partial s} \tag{3.01.3}$$

where $\partial f/\partial s$ is the rate of change of f in the direction PP'. In this connection, it may be helpful to consider f as defined on some curve

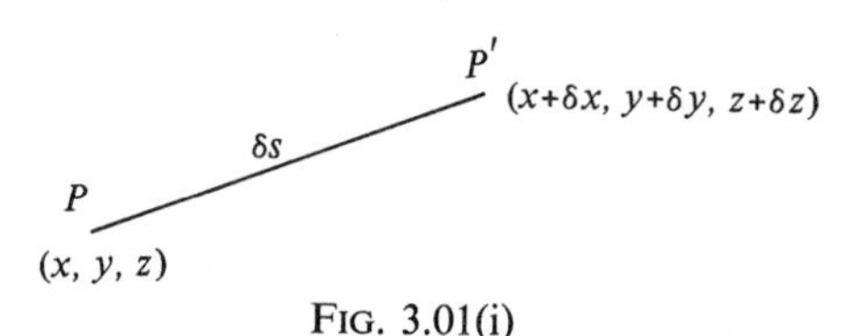

FIG. 3.01(i)

which runs through P and P', P being an arc distance s from some arbitrary point on the curve P_0 and P' an arc distance $s + \delta s$ from P_0. An alternative expression for $f(P')$ is

$$f(P') = f(P) + \delta x \frac{\partial f}{\partial x} + \delta y \frac{\partial f}{\partial y} + \delta z \frac{\partial f}{\partial z}. \tag{3.01.4}$$

Thus

$$\delta s \frac{\partial f}{\partial s} = \delta x \frac{\partial f}{\partial x} + \delta y \frac{\partial f}{\partial y} + \delta z \frac{\partial f}{\partial z} \tag{3.01.5}$$

and using the relations (3.01.2)

$$\frac{\partial f}{\partial s} = \lambda \frac{\partial f}{\partial x} + \mu \frac{\partial f}{\partial y} + \nu \frac{\partial f}{\partial z}. \tag{3.01.6}$$

In exactly the same way, if $\mathbf{A}(\mathbf{r})$ is a vector point function

$$\frac{\partial \mathbf{A}}{\partial s} = \lambda \frac{\partial \mathbf{A}}{\partial x} + \mu \frac{\partial \mathbf{A}}{\partial y} + \nu \frac{\partial \mathbf{A}}{\partial z}. \tag{3.01.7}$$

Thus the rate of change of a quantity in the direction PP' is obtained by means of an operator

$$\frac{\partial}{\partial s} = \lambda \frac{\partial}{\partial x} + \mu \frac{\partial}{\partial y} + \nu \frac{\partial}{\partial z} \tag{3.01.8}$$

which may be written formally as

$$\frac{\partial}{\partial s} = (\lambda\mathbf{i} + \mu\mathbf{j} + \nu\mathbf{k}).\left(\mathbf{i}\frac{\partial}{\partial x} + \mathbf{j}\frac{\partial}{\partial y} + \mathbf{k}\frac{\partial}{\partial z}\right) \tag{3.01.9}$$

$$= \hat{\mathbf{s}}.\boldsymbol{\nabla}$$

where

$$\boldsymbol{\nabla} = \mathbf{i}\frac{\partial}{\partial x} + \mathbf{j}\frac{\partial}{\partial y} + \mathbf{k}\frac{\partial}{\partial z}. \tag{3.01.10}$$

$\boldsymbol{\nabla}$ is an operator. That is, it acts on a quantity, converting it into some other quantity. $\boldsymbol{\nabla}$ is pronounced del (or sometimes nabla). It possesses properties which are analogous in many ways to those of ordinary vectors. It should be noticed though, and this will be borne out later, that the order of symbols in an expression involving $\boldsymbol{\nabla}$ is important.

3.1. Gradient

Let $f(\mathbf{r})$ be some scalar point function. It is clear from section 3.01 that $(\hat{\mathbf{s}}.\boldsymbol{\nabla})f$ is $\partial f/\partial s$, the rate of change of f in the direction of f increasing.

Thus

$$\left.\begin{aligned} \frac{\partial f}{\partial x} &= (\mathbf{i}.\boldsymbol{\nabla})f \\ \frac{\partial f}{\partial y} &= (\mathbf{j}.\boldsymbol{\nabla})f \\ \frac{\partial f}{\partial z} &= (\mathbf{k}.\boldsymbol{\nabla})f \end{aligned}\right\} \tag{3.1.1}$$

The vector quantity

$$\boldsymbol{\nabla}f = \mathbf{i}\frac{\partial f}{\partial x} + \mathbf{j}\frac{\partial f}{\partial y} + \mathbf{k}\frac{\partial f}{\partial z} \tag{3.1.2}$$

is termed the gradient of f. If two quantities f_1 and f_2 differ by a constant their gradients are clearly the same.

If $\mathbf{E} = \boldsymbol{\nabla}f$, f is said to be the scalar potential of $\mathbf{E}$. From what has just been said, it is indefinite with respect to an arbitrary constant.

It will be seen that

$$\hat{\mathbf{s}}.(\boldsymbol{\nabla}f) = (\lambda, \mu, \nu).\left(\frac{\partial f}{\partial x}, \frac{\partial f}{\partial y}, \frac{\partial f}{\partial z}\right)$$

$$= \lambda\frac{\partial f}{\partial x} + \mu\frac{\partial f}{\partial y} + \nu\frac{\partial f}{\partial z} = \frac{\partial f}{\partial s}$$

$$= (\hat{\mathbf{s}}.\boldsymbol{\nabla})f \tag{3.1.3}$$

Thus the brackets are not essential and the symbolism $\hat{\mathbf{s}}.\nabla f$ is unambiguous. It should be noted, however, that although by the properties of the scalar product,

$$\hat{\mathbf{s}}.\nabla f = \hat{\mathbf{s}}.(\nabla f) = (\nabla f).\hat{\mathbf{s}} \tag{3.1.4}$$

$$(\nabla f).\hat{\mathbf{s}} \neq \nabla.\{f\hat{\mathbf{s}}\}$$

and $\nabla f.\hat{\mathbf{s}}$ would be ambiguous.

An alternative approach is as follows. Let $V(\mathbf{r})$ be a scalar field. The points defined by

$$V(\mathbf{r}) = C \tag{3.1.5}$$

where C is a constant, form a surface in space and as C varies, a family of surfaces is generated. Consider now two neighbouring surfaces of this family.

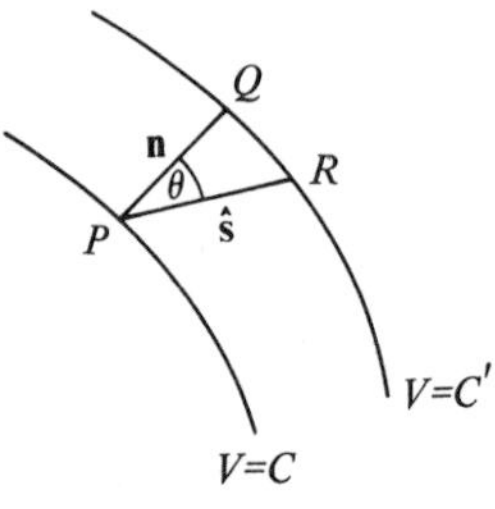

FIG. 3.1(i)

Let P be on $V = C$, Q and R on $V = C'$. Let PQ be in the direction of the normal to the surface $V = C$ and PR in some other direction defined by a unit vector $\hat{\mathbf{s}}$.

$$\overrightarrow{PQ} = \mathbf{n}\,\delta s_n, \qquad \overrightarrow{PR} = \hat{\mathbf{s}}\,\delta s \tag{3.1.6}$$

$$\delta s_n = \delta s \cos\theta = \delta s(\hat{\mathbf{s}}.\mathbf{n}) \tag{3.1.7}$$

$\mathbf{n}$ is the unit normal to the surface $V = C$ at P. Then $\partial V/\partial s$ is the rate of change of V in the direction PR, and $\partial V/\partial s_n$, usually written $\partial V/\partial n$, is the rate of change in the direction PQ, that is in the direction of normal to $B = C$ at P. Now

$$\frac{\partial V}{\partial s} = \cos\theta \frac{\partial V}{\partial n} \tag{3.1.8}$$

$$= (\hat{\mathbf{s}}.\mathbf{n})\frac{\partial V}{\partial n} = \hat{\mathbf{s}}.\left(\mathbf{n}\frac{\partial V}{\partial n}\right). \tag{3.1.9}$$

The quantity $\mathbf{n}(\partial V/\partial n)$ is thus the gradient of V.

Putting $\hat{\mathbf{s}} = \mathbf{i}, \mathbf{j}, \mathbf{k}$ in turn, it follows that

$$\mathbf{i}.\left(\mathbf{n}\frac{\partial V}{\partial n}\right) = \frac{\partial V}{\partial x}, \text{ etc.} \tag{3.1.10}$$

and so

$$\mathbf{n}\frac{\partial V}{\partial n} = \mathbf{i}\frac{\partial V}{\partial x} + \mathbf{j}\frac{\partial V}{\partial y} + \mathbf{k}\frac{\partial V}{\partial z} = \boldsymbol{\nabla} V. \tag{3.1.11}$$

It follows that if $\mathbf{n}$ is the unit normal at a point on the surface $V(\mathbf{r}) = C$

$$\mathbf{n} = \frac{\boldsymbol{\nabla} V}{|\boldsymbol{\nabla} V|}, \qquad \mathbf{n}.\boldsymbol{\nabla} V = \frac{\partial V}{\partial n}. \tag{3.1.12}$$

The second method of definition of $\boldsymbol{\nabla} V$ does not involve any coordinate system. Equations (3.1.10) and (3.1.11) depend only upon the vectors $\mathbf{i}, \mathbf{j}, \mathbf{k}$ being orthogonal. If $\mathbf{i}_1$, $\mathbf{i}_2$, $\mathbf{i}_3$ are unit vectors which form an orthogonal set and if δs_1, δs_2, δs_3 are the respective elements of distance parallel to these

$$\boldsymbol{\nabla} = \mathbf{i}_1\frac{\partial V}{\partial s_1} + \mathbf{i}_2\frac{\partial V}{\partial s_2} + \mathbf{i}_3\frac{\partial V}{\partial s_3} \tag{3.1.13}$$

and

$$\frac{\partial V}{\partial s_r} = \mathbf{i}_r.\boldsymbol{\nabla} V. \tag{3.1.14}$$

For cylindrical polar coordinates

$$\boldsymbol{\nabla} V = \hat{\boldsymbol{\rho}}\frac{\partial V}{\partial \rho} + \hat{\boldsymbol{\varphi}}\frac{1}{\rho}\frac{\partial V}{\partial \phi} + \mathbf{k}\frac{\partial V}{\partial z} \tag{3.1.15}$$

and for spherical polar coordinates

$$\boldsymbol{\nabla} V = \hat{\mathbf{r}}\frac{\partial V}{\partial r} + \hat{\boldsymbol{\theta}}\frac{1}{r}\frac{\partial V}{\partial \theta} + \hat{\boldsymbol{\phi}}\frac{1}{r\sin\theta}\frac{\partial V}{\partial \phi}. \tag{3.1.16}$$

The proofs of the statements

$$\boldsymbol{\nabla}(U + V) = \boldsymbol{\nabla} U + \boldsymbol{\nabla} V \tag{3.1.17}$$

$$\boldsymbol{\nabla}(UV) = V\boldsymbol{\nabla} U + U\boldsymbol{\nabla} V \tag{3.1.18}$$

$$\boldsymbol{\nabla}\left(\frac{U}{V}\right) = \frac{V\boldsymbol{\nabla} U - U\boldsymbol{\nabla} V}{V^2} \quad V \neq 0 \tag{3.1.19}$$

are trivial and are left to the reader. Also, if $F = F(U, V, \ldots)$ U, $V\ldots$ being point functions

$$\boldsymbol{\nabla} F = \frac{\partial F}{\partial U}\boldsymbol{\nabla} U + \frac{\partial F}{\partial V}\boldsymbol{\nabla} V + \cdots. \tag{3.1.20}$$

If $\mathbf{A}$ is an arbitrary vector, the quantity

$$(\mathbf{A}.\boldsymbol{\nabla})V = \mathbf{A}(\hat{\mathbf{A}}.\boldsymbol{\nabla})V = A\frac{\partial V}{\partial s_A} = \mathbf{A}.(\boldsymbol{\nabla}V) \qquad (3.1.21)$$

where $\partial/\partial s_A$ is differentiation in the direction associated with $\hat{\mathbf{A}}$. Similarly

$$(\mathbf{A}.\boldsymbol{\nabla})\mathbf{P} = A\frac{\partial \mathbf{P}}{\partial s_A} \qquad (3.1.22)$$

and in particular

$$(\mathbf{A}.\boldsymbol{\nabla})\mathbf{r} = \mathbf{A}. \qquad (3.1.23)$$

3.11. *Special gradients*

Let $\mathbf{r}$ be the position vector of $\mathbf{P}$.

$$\begin{aligned}\boldsymbol{\nabla}r &= \left(\mathbf{i}\frac{\partial}{\partial x} + \mathbf{j}\frac{\partial}{\partial y} + \mathbf{k}\frac{\partial}{\partial z}\right)(x^2+y^2+z^2)^{1/2} \\ &= \frac{\mathbf{i}x+\mathbf{j}y+\mathbf{k}z}{(x^2+y^2+z^2)^{1/2}} = \frac{\mathbf{r}}{r} \\ &= \hat{\mathbf{r}}. \end{aligned} \qquad (3.11.1)$$

Generalizing this,

$$\begin{aligned}\boldsymbol{\nabla}f(r) &= \left(\mathbf{i}\frac{\partial}{\partial x} + \mathbf{j}\frac{\partial}{\partial y} + \mathbf{k}\frac{\partial}{\partial z}\right)f(r) \\ &= \left(\mathbf{i}\frac{\partial r}{\partial x} + \mathbf{j}\frac{\partial r}{\partial y} + \mathbf{k}\frac{\partial r}{\partial z}\right)f'(r) \\ &= \hat{\mathbf{r}}f'(r). \end{aligned} \qquad (3.11.2)$$

If $\mathbf{A}$ is a constant vector,

$$\begin{aligned}\boldsymbol{\nabla}(\mathbf{A}.\mathbf{r}) &= \left(\mathbf{i}\frac{\partial}{\partial x} + \mathbf{j}\frac{\partial}{\partial y} + \mathbf{k}\frac{\partial}{\partial y}\right)(A_x x + A_y y + A_z z) \\ &= \mathbf{i}A_x + \mathbf{j}A_y + \mathbf{k}A_z = \mathbf{A}. \end{aligned} \qquad (3.11.3)$$

The evaluation of the gradients of more general quantities such as $\mathbf{A}.\mathbf{B}$ will be considered later.

3.12. *The operator* $\boldsymbol{\nabla}'$

Let P' be the point with position vector $\mathbf{r}'$, let $\overrightarrow{P'P} = \mathbf{R} = \mathbf{r} - \mathbf{r}'$ and consider the function $f(R)$. This is effectively the same as considering $f(r)$ when there has been a change of origin. If P be regarded as the field point, that is the point at which a quantity is measured and P' as a source point, $f(R)$ represents a field symmetrical about a source at P', and $f(r)$ a field symmetrical about a source at O.

Let

$$\boldsymbol{\nabla}' = \mathbf{i}\frac{\partial}{\partial x'} + \mathbf{j}\frac{\partial}{\partial y'} + \mathbf{k}\frac{\partial}{\partial z'}. \tag{3.12.1}$$

This is in effect a differentiation, as the position of the source point is altered.

$$\boldsymbol{\nabla}' f(R) = \left(\mathbf{i}\frac{\partial R}{\partial x'} + \mathbf{j}\frac{\partial R}{\partial y'} + \mathbf{k}\frac{\partial R}{\partial z'}\right) f'(R)$$

as in the derivation of equation (3.11.2).

Now

$$R^2 = (x-x')^2 + (y-y')^2 + (z-z')^2$$

and so

$$\mathbf{i}\frac{\partial R}{\partial x'} + \mathbf{j}\frac{\partial R}{\partial y'} + \mathbf{k}\frac{\partial R}{\partial z'} = \frac{\mathbf{i}(x'-x) + \mathbf{j}(y'-y) + \mathbf{k}(z'-z)}{\{(x-x')^2 + (y-y')^2 + (z-z')^2\}^{1/2}} = -\frac{\mathbf{R}}{R}.$$

Thus

$$\boldsymbol{\nabla}' f(R) = -\hat{\mathbf{R}} f'(R). \tag{3.12.2}$$

Similarly,

$$\boldsymbol{\nabla} f(R) = f'(R)\hat{\mathbf{R}} \tag{3.12.3}$$

and so

$$\boldsymbol{\nabla} f(R) + \boldsymbol{\nabla}' f(R) = 0. \tag{3.12.4}$$

Extending the result of equation (3.11.3)

$$\begin{aligned} \boldsymbol{\nabla}(\mathbf{A}.\mathbf{R}) &= \boldsymbol{\nabla}(\mathbf{A}.\mathbf{r}) = \mathbf{A} \\ \boldsymbol{\nabla}'(\mathbf{A}.\mathbf{R}) &= -\boldsymbol{\nabla}'(\mathbf{A}.\mathbf{r}') = -\mathbf{A} = -\boldsymbol{\nabla}(\mathbf{A}.\mathbf{R}). \end{aligned} \tag{3.12.5}$$

3.13. *Differentiation with respect to a vector*

By the definition of gradient, if V is a point function

$$\delta V = (\boldsymbol{\nabla} V).\delta\mathbf{r}. \tag{3.13.1}$$

This can be written formally in analogy with well-known formulae of differential calculus as

$$\delta V = \frac{\partial V}{\partial \mathbf{r}}.\delta\mathbf{r} \tag{3.13.2}$$

and, in a sense, taking the gradient is equivalent to differentiating with respect to the position vector $\mathbf{r}$. Indeed in some works the notation $\partial/\partial\mathbf{r}$ is used instead of $\boldsymbol{\nabla}$.

By extension therefore, if U is some scalar quantity which depends upon some vector $\mathbf{q}$, the change in U due to a change in $\mathbf{q}$ may be written as

$$\delta U = \frac{\partial U}{\partial q_x}\delta q_x + \frac{\partial U}{\partial q_y}\delta q_y + \frac{\partial U}{\partial q_z}\delta q_z$$

$$= \left[\left(\mathbf{i}\frac{\partial}{\partial q_x} + \mathbf{j}\frac{\partial}{\partial q_y} + \mathbf{k}\frac{\partial}{\partial q_z}\right)U\right].\delta\mathbf{q} \tag{3.13.3}$$

$$= \frac{\partial U}{\partial \mathbf{q}}.\delta\mathbf{q} \tag{3.13.4}$$

where

$$\frac{\partial}{\partial \mathbf{q}} = \mathbf{i}\frac{\partial}{\partial q_x} + \mathbf{j}\frac{\partial}{\partial q_y} + \mathbf{k}\frac{\partial}{\partial q_z} \tag{3.13.5}$$

the right-hand side of this defining differentiation with respect to the vector $\mathbf{q}$.

3.14. *Differentiation following the fluid*

In many physical problems it is of interest to find the time derivative of some quantity which is associated with a moving point in space, as for example a particle of a fluid in motion. This rate of change will not, in general, be the same as that which is measured at some fixed point. The former may be termed differentiation following the fluid, or full differentiation, whereas the other may be termed local differentiation, or partial differentiation. (An analogous problem is discussed in example 17 of chapter 2, where local differentiation corresponds to the time derivative in the moving frame and full differentiation corresponds to the time derivative in the fixed frame.)

Consider now some quantity U which depends on position in space and time, for example the temperature in a moving fluid.

$$U = U(\mathbf{r}, t) = U(x, y, z, t). \tag{3.14.1}$$

A small variation in U is given by

$$\delta U = \frac{\partial U}{\partial t}\delta t + \frac{\partial U}{\partial x}\delta x + \frac{\partial U}{\partial y}\delta y + \frac{\partial U}{\partial z}\delta z \tag{3.14.2}$$

$$= \frac{\partial U}{\partial t}\delta t + (\delta\mathbf{r}.\boldsymbol{\nabla})\,U.$$

$\partial U/\partial t$ represents the rate of change of U when t only is varied and is thus the local variation. Equation (3.14.2) represents the change in

U due to arbitrary variations δt, $\delta\mathbf{r}$. Suppose now that we wish to compare say the temperature of, or the pressure of, some element of a moving fluid at times t and $t+\delta t$. This will change with time and also with the position in the fluid at which it is measured. Suppose that the velocity of the fluid is $\mathbf{v}$. Then the portion of fluid which, loosely speaking, has position vector $\mathbf{r}$ at time t, has position vector $\mathbf{r}+\mathbf{v}\,\delta t$ at time $t+\delta t$. The change in U is the difference between the value of U at the point with position vector $\mathbf{r}+\mathbf{v}\,\delta t$ at time $t+\delta t$ and the value of U at the point with position vector $\mathbf{r}$ at time t.

Then

$$\delta r = \mathbf{v}\,\delta t \tag{3.14.3}$$

and

$$\delta U = \frac{\partial U}{\partial t}\delta t + \delta t(\mathbf{v}\,.\,\boldsymbol{\nabla})\,U. \tag{3.14.4}$$

It is this variation over time when measured at the moving point which corresponds to full differentiation and

$$\delta U = \frac{dU}{dt}\delta t. \tag{3.14.5}$$

Thus

$$\frac{dU}{dt} = \frac{\partial U}{\partial t} + (\mathbf{v}\,.\,\boldsymbol{\nabla})\,U \tag{3.14.6}$$

where dU/dt is full differentiation, or differentiation following the fluid, and $\partial U/\partial t$ is local differentiation.

It follows immediately that

$$\frac{d\mathbf{A}}{dt} = \frac{\partial \mathbf{A}}{\partial t} + (\mathbf{v}\,.\,\boldsymbol{\nabla})\,\mathbf{A} \tag{3.14.7}$$

$\mathbf{A}$ being a vector quantity which depends upon position and time, such as for example the velocity of a moving fluid.

In particular, the velocity of a particle is given by

$$\frac{d\mathbf{r}}{dt} = \frac{\partial \mathbf{r}}{\partial t} + (\mathbf{v}\,.\,\boldsymbol{\nabla})\,\mathbf{r} \tag{3.14.8}$$

and because t, $\mathbf{r}$ are independent variables

$$\frac{\partial \mathbf{r}}{\partial t} = 0$$

and

$$\frac{d\mathbf{r}}{dt} = (\mathbf{v}\,.\,\boldsymbol{\nabla})\,\mathbf{r} = \mathbf{v} \tag{3.14.9}$$

by equation (3.1.22). This result is of course to be expected.

The acceleration is given by

$$\boldsymbol{\alpha} = \frac{d\mathbf{v}}{dt} = \frac{\partial \mathbf{v}}{\partial t} + (\mathbf{v}.\boldsymbol{\nabla})\,\mathbf{v}. \qquad (3.14.10)$$

For these two results, the moving frame is one following the particle.

3.2. Divergence of a vector

The divergence of a vector $\mathbf{A}$ is defined by

$$\boldsymbol{\nabla}.\mathbf{A} = \left(\mathbf{i}\frac{\partial}{\partial x} + \mathbf{j}\frac{\partial}{\partial y} + \mathbf{k}\frac{\partial}{\partial z}\right).(A_x\mathbf{i} + A_y\mathbf{j} + A_z\mathbf{k})$$

$$= \frac{\partial A_x}{\partial x} + \frac{\partial A_y}{\partial y} + \frac{\partial A_z}{\partial z}. \qquad (3.2.1)$$

It may be regarded as the scalar product of the vector operator $\boldsymbol{\nabla}$ and the vector $\mathbf{A}$. It should be observed that

$$\mathbf{A}.\boldsymbol{\nabla} \neq \boldsymbol{\nabla}.\mathbf{A}. \qquad (3.2.2)$$

This is because multiplication involving operators is not of necessity commutative. Vectors whose divergence is zero are said to be solenoidal.

It is easy to see that

$$\boldsymbol{\nabla}.(\mathbf{A} + \mathbf{B}) = \boldsymbol{\nabla}.\mathbf{A} + \boldsymbol{\nabla}.\mathbf{B}. \qquad (3.2.3)$$

The divergence of the vector formed by the product of a scalar and a vector can easily be found

$$\boldsymbol{\nabla}.(\phi\mathbf{A}) = \left(\mathbf{i}\frac{\partial}{\partial x} + \mathbf{j}\frac{\partial}{\partial y} + \mathbf{k}\frac{\partial}{\partial z}\right).(\phi A_x\mathbf{i} + \phi A_y\mathbf{j} + \phi A_z\mathbf{k})$$

$$= \frac{\partial}{\partial x}(\phi A_x) + \frac{\partial}{\partial y}(\phi A_y) + \frac{\partial}{\partial z}(\phi A_z) \qquad (3.2.4)$$

$$= \frac{\partial \phi}{\partial x}A_x + \frac{\partial \phi}{\partial y}A_y + \frac{\partial \phi}{\partial z}A_z$$

$$+ \phi\left(\frac{\partial A_x}{\partial x} + \frac{\partial A_y}{\partial y} + \frac{\partial A_z}{\partial z}\right)$$

$$= (\boldsymbol{\nabla}\phi).\mathbf{A} + \phi\boldsymbol{\nabla}.\mathbf{A}. \qquad (3.2.5)$$

3.21. *Special divergences*

With the usual notation

$$\boldsymbol{\nabla}.\mathbf{r} = \boldsymbol{\nabla}.\mathbf{R} = 3 \tag{3.21.1}$$

$$\begin{aligned}\boldsymbol{\nabla}.\{f(r)\hat{\mathbf{r}}\} &= \boldsymbol{\nabla}.\left\{\mathbf{r}\frac{f(r)}{r}\right\} \\ &= \boldsymbol{\nabla}.\mathbf{r}\frac{f(r)}{r} + \mathbf{r}.\boldsymbol{\nabla}\frac{f(r)}{r} \\ &= \frac{3f(r)}{r} + \mathbf{r}.\hat{\mathbf{r}}\frac{d}{dr}\left\{\frac{f(r)}{r}\right\} \\ &= \frac{1}{r^2}\frac{d}{dr}\{r^2 f(r)\}.\end{aligned} \tag{3.21.2}$$

3.22. *Laplacian operator*

Let ϕ be some scalar depending upon position. Then, if $\mathbf{P}$ is defined by

$$\mathbf{P} = \boldsymbol{\nabla}\phi \tag{3.22.1}$$

$$\boldsymbol{\nabla}.\mathbf{P} = \boldsymbol{\nabla}.(\boldsymbol{\nabla}\phi) \tag{3.22.2}$$

$$\begin{aligned}&= \left(\mathbf{i}\frac{\partial}{\partial x} + \mathbf{j}\frac{\partial}{\partial y} + \mathbf{k}\frac{\partial}{\partial z}\right)\cdot\left(\mathbf{i}\frac{\partial\phi}{\partial x} + \mathbf{j}\frac{\partial\phi}{\partial y} + \mathbf{k}\frac{\partial\phi}{\partial z}\right) \\ &= \frac{\partial^2\phi}{\partial x^2} + \frac{\partial^2\phi}{\partial y^2} + \frac{\partial^2\phi}{\partial z^2}.\end{aligned} \tag{3.22.3}$$

It is convenient to write this as $\nabla^2\phi$ with an obvious notation. ∇^2 is an operator, termed the Laplacian operator. A function ϕ such that

$$\nabla^2\phi = 0 \tag{3.22.4}$$

is said to be harmonic. The Laplacian operator may be applied to a vector also with the meaning

$$\nabla^2\mathbf{A} = i\boldsymbol{\nabla}^2 A_x + \mathbf{j}\nabla^2 A_y + \mathbf{k}\nabla^2 A_z. \tag{3.22.5}$$

3.3. **Curl of a vector**

The curl of a vector $\mathbf{A}$ is defined by

$$\begin{aligned}\boldsymbol{\nabla}\times\mathbf{A} &= \left(\mathbf{i}\frac{\partial}{\partial x} + \mathbf{j}\frac{\partial}{\partial y} + \mathbf{k}\frac{\partial}{\partial z}\right)\times(A_x\mathbf{i} + A_y\mathbf{j} + A_z\mathbf{k}) \\ &= \mathbf{i}\left(\frac{\partial A_z}{\partial y} - \frac{\partial A_y}{\partial z}\right) + \mathbf{j}\left(\frac{\partial A_x}{\partial z} - \frac{\partial A_z}{\partial x}\right) + \mathbf{k}\left(\frac{\partial A_y}{\partial x} - \frac{\partial A_x}{\partial y}\right).\end{aligned} \tag{3.3.1}$$

An alternative way of writing this is

$$\nabla \times \mathbf{A} = \begin{vmatrix} \mathbf{i} & \mathbf{j} & \mathbf{k} \\ \dfrac{\partial}{\partial x} & \dfrac{\partial}{\partial y} & \dfrac{\partial}{\partial z} \\ A_x & A_y & A_z \end{vmatrix} \tag{3.3.2}$$

The curl may be regarded as the vector product of the vector operator ∇ and the vector **A**. Vectors whose curl vanishes are termed irrotational.

The curl of a polar vector is an axial vector and the curl of an axial vector is a polar vector. The proof is almost identical with the proof that the vector product of a polar vector and a polar vector is an axial vector and that the vector product of a polar vector and an axial vector is a polar vector.

It is easy to see that

$$\nabla \times (\mathbf{A} + \mathbf{B}) = \nabla \times \mathbf{A} + \nabla \times \mathbf{B} \tag{3.3.3}$$

and to prove that

$$\nabla \times (\phi \mathbf{A}) = (\nabla \phi) \times \mathbf{A} + \phi (\nabla \times \mathbf{A}). \tag{3.3.4}$$

3.31. *Special curls*

With the usual notation

$$\nabla \times \mathbf{r} = \mathbf{i}\left(\frac{\partial z}{\partial y} - \frac{\partial y}{\partial z}\right) + \mathbf{j}\left(\frac{\partial x}{\partial z} - \frac{\partial z}{\partial x}\right) + \mathbf{k}\left(\frac{\partial y}{\partial x} - \frac{\partial x}{\partial y}\right)$$

$$= \mathbf{0}. \tag{3.31.1}$$

Similarly

$$\nabla \times \mathbf{R} = \nabla' \times \mathbf{R} = \mathbf{0} \tag{3.31.2}$$

where $\mathbf{R} = \mathbf{r} - \mathbf{r}'$.

It is sometimes convenient when considering the curl of a vector to break down the formula into two components, one parallel to a certain direction, and the other perpendicular to it.

Suppose that this is the z direction. Let

$$\mathbf{A} = \mathbf{A}_\perp + A_z \mathbf{k} \tag{3.31.3}$$

$$\nabla = \nabla_\perp + \mathbf{k}\frac{\partial}{\partial z}. \tag{3.31.4}$$

Then

$$\nabla \times \mathbf{A} = \left(\nabla_\perp + \mathbf{k}\frac{\partial}{\partial z}\right) \times (\mathbf{A}_\perp + A_z \mathbf{k})$$

$$= \nabla_\perp \times \mathbf{A}_\perp + \mathbf{k} \times \frac{\partial \mathbf{A}_\perp}{\partial z} + \nabla_\perp \times (A_z \mathbf{k})$$

because $\mathbf{k} \times \mathbf{k}$ vanishes

$$= \nabla_{\perp} \times \mathbf{A}_{\perp} + \mathbf{k} \times \left(\frac{\partial \mathbf{A}_{\perp}}{\partial z} - \nabla_{\perp} A_z \right).$$

Now

$$\mathbf{k} \times \left(\frac{\partial \mathbf{A}_{\perp}}{\partial z} - \nabla_{\perp} A_z \right)$$

is perpendicular to $\mathbf{k}$ and it may be verified that $\nabla_{\perp} \times \mathbf{A}_{\perp}$ is in fact parallel to $\mathbf{k}$. Thus the required resolution into two components has been accomplished.

3.32. *Second-order differential formulae*

There are a number of formulae in the vector calculus which involve combinations of gradient, divergence and curl. These formulae are analogous with those involving products of vectors and may be obtained by expansion into rectangular cartesian components. There are in particular two formulae of great importance which are analogous with the vanishing of $\mathbf{a}.(\mathbf{a} \times \mathbf{b})$ and $\mathbf{a} \times (\mathbf{a}\phi)$. Now

$$\nabla \times (\nabla \phi) = \begin{vmatrix} \mathbf{i} & \mathbf{j} & \mathbf{k} \\ \frac{\partial}{\partial x} & \frac{\partial}{\partial y} & \frac{\partial}{\partial z} \\ \frac{\partial \phi}{\partial x} & \frac{\partial \phi}{\partial y} & \frac{\partial \phi}{\partial z} \end{vmatrix}$$

$$= \mathbf{i} \left(\frac{\partial^2 \phi}{\partial y \, \partial z} - \frac{\partial^2 \phi}{\partial z \, \partial y} \right) + \mathbf{j} \left(\frac{\partial^2 \phi}{\partial z \, \partial x} - \frac{\partial^2 \phi}{\partial x \, \partial z} \right) + \mathbf{k} \left(\frac{\partial^2 \phi}{\partial x \, \partial y} - \frac{\partial^2 \phi}{\partial y \, \partial x} \right)$$

and

$$\nabla.(\nabla \times \mathbf{A}) = \left(\frac{\partial}{\partial x}, \frac{\partial}{\partial y}, \frac{\partial}{\partial z} \right).$$

$$\left(\frac{\partial A_y}{\partial z} - \frac{\partial A_z}{\partial y}, \frac{\partial A_z}{\partial x} - \frac{\partial A_x}{\partial z}, \frac{\partial A_x}{\partial y} - \frac{\partial A_y}{\partial x} \right)$$

$$= \frac{\partial^2 A_y}{\partial x \, \partial z} - \frac{\partial^2 A_y}{\partial z \, \partial x} + \frac{\partial^2 A_z}{\partial y \, \partial x} - \frac{\partial^2 A_z}{\partial x \, \partial y} + \frac{\partial^2 A_x}{\partial z \, \partial y} - \frac{\partial^2 A_x}{\partial y \, \partial z}.$$

Now in general

$$\frac{\partial^2 U}{\partial \alpha \, \partial \beta} = \frac{\partial^2 U}{\partial \beta \, \partial \alpha}$$

(see e.g., E. G. Phillips, *A Course in Analysis*, Cambridge Univ. Press (1939), p. 225), and so it follows that

$$\nabla \times (\nabla \phi) = \mathbf{0} \tag{3.32.1}$$

and

$$\nabla . (\nabla \times \mathbf{A}) = 0. \tag{3.32.2}$$

Both these formulae are of great importance.

It follows from equation (3.32.1) that if $\mathbf{B} = \nabla \times \mathbf{A}$, then $\mathbf{A}$ can be replaced by $\mathbf{A} + \nabla \psi$ where ψ is arbitrary. $\mathbf{A}$ is in this case the vector potential of $\mathbf{B}$, but is indefinite with respect to an arbitrary gradient.

Consider now the repeated curl $\nabla \times (\nabla \times \mathbf{A})$. This may obviously be evaluated by expansion. However, it is not necessary to evaluate all the terms in the expansion because if the x component, say, is evaluated, the others follow from symmetry.

Let

$$\mathbf{B} = \nabla \times \mathbf{A}. \tag{3.32.3}$$

The x component of $\nabla \times (\nabla \times \mathbf{A})$ is $\mathbf{i} . (\nabla \times \mathbf{B})$, which is

$$\frac{\partial B_y}{\partial z} - \frac{\partial B_z}{\partial y} = \frac{\partial}{\partial z}\left(\frac{\partial A_x}{\partial x} - \frac{\partial A_x}{\partial z}\right) - \frac{\partial}{\partial y}\left(\frac{\partial A_x}{\partial y} - \frac{\partial A_y}{\partial x}\right)$$

$$= \frac{\partial}{\partial x}\left(\frac{\partial A_x}{\partial x} + \frac{\partial A_y}{\partial y} + \frac{\partial A_z}{\partial z}\right) - \left(\frac{\partial^2}{\partial x^2} + \frac{\partial^2}{\partial y^2} + \frac{\partial^2}{\partial z^2}\right) A_x. \tag{3.32.4}$$

A term $\partial^2 A_x/\partial x^2$ has been added in and subtracted out, and the usual assumption is made that differentiation with respect to various variables is commutative.

Thus

$$\mathbf{i} . \{\nabla \times (\nabla \times \mathbf{A})\} = \frac{\partial}{\partial x}(\nabla . \mathbf{A}) - \nabla^2 A_x$$

whence, on considering the three components,

$$\nabla \times (\nabla \times \mathbf{A}) = \nabla(\nabla . \mathbf{A}) - \nabla^2 \mathbf{A} \tag{3.32.5}$$

where $\nabla^2 \mathbf{A}$ is interpreted as in (3.22.5). This may be compared with the formula

$$\mathbf{a} \times (\mathbf{a} \times \mathbf{b}) = \mathbf{a}(\mathbf{a} . \mathbf{b}) - a^2 \mathbf{b}.$$

3.33. *Expansion formulae*

A number of expansion formulae have already been discussed. There remain, however, formulae which involve the application of a differential operator to a product of vectors.

$$\nabla.(\mathbf{A}\times\mathbf{B}) = \mathbf{B}.(\nabla\times\mathbf{A}) - \mathbf{A}.(\nabla\times\mathbf{B}) \tag{3.33.1}$$

$$\nabla\times(\mathbf{A}\times\mathbf{B}) = (\mathbf{B}.\nabla)\mathbf{A} - (\mathbf{A}.\nabla)\mathbf{B} + \mathbf{A}(\nabla.\mathbf{B}) - \mathbf{B}(\nabla.\mathbf{A}) \tag{3.33.2}$$

$$\nabla(\mathbf{A}.\mathbf{B}) = (\mathbf{B}.\nabla)\mathbf{A} + (\mathbf{A}.\nabla)\mathbf{B} + \mathbf{B}\times(\nabla\times\mathbf{A}) + \mathbf{A}\times(\nabla\times\mathbf{B}). \tag{3.33.3}$$

It will be noted that formulae (3.33.1) and (3.33.2) which involve the vector product $\mathbf{A}\times\mathbf{B}$ are antisymmetrical in $\mathbf{A}$ and $\mathbf{B}$, whereas formula (3.33.3) which involves the scalar product $\mathbf{A}.\mathbf{B}$ is symmetrical in $\mathbf{A}$ and $\mathbf{B}$. These formulae can be proved by expansion into the rectangular cartesian components.
For example, if $\mathbf{A}\times\mathbf{B} = \mathbf{C}$

$$\begin{aligned}
\mathbf{i}.(\nabla\times\mathbf{C}) &= \frac{\partial C_z}{\partial y} - \frac{\partial C_y}{\partial z} \\
&= \frac{\partial}{\partial y}(A_x B_y - A_y B_x) - \frac{\partial}{\partial z}(A_z B_x - A_x B_z) \\
&= B_x\frac{\partial A_x}{\partial x} + B_y\frac{\partial A_x}{\partial y} + B_z\frac{\partial A_x}{\partial z} \\
&\quad - A_x\frac{\partial B_x}{\partial x} - A_y\frac{\partial B_x}{\partial y} - A_z\frac{\partial B_x}{\partial z} \\
&\quad + A_x\left(\frac{\partial B_x}{\partial x} + \frac{\partial B_y}{\partial y} + \frac{\partial B_z}{\partial z}\right) \\
&\quad - B_x\left(\frac{\partial A_x}{\partial x} + \frac{\partial A_y}{\partial y} + \frac{\partial A_z}{\partial z}\right)
\end{aligned}$$

on adding certain terms in and taking them away again

$$= (\mathbf{B}.\nabla)A_x - (\mathbf{A}.\nabla)B_x + A_x(\nabla.\mathbf{B}) - B_x(\nabla.\mathbf{A}) \tag{3.33.4}$$

whence by considering each of the components in turn, and adding, equation (3.33.2) follows.

An alternative method of obtaining the expansion formulae follows.

∇ is written in the form $\nabla = \nabla_{\mathrm{A}} + \nabla_{\mathrm{B}}$, where ∇_{A} operates only on $\mathbf{A}$, and ∇_{B} operates only on $\mathbf{B}$. ∇_{A} is a mere vector as regards $\mathbf{B}$ and ∇_{B} as regards $\mathbf{A}$.

Consider

$$\nabla(\mathbf{A}.\mathbf{B}) = \nabla_{\mathrm{A}}(\mathbf{A}.\mathbf{B}) + \nabla_{\mathrm{B}}(\mathbf{A}.\mathbf{B}). \tag{3.33.5}$$

Were $\boldsymbol{\nabla}_{\mathbf{A}}$ a mere vector, $\boldsymbol{\nabla}_{\mathbf{A}}(\mathbf{A}.\mathbf{B})$ would be part of the triple vector product expansion and so

$$\begin{aligned}\boldsymbol{\nabla}_{\mathbf{A}}(\mathbf{A}.\mathbf{B}) &= (\mathbf{B}.\boldsymbol{\nabla}_{\mathbf{A}})\mathbf{A} + \mathbf{B} \times (\boldsymbol{\nabla}_{\mathbf{A}} \times \mathbf{A}) \\ &= (\mathbf{B}.\boldsymbol{\nabla})\mathbf{A} + \mathbf{B} \times (\boldsymbol{\nabla} \times \mathbf{A}) \qquad (3.33.6)\end{aligned}$$

because $\boldsymbol{\nabla}_{\mathbf{A}}$ acting on $\mathbf{A}$ has exactly the same form as $\boldsymbol{\nabla}$ acting on $\mathbf{A}$. Similarly

$$\begin{aligned}\boldsymbol{\nabla}_{\mathbf{B}}(\mathbf{A}.\mathbf{B}) &= \boldsymbol{\nabla}_{\mathbf{B}}(\mathbf{B}.\mathbf{A}) \\ &= (\mathbf{A}.\boldsymbol{\nabla})\mathbf{B} + \mathbf{A} \times (\boldsymbol{\nabla} \times \mathbf{B}) \qquad (3.33.7)\end{aligned}$$

and it follows that

$$(\mathbf{A}.\mathbf{B}) = (\mathbf{B}.\boldsymbol{\nabla})\mathbf{A} + (\mathbf{A}.\boldsymbol{\nabla})\mathbf{B} + \mathbf{B} \times (\boldsymbol{\nabla} \times \mathbf{A}) + \mathbf{A} \times (\boldsymbol{\nabla} \times \mathbf{B}). \qquad (3.33.8)$$

3.34. *Beltrami fields*

A vector field $\mathbf{F}$ which satisfies the relation

$$\mathbf{F} \times (\boldsymbol{\nabla} \times \mathbf{F}) = \mathbf{0} \qquad (\boldsymbol{\nabla} \times \mathbf{F} \neq \mathbf{0}) \qquad (3.34.1)$$

is termed a Beltrami field.

It follows that

$$\boldsymbol{\nabla} \times \mathbf{F} = \alpha\mathbf{F} \qquad (3.34.2)$$

where α is arbitrary, and that

$$\begin{aligned}0 = \boldsymbol{\nabla}.(\boldsymbol{\nabla} \times \mathbf{F}) &= \boldsymbol{\nabla}.(\alpha\mathbf{F}) \\ &= (\mathbf{F}.\boldsymbol{\nabla})\,\alpha + \alpha(\boldsymbol{\nabla}.\mathbf{F}).\end{aligned}$$

It follows that if α is a constant

$$\boldsymbol{\nabla}.\mathbf{F} = 0. \qquad (3.34.3)$$

From equation (3.34.1)

$$\boldsymbol{\nabla} \times (\boldsymbol{\nabla} \times \mathbf{F}) = \alpha(\boldsymbol{\nabla} \times \mathbf{F})$$

which can be rewritten as

$$-\nabla^2\mathbf{F} + \boldsymbol{\nabla}(\boldsymbol{\nabla}.\mathbf{F}) = \alpha^2\mathbf{F}$$

and using equation (3.34.3) it follows that

$$\nabla^2\mathbf{F} + \alpha^2\mathbf{F} = 0. \qquad (3.34.4)$$

Not all solutions of equation (3.34.4) are, however, solutions of (3.34.2). Because α occurs as α^2 in equation (3.34.4), it follows that solutions of the congugate equation $\boldsymbol{\nabla} \times \mathbf{F} = -\alpha\mathbf{F}$ are also solutions of (3.34.4). It is fairly easy to verify that the curl of a Beltrami field is also a Beltrami field.

3.4. Expressions involving $\boldsymbol{\nabla}$ in orthogonal curvilinear coordinates

Suppose that u_1, u_2, u_3 form a system of orthogonal curvilinear coordinates as discussed in section 2.9.

The vector operator $\boldsymbol{\nabla}$ may be defined through

$$f(\mathbf{r}+\delta\mathbf{r}) = f(\mathbf{r}) + \delta\mathbf{r}\,.\,\boldsymbol{\nabla} f. \tag{3.4.1}$$

If now the quantity $f(\mathbf{r})$ be expressed, not in terms of the position vector, but in terms of u_1, u_2, u_3, this takes the form

$$f(u_1+\delta u_1, u_2+\delta u_2, u_3+\delta u_3) = f(u_1, u_2, u_3) + (\delta\mathbf{r}\,.\,\boldsymbol{\nabla}) f \tag{3.4.2}$$

$$= f(u_1, u_2, u_3) + \delta u_1 \frac{\partial f}{\partial u_1} + \delta u_2 \frac{\partial f}{\partial u_2} + \delta u_3 \frac{\partial f}{\partial u_3} \tag{3.4.3}$$

by Taylor's theorem.

It follows that

$$(\delta\mathbf{r}\,.\,\boldsymbol{\nabla}) f = u_1 \frac{\partial f}{\partial u_1} + \delta u_2 \frac{\partial f}{\partial u_2} + \delta u_3 \frac{\partial f}{\partial u_3}. \tag{3.4.4}$$

Now in orthogonal curvilinear coordinates

$$\delta\mathbf{r} = h_1\,\delta u_1\,\mathbf{i}_1 + h_2\,\delta u_2\,\mathbf{i}_2 + h_3\,\delta u_3\,\mathbf{i}_3 \tag{3.4.5}$$

and if

$$\boldsymbol{\nabla} = \mathbf{i}_1\,\partial_1 + \mathbf{i}_2\,\partial_2 + \mathbf{i}_3\,\partial_3 \tag{3.4.6}$$

where ∂_1, ∂_2, ∂_3 are operators

$$(h_1\,\delta u_1\,\partial_1 + h_2\,\delta u_2\,\partial_2 + h_3\,\delta u_3\,\partial_3) f = \delta u_1 \frac{\partial f}{\partial u_1} + \delta u_2 \frac{\partial f}{\partial u_2} + \delta u_3 \frac{\partial f}{\partial u_3} \tag{3.4.7}$$

Now δu_1, δu_2, δu_3 are arbitrary and so

$$h_1\,\delta u_1\,\partial_1 f = \delta u_1 \frac{\partial f}{\partial u_1}$$

giving

$$\partial_1 f = \frac{1}{h_1}\frac{\partial f}{\partial u_1}, \text{ etc.} \tag{3.4.8}$$

and so

$$\boldsymbol{\nabla} = \mathbf{i}_1 \frac{1}{h_1}\frac{\partial}{\partial u_1} + \mathbf{i}_2 \frac{1}{h_2}\frac{\partial}{\partial u_2} + \mathbf{i}_3 \frac{1}{h_3}\frac{\partial}{\partial u_3}. \tag{3.4.9}$$

It must be remembered that $\mathbf{i}_1$, $\mathbf{i}_2$, $\mathbf{i}_3$, although unit vectors, are not constant vectors. Accordingly the unit vectors must be written before the differential operators.

It follows immediately that

$$\boldsymbol{\nabla} f = \mathbf{i}_1 \frac{1}{h_1}\frac{\partial f}{\partial u_1} + \mathbf{i}_2 \frac{1}{h_2}\frac{\partial f}{\partial u_2} + \mathbf{i}_3 \frac{1}{h_3}\frac{\partial f}{\partial u_3}. \tag{3.4.10}$$

The gradients of the coordinates themselves take a particularly simple form

$$\nabla u_1 = \mathbf{i}_1 \frac{1}{h_1}, \text{ etc.} \tag{3.4.11}$$

Before proceeding to obtaining expressions for divergence and curl in orthogonal coordinates, it is necessary to evaluate the divergence and curl of the unit vectors. This is because, as has been mentioned, they are not constants and differentiating them does not give rise to zero. The evaluation of these is somewhat indirect and is done as follows

$$\begin{aligned} \mathbf{0} = \nabla \times (\nabla u_1) &= \nabla \times \left(\frac{\mathbf{i}_1}{h_1}\right) \\ &= \frac{1}{h_1}(\nabla \times \mathbf{i}_1) - \mathbf{i}_1 \times \nabla\left(\frac{1}{h_1}\right) \end{aligned}$$

whence

$$\nabla \times \mathbf{i}_1 = h_1 \mathbf{i}_1 \times \nabla\left(\frac{1}{h_1}\right) \tag{3.4.12}$$

$$= \frac{1}{h_1}\left(\mathbf{i}_2 \frac{1}{h_3}\frac{\partial h_1}{\partial u_3} - \mathbf{i}_3 \frac{1}{h_2}\frac{\partial h_1}{\partial u_2}\right). \tag{3.4.13a}$$

Similarly

$$\nabla \times \mathbf{i}_2 = \frac{1}{h_2}\left(\mathbf{i}_3 \frac{1}{h_1}\frac{\partial h_2}{\partial u_1} - \mathbf{i}_1 \frac{1}{h_3}\frac{\partial h_2}{\partial u_3}\right) \tag{3.4.13b}$$

and

$$\nabla \times \mathbf{i}_3 = \frac{1}{h_3}\left(\mathbf{i}_1 \frac{1}{h_2}\frac{\partial h_3}{\partial u_2} - \mathbf{i}_2 \frac{1}{h_1}\frac{\partial h_3}{\partial u_1}\right). \tag{3.4.13c}$$

It follows from equations (3.4.13) that

$$\nabla . \mathbf{i}_1 = \nabla . (\mathbf{i}_2 \times \mathbf{i}_3) = \mathbf{i}_3 . (\nabla \times \mathbf{i}_2) - \mathbf{i}_2 . (\nabla \times \mathbf{i}_3)$$

$$= \frac{1}{h_1 h_2 h_3}\frac{\partial}{\partial u_1}(h_2 h_3). \tag{3.4.14a}$$

Similarly

$$\nabla . \mathbf{i}_2 = \frac{1}{h_1 h_2 h_3}\frac{\partial}{\partial u_2}(h_3 h_1) \tag{3.4.14b}$$

$$\nabla . \mathbf{i}_3 = \frac{1}{h_1 h_2 h_3}\frac{\partial}{\partial u_3}(h_1 h_2). \tag{3.4.14c}$$

Now

$$\nabla \times \mathbf{F} = \nabla \times (F_1 \mathbf{i}_1 + F_2 \mathbf{i}_2 + F_3 \mathbf{i}_3)$$

where F_α is the component of $\mathbf{F}$ in the direction of unit vector $\mathbf{i}_\alpha$, that is $\mathbf{F}.\mathbf{i}_\alpha$

$$\begin{aligned}\nabla \times (\mathbf{F}_1 \mathbf{i}_1) &= F_1(\nabla \times \mathbf{i}_1) - \mathbf{i}_1 \times \nabla F_1 \\ &= \frac{F_1}{h_1}\left(\mathbf{i}_2 \frac{\partial h_1}{\partial u_3} - \mathbf{i}_3 \frac{\partial h_1}{\partial u_2}\right) \\ &\quad + \mathbf{i}_2 \frac{1}{h_3} \frac{\partial F_1}{\partial u_3} - \mathbf{i}_3 \frac{1}{h_2} \frac{\partial F_1}{\partial u_2} \\ &= \frac{1}{h_1}\left\{\mathbf{i}_2 \frac{1}{h_3} \frac{\partial}{\partial u_3}(h_1 F_1) - \mathbf{i}_3 \frac{1}{h_2} \frac{\partial}{\partial u_2}(h_1 F_1)\right\}.\end{aligned}$$

Thus

$$\begin{aligned}\nabla \times \mathbf{F} &= \frac{\mathbf{i}_1}{h_2 h_3}\left\{\frac{\partial}{\partial u_2}(h_3 F_3) - \frac{\partial}{\partial u_3}(h_2 F_2)\right\} \\ &\quad + \frac{\mathbf{i}_2}{h_3 h_1}\left\{\frac{\partial}{\partial u_3}(h_1 F_1) - \frac{\partial}{\partial u_1}(h_3 F_3)\right\} \\ &\quad + \frac{\mathbf{i}_3}{h_1 h_2}\left\{\frac{\partial}{\partial u_1}(h_2 F_2) - \frac{\partial}{\partial u_2}(h_1 F_1)\right\} \qquad (3.4.15)\end{aligned}$$

$$= \frac{1}{h_1 h_2 h_3}\begin{vmatrix} h_1 \mathbf{i}_1 & h_2 \mathbf{i}_2 & h_3 \mathbf{i}_3 \\ \dfrac{\partial}{\partial u_1} & \dfrac{\partial}{\partial u_2} & \dfrac{\partial}{\partial u_3} \\ h_1 F_1 & h_2 F_2 & h_3 F_3 \end{vmatrix} \qquad (3.4.16)$$

For the divergence

$$\nabla.\mathbf{F} = \nabla.(F_1 \mathbf{i}_1 + F_2 \mathbf{i}_2 + F_3 \mathbf{i}_3)$$

$$\begin{aligned}\nabla.(F_1 \mathbf{i}_1) &= (\mathbf{i}_1 . \nabla) F_1 + F_1(\nabla . \mathbf{i}_1) \\ &= \frac{1}{h_1} \frac{\partial F_1}{\partial u_1} + F_1 \frac{1}{h_1 h_2 h_3} \frac{\partial}{\partial u_1}(h_2 h_3) \\ &= \frac{1}{h_1 h_2 h_3} \frac{\partial}{\partial u_1}(h_2 h_3 F_1).\end{aligned}$$

It follows by symmetry that

$$\nabla.\mathbf{F} = \frac{1}{h_1 h_2 h_3}\left\{\frac{\partial}{\partial u_1}(h_2 h_3 F_1) + \frac{\partial}{\partial u_2}(h_3 h_1 F_2) + \frac{\partial}{\partial u_3}(h_1 h_2 F_3)\right\}. \quad (3.4.17)$$

The Laplacian is given by

$$\nabla^2 f = \boldsymbol{\nabla}.(\boldsymbol{\nabla} f)$$

$$= \frac{1}{h_1 h_2 h_3}\left\{\frac{\partial}{\partial u_1}\left(\frac{h_2 h_3}{h_1}\frac{\partial f}{\partial u_1}\right) + \frac{\partial}{\partial u_2}\left(\frac{h_3 h_1}{h_2}\frac{\partial f}{\partial u_2}\right) + \frac{\partial}{\partial u_3}\left(\frac{h_1 h_2}{h_3}\frac{\partial f}{\partial u_3}\right)\right\} \tag{3.4.18}$$

The operator ∇^2 when applied to a vector $\mathbf{F}$ can be expressed in the same way. However, unless u_1, u_2, u_3 are cartesian coordinates, $\mathbf{i}_1$, $\mathbf{i}_2$, $\mathbf{i}_3$ are all functions of u_1, u_2, u_3 and so in general

$$\begin{aligned} \nabla^2 \mathbf{F} &= \nabla^2(F_1 \mathbf{i}_1 + F_2 \mathbf{i}_2 + F_3 \mathbf{i}_3) \\ &\neq \mathbf{i}_1 \nabla^2 F_1 + \mathbf{i}_2 \nabla^2 F_2 + \mathbf{i}_3 \nabla^2 F_3 \end{aligned} \tag{3.4.19}$$

It is possible, however, to write

$$\nabla^2 \mathbf{F} = \mathbf{i}\nabla^2 F_x + \mathbf{j}\nabla^2 F_y + \mathbf{k}\nabla^2 F_2 \tag{3.4.20}$$

where ∇^2 expressed in orthogonal curvilinear coordinates is applied to the rectangular cartesian components which are, of course, scalar quantities. It is worth noting that it is possible to find ∇^2 in any system of coordinates (whether orthogonal or not) by means of a transformation of variables in the differential operator

$$\frac{\partial^2}{\partial x^2} + \frac{\partial^2}{\partial y^2} + \frac{\partial^2}{\partial z^2}.$$

3.41. $\boldsymbol{\nabla}$ *in cylindrical and spherical coordinate systems*

(*a*) **Cylindrical polars.** Let $\rho = u_1$, $\phi = u_2$, $z = u_3$, where $x = \rho\cos\phi$, $y = \rho\sin\phi$

$$\begin{aligned} (\delta s)^2 &= h_1^2(\delta u_1)^2 + h_2^2(\delta u_2)^2 + h_3^2(\delta u_3)^2 \\ &= (\delta\rho)^2 + \rho^2(\delta\phi)^2 + (\delta z)^2 \end{aligned} \tag{3.41.1}$$

Thus $h_1 = 1$, $h_2 = \rho$, $h_3 = 1$, and so

$$\boldsymbol{\nabla} = \hat{\boldsymbol{\rho}}\frac{\partial}{\partial\rho} + \hat{\boldsymbol{\varphi}}\frac{1}{\rho}\frac{\partial}{\partial\phi} + \mathbf{k}\frac{\partial}{\partial z} \tag{3.41.2}$$

$$\boldsymbol{\nabla}\psi = \hat{\boldsymbol{\rho}}\frac{\partial\psi}{\partial\rho} + \hat{\boldsymbol{\varphi}}\frac{1}{\rho}\frac{\partial\psi}{\partial\phi} + \mathbf{k}\frac{\partial\psi}{\partial z} \tag{3.41.3}$$

and if

$$\mathbf{F} = \hat{\boldsymbol{\rho}} F_\rho + \hat{\boldsymbol{\varphi}} F_\phi + \mathbf{k} F_z$$

$$\nabla . \mathbf{F} = \frac{1}{\rho}\frac{\partial}{\partial \rho}(\rho F_\rho) + \frac{1}{\rho}\frac{\partial F_\phi}{\partial \phi} + \frac{\partial F_z}{\partial z} \tag{3.41.4}$$

$$\nabla \times \mathbf{F} = \left(\frac{1}{\rho}\frac{\partial F_z}{\partial \phi} - \frac{\partial F_\phi}{\partial z}\right)\hat{\boldsymbol{\rho}} + \left(\frac{\partial F_\rho}{\partial z} - \frac{\partial F_z}{\partial \rho}\right)\hat{\boldsymbol{\varphi}} \tag{3.41.5}$$

$$+ \left(\frac{1}{\rho}\frac{\partial}{\partial \rho}(\rho F_\phi) - \frac{1}{\rho}\frac{\partial F_\rho}{\partial \phi}\right)\mathbf{k}$$

$$\nabla^2 \psi = \frac{1}{\rho}\frac{\partial}{\partial \rho}\left(\rho \frac{\partial \psi}{\partial \rho}\right) + \frac{1}{\rho^2}\frac{\partial^2 \psi}{\partial \phi^2} + \frac{\partial^2 \psi}{\partial z^2}. \tag{3.41.6}$$

(*b*) **Spherical polars.** Let $r = u_1$, $\theta = u_2$, $\phi = u_3$, where

$$x = r\sin\theta\cos\phi, \qquad y = r\sin\theta\sin\phi, \qquad x = r\cos\theta.$$

$$(\delta s)^2 = h_1^2(\delta u_1)^2 + h_2^2(\delta u_2)^2 + h_3^2(\delta u_3)^2$$

$$= (\delta r)^2 + r^2(\delta\theta)^2 + r^2\sin^2\theta(\delta\phi)^2. \tag{3.41.7}$$

Thus $h_1 = 1$, $h_2 = r$, $h_3 = r\sin\theta$, and so

$$\nabla = \hat{\mathbf{r}}\frac{\partial}{\partial r} + \hat{\boldsymbol{\theta}}\frac{1}{r}\frac{\partial}{\partial \theta} + \hat{\boldsymbol{\varphi}}\frac{1}{r\sin\theta}\frac{\partial}{\partial \phi} \tag{3.41.8}$$

$$\nabla\psi = \hat{\mathbf{r}}\frac{\partial \psi}{\partial r} + \hat{\boldsymbol{\theta}}\frac{1}{r}\frac{\partial \psi}{\partial \theta} + \hat{\boldsymbol{\varphi}}\frac{1}{r\sin\theta}\frac{\partial \psi}{\partial \phi} \tag{3.41.9}$$

$$\nabla . \mathbf{F} = \frac{1}{r^2}\frac{\partial}{\partial r}(r^2 F_r) + \frac{1}{r\sin\theta}\frac{\partial}{\partial \theta}(\sin\theta F_\theta) + \frac{1}{r\sin\theta}\frac{\partial F_\phi}{\partial \phi} \tag{3.41.10}$$

$$\nabla \times \mathbf{F} = \frac{1}{r\sin\theta}\left\{\frac{\partial}{\partial \theta}(\sin\theta F_\phi) - \frac{\partial F_\theta}{\partial \phi}\right\}\hat{\mathbf{r}}$$

$$+ \frac{1}{r}\left\{\frac{1}{\sin\theta}\frac{\partial F_r}{\partial \phi} - \frac{\partial}{\partial r}(rF_\phi)\right\}\hat{\boldsymbol{\theta}}$$

$$+ \frac{1}{r}\left\{\frac{\partial}{\partial r}(rF_\theta) - \frac{\partial F_r}{\partial \theta}\right\}\hat{\boldsymbol{\varphi}} \tag{3.41.11}$$

$$\nabla^2 \psi = \frac{1}{r^2}\frac{\partial}{\partial r}\left(r^2\frac{\partial \psi}{\partial r}\right) + \frac{1}{r^2\sin\theta}\frac{\partial}{\partial \theta}\left(\sin\theta\frac{\partial \psi}{\partial \theta}\right) + \frac{1}{r^2\sin^2\theta}\frac{\partial^2 \psi}{\partial \phi^2}. \tag{3.41.12}$$

Worked examples

1. If $\psi(x,y,z) = 2ax^2 - 3axy + 2z^3$, find $\nabla\psi$ at the point $(a, -a, 2a)$.

$$\nabla\psi = \left(\mathbf{i}\frac{\partial}{\partial x} + \mathbf{j}\frac{\partial}{\partial y} + \mathbf{k}\frac{\partial}{\partial z}\right)(2ax^2 - 3axy + 2z^3)$$

$$= \mathbf{i}\{4ax - 3ay\} + \mathbf{j}\{-3ax\} + \mathbf{k}\,.\,6z^2$$

$$= a^2\{7\mathbf{i} - 3\mathbf{j} + 24\mathbf{k}\}$$

at the given point.

2. Find the unit normal to the surface

$$\frac{x^2}{a^2} + \frac{y^2}{b^2} + \frac{z^2}{c^2} - 1 = 0$$

at the point x, y, z and the length of the perpendicular from the origin to the tangent plane there.

The unit normal to the surface $\psi(\mathbf{r}) = 0$ is given by

$$\mathbf{n} = \frac{\nabla\psi}{|\nabla\psi|}$$

$$\psi = \frac{x^2}{a^2} + \frac{y^2}{b^2} + \frac{z^2}{c^2} - 1$$

$$\nabla\psi = \frac{2x\mathbf{i}}{a^2} + \frac{2y\mathbf{j}}{b^2} + \frac{2z\mathbf{k}}{c^2}$$

and so

$$\mathbf{n} = \frac{\dfrac{x\mathbf{i}}{a^2} + \dfrac{y\mathbf{j}}{b^2} + \dfrac{z\mathbf{k}}{c^2}}{\left(\dfrac{x^2}{a^4} + \dfrac{y^2}{b^4} + \dfrac{z^2}{c^4}\right)^{1/2}}.$$

The length of the perpendicular from the origin to the tangent plane is

$$p = \mathbf{n}\,.\,\mathbf{r} = \frac{\dfrac{x^2}{a^2} + \dfrac{y^2}{b^2} + \dfrac{z^2}{c^2}}{\left(\dfrac{x^2}{a^4} + \dfrac{y^2}{b^4} + \dfrac{z^2}{c^4}\right)^{1/2}}$$

$$= \frac{1}{\left(\dfrac{x^2}{a^4} + \dfrac{y^2}{b^4} + \dfrac{z^2}{c^4}\right)^{1/2}}$$

because x, y, z is on the given ellipsoid.

3. Find the equation of the tangent plane to the surface

$$x^4 - 3xyz + z^2 + 1 = 0$$

at the point $(1, 1, 1)$.

$$\nabla(x^3 - 3xyz + z^3 + 1) = \mathbf{i}\{4x^3 - 3yz\} + \mathbf{j}\{-3xz\} + \mathbf{k}\{-3xy + 2z\}.$$

At the point of interest this is

$$\mathbf{i} - 3\mathbf{j} - \mathbf{k}$$

and the unit normal there is

$$\frac{(1, -3, -1)}{\{1^2 + (-3)^2 + (-1)^2\}} = \frac{1}{\sqrt{11}}(1, -3, -1).$$

The equation of the plane through the point $\mathbf{r}_0$ with unit normal $\mathbf{n}$ is given by

$$\mathbf{n}.(\mathbf{r} - \mathbf{r}_0) = 0$$

and in this particular problem this becomes

$$\frac{1}{\sqrt{11}}(1, -3, -1).(x - 1, y - 1, z - 1) = 0$$

which is $x - 3y - z + 3 = 0$. It may be verified that this passes through $(1, 1, 1)$.

4. Evaluate ∇V, where $V = |\mathbf{c} \times \mathbf{r}|^2$, where $\mathbf{c}$ is a constant vector and $\mathbf{r}$ is the position vector.

It is possible to write $V = \mathbf{A}.\mathbf{B}$ and calculate the result by means of the formulae for $\nabla(\mathbf{A}.\mathbf{B})$ with $\mathbf{A} = \mathbf{B} = \mathbf{c} \times \mathbf{r}$. There is, however, a method of proceeding which avoids this and has recourse only to the definition of ∇ and the properties of products.

$$\begin{aligned}\delta V &= \delta(\mathbf{c} \times \mathbf{r}).(\mathbf{c} \times \mathbf{r}) = 2(\mathbf{c} \times \mathbf{r}).(\mathbf{c} \times \delta\mathbf{r}) \\ &= 2[(\mathbf{c} \times \mathbf{r}) \times \mathbf{c}].\delta\mathbf{r}\end{aligned}$$

and so

$$\nabla V = 2(\mathbf{c} \times \mathbf{r}) \times \mathbf{c}.$$

5. Find the directional derivative of $\psi = 2x^3 - 3yz$ at the point $(2, 1, 3)$ in the direction parallel to $(2, 1, -2)$.

The unit vector in the direction $(2, 1, -2)$ is given by

$$\frac{(2, 1, -2)}{\{2^2 + 1^2 + (-2)^2\}} = \tfrac{1}{3}(2, 1, -2) = \mathbf{n}$$

$$\nabla\psi = 6x^2\mathbf{i} - 3z\mathbf{j} - 3y\mathbf{k}$$

$\nabla\psi$ at the given point is

$$\mathbf{A} = 24\mathbf{i} - 9\mathbf{j} - 3\mathbf{k}$$

and

$$\begin{aligned}\mathbf{A}.\mathbf{n} &= \{2.24 + 1(-9) + (-2)(-3)\}\tfrac{1}{3} \\ &= 15.\end{aligned}$$

6. Find

$$\nabla\left(\frac{e^{\lambda r}}{r}\right)$$

and hence

$$\nabla^2\left(\frac{e^{\lambda r}}{r}\right)$$

$$\nabla\left(\frac{e^{\lambda r}}{r}\right)=\hat{\mathbf{r}}\frac{d}{dr}\left(\frac{e^{\lambda r}}{r}\right).$$

This is a full derivative in r because θ and ϕ are not involved

$$=\hat{\mathbf{r}}\frac{e^{\lambda r}}{r}\left(\lambda-\frac{1}{r}\right)$$

$$\nabla^2\left(\frac{e^{\lambda r}}{r}\right)=\frac{1}{r^2}\frac{d}{dr}\left[r^2\frac{d}{dr}\left(\frac{e^{\lambda r}}{r}\right)\right]$$

$$=\frac{1}{r^2}\frac{d}{dr}(\lambda r e^{\lambda r}-e^{\lambda r})$$

$$=\lambda^2\frac{e^{\lambda r}}{r}$$

Similarly

$$\nabla^2\left(\frac{e^{-\lambda r}}{r}\right)=\lambda^2\left(\frac{e^{-\lambda r}}{r}\right)$$

It follows from this that, if $\psi=e^{\lambda r}/r$ or $e^{-\lambda r}/r$ or any linear combination thereof, such as $\cosh\lambda r/r$, $\sinh\lambda r/r$, obeys the differential equation

$$(\nabla^2-\lambda^2)\psi=0$$

except when $r=0$. At $r=0$, these functions are singular and therefore cannot obey any differential equation there. (However, it is possible to overcome this difficulty by means of the so-called delta functions. This will be referred to later.)

Similarly, $e^{\lambda R}/R$, etc. obey the differential equation

$$(\nabla^2-\lambda^2)\psi=0$$

except at $\mathbf{r}=\mathbf{r}'$. This follows by a change of origin.

7. Prove that the angle between the normal at a point on a paraboloid of revolution, and the axis of the paraboloid, is equal to the angle between the normal and the line drawn to the focus.

In cylindrical coordinates, the equation of the paraboloid may be expressed as $\rho^2=4az$, the focus S being the point $\rho=0$, $z=a$.

Let $P'(\rho',\phi',z')$ be a point on the paraboloid. Writing the equation of the paraboloid in the form

$$\psi \equiv \rho^2 - 4az = 0$$

the unit normal $\mathbf{n}$ is given by

$$\frac{2\rho'\hat{\boldsymbol{\rho}} - 4a\mathbf{k}}{\{4\rho'^2 + 16a^2\}^{1/2}}.$$

Now

$$\overrightarrow{P'S} = \overrightarrow{OS} - \overrightarrow{OP'}$$
$$= a\mathbf{k} - (\rho'\hat{\boldsymbol{\rho}} + z'\mathbf{k})$$

and so

$$\widehat{\overrightarrow{P'S}} = \frac{-\rho'\hat{\boldsymbol{\rho}} + (a-z')\mathbf{k}}{\{\rho^{12} + (a-z')^2\}^{1/2}}$$

The two angles are equal if $\mathbf{n}.\mathbf{k} = \mathbf{n}.\widehat{\overrightarrow{P'S}}$. That is if

$$\frac{-4a}{\{4\rho'^2 + 16a^2\}^{1/2}} = \frac{-2\rho'^2 - 4a(a-z')}{(4\rho'^2 + 16a^2)^{1/2}\{\rho'^2 + (a-z')^2\}^{1/2}}$$

Substituting $\rho'^2 = 4az'$, the condition is equivalent to

$$4a = \frac{8az' + 4a(a-z')}{\{4az' + (a-z')^2\}^{1/2}}$$

which is so.

8. Prove that

$$\nabla \times (\mathbf{G} \times \mathbf{r}) = 2\mathbf{G} + (\mathbf{r}.\nabla)\mathbf{G}$$

if

$$\nabla.\mathbf{G} = 0$$
$$\nabla \times (\mathbf{G} \times \mathbf{r}) = \mathbf{G}(\nabla.\mathbf{r}) - \mathbf{r}(\nabla.\mathbf{G}) + (\mathbf{r}.\nabla)\mathbf{G} - (\mathbf{G}.\nabla)\mathbf{r}$$
$$= 3\mathbf{G} + (\mathbf{r}.\nabla)\mathbf{G} - \mathbf{G}$$
$$= 2\mathbf{G} + (\mathbf{r}.\nabla)\mathbf{G}.$$

9. Prove that, if $\nabla.\mathbf{F} = 0$, then $\mathbf{F}$ can be written in the form

$$\mathbf{F} = \nabla \times \mathbf{A}$$

where

$$\mathbf{A} = \mathbf{G} \times \mathbf{r}$$

and

$$\mathbf{G} = \int_0^1 \mathbf{F}(\lambda\mathbf{r})\,\lambda\,d\lambda$$

provided that $\lim \lambda \to 0\{\lambda^2\mathbf{F}(\lambda\mathbf{r})\} = \mathbf{0}$. (Leicester University.)

$$\nabla . \mathbf{G} = \nabla . \int_0^1 \mathbf{F}(\lambda \mathbf{r})\, \lambda\, d\lambda$$

$$= \int_0^1 \lambda^2 \nabla' . \mathbf{F}(\mathbf{r}')\, d\lambda$$

where $\mathbf{r}' = \lambda \mathbf{r}, \quad \nabla' = \dfrac{\partial}{\partial \mathbf{r}'}$

$= 0$ because $\nabla . \mathbf{F} = 0$.

Applying example 8, it follows that

$$\nabla \times (\mathbf{G} \times \mathbf{r}) = 2\mathbf{G} + (\mathbf{r} . \nabla)\mathbf{G}$$

$$= 2\int_0^1 \mathbf{F}(\lambda \mathbf{r})\, \lambda\, d\lambda + (\mathbf{r} . \nabla) \int_0^1 \lambda \mathbf{F}(\lambda \mathbf{r})\, d\lambda$$

$$= 2\int_0^1 \mathbf{F}(\mathbf{r}')\, \lambda\, d\lambda + \int_0^1 (\mathbf{r}' . \nabla')\, \lambda \mathbf{F}(\mathbf{r}')\, d\lambda$$

$$= \Big[\mathbf{F}(\mathbf{r}')\, \lambda^2\Big]_0^1 - \int_0^1 \frac{\partial}{\partial \lambda}\{\mathbf{F}(\mathbf{r}')\}\, \lambda^2\, d\lambda$$

$$+ \int_0^1 (\mathbf{r}' . \nabla')\, \lambda \mathbf{F}(\mathbf{r}')\, d\lambda$$

$$= \mathbf{F}(\mathbf{r}) - \int_0^1 (\mathbf{r} . \nabla')\, \mathbf{F}(\mathbf{r}')\, \lambda^2\, d\lambda + \int_0^1 (\mathbf{r}' . \nabla')\, \mathbf{F}(\mathbf{r}')\, \lambda\, d\lambda$$

using the condition at $\lambda = 0$.

$$= \mathbf{F}(\mathbf{r})$$

which gives the required results.

10. Prove that

$$(\mathbf{m}_1 . \nabla)(\mathbf{m}_2 . \nabla)\frac{1}{R} = \frac{3(\mathbf{m}_1 . \mathbf{R})(\mathbf{m}_2 . \mathbf{R})}{R^5} - \frac{\mathbf{m}_1 . \mathbf{m}_2}{R^3}$$

$$\nabla\left(\frac{1}{R}\right) = -\frac{1}{R^2}\hat{\mathbf{R}} = -\frac{\mathbf{R}}{R^3}$$

$$\mathbf{m}_2 . \nabla\left(\frac{1}{R}\right) = -\frac{\mathbf{m}_2 . \mathbf{R}}{R^3}$$

$$\nabla\left\{(\mathbf{m}_2 . \nabla)\left(\frac{1}{R}\right)\right\} = -(\mathbf{m}_2 . \mathbf{R})\left(-\frac{3\hat{\mathbf{R}}}{R^4}\right) - \frac{\mathbf{m}_2}{R^3}$$

$$= (\mathbf{m}_2 . \mathbf{R})\frac{3\mathbf{R}}{R^5} - \frac{\mathbf{m}_2}{R^3}$$

whence the result follows.

This result is of considerable importance in electrostatics and hydrodynamics. Dimensional considerations apart, $1/R$ represents the field at the point with position vector $\mathbf{r}$ due to a point charge or source at the point with position vector $\mathbf{r}'$, $\mathbf{m}_2 . \nabla(1/R)$ represents the field due to a dipole of moment $\mathbf{m}_2$, and $(\mathbf{m}_1 . \nabla)(\mathbf{m}_2 . \nabla) 1/R$ the mutual energy of dipoles, moments $\mathbf{m}_1$, $\mathbf{m}_2$, situated at the points with position vectors at $\mathbf{r}$ and $\mathbf{r}'$.

11. Parabolic coordinates are defined by

$$x = auv \cos\phi, \qquad y = auv \sin\phi, \qquad z = \frac{a}{2}(u^2 - v^2)$$

where $u > 0$, $v > 0$, $-\pi < \phi < \pi$.

Find $\nabla^2 \psi$

$$\begin{aligned} \delta x &= a\,\delta u v \cos\phi + au\,\delta v \cos\phi - auv \sin\phi\,\delta\phi \\ \delta y &= a\,\delta u v \sin\phi + au\,\delta v \sin\phi + auv \cos\phi\,\delta\phi \\ \delta z &= a(u\,\delta u - v\,\delta v) \end{aligned}$$

$$(\delta x)^2 + (\delta y)^2 + (\delta z)^2 = a^2(u^2 + v^2)\{(\delta u)^2 + (\delta v)^2\} + a^2 u^2 v^2 (\delta\phi)^2.$$

Thus

$$h_u = a(u^2 + v^2)^{1/2}, \qquad h_v = a(u^2 + v^2)^{1/2}, \qquad h_\phi = auv.$$

Substituting these in equation (3.4.18)

$$\nabla^2 \psi = \frac{1}{a^2(u^2 + v^2)}\left\{\frac{1}{u}\frac{\partial}{\partial u}\left(u\frac{\partial\psi}{\partial u}\right) + \frac{1}{v}\frac{\partial}{\partial v}\left(v\frac{\partial\psi}{\partial v}\right) + \left(\frac{1}{u^2} + \frac{1}{v^2}\right)\frac{\partial^2\psi}{\partial\phi^2}\right\}.$$

12. Maxwell's equations of the electromagnetic field are defined for a homogeneous medium by

$$\nabla \times \mathbf{H} - \epsilon\frac{\partial \mathbf{E}}{\partial t} = \mathbf{J}, \qquad \nabla \times \mathbf{E} + \mu\frac{\partial \mathbf{H}}{\partial t} = \mathbf{0}$$

$$\nabla . \mathbf{E} = \rho/\epsilon, \qquad \nabla . \mathbf{H} = 0.$$

(For interpretations of the various physical quantities involved, see for example, A. Sommerfield, *Electrodynamics*, Academic Press, New York (1964).)

Find differential equations expressing $\mathbf{E}$ and $\mathbf{H}$ in terms of J and ρ.

$$\nabla \times (\nabla \times \mathbf{H}) - \epsilon\frac{\partial}{\partial t}(\nabla \times \mathbf{E}) = \nabla \times \mathbf{J}$$

$$+ \nabla(\nabla . \mathbf{H}) - \nabla^2 \mathbf{H} + \mu\epsilon\frac{\partial^2 \mathbf{H}}{\partial t^2} = \nabla \times \mathbf{J}$$

whence

$$\nabla^2 \mathbf{H} - \mu\epsilon\frac{\partial^2 \mathbf{H}}{\partial t^2} = -\nabla \times \mathbf{J}.$$

Similarly

$$\nabla \times (\nabla \times \mathbf{E}) + \mu \frac{\partial}{\partial t}(\nabla \times \mathbf{H}) = \mathbf{0}$$

$$+ \nabla(\nabla . \mathbf{E}) - \nabla^2 \mathbf{E} + \mu\epsilon \frac{\partial^2 \mathbf{E}}{\partial t^2} + \mu \frac{\partial \mathbf{J}}{\partial t} = \mathbf{0}$$

$$\frac{\nabla \rho}{\epsilon} - \nabla^2 \mathbf{E} + \mu\epsilon \frac{\partial^2 \mathbf{E}}{\partial t^2} + \mu \frac{\partial \mathbf{J}}{\partial t} = \mathbf{0}$$

where

$$\nabla^2 \mathbf{E} - \mu\epsilon \frac{\partial^2 \mathbf{E}}{\partial t^2} = \frac{\nabla \rho}{\epsilon} + \mu \frac{\partial \mathbf{J}}{\partial t}.$$

It is assumed in this book that the medium is homogeneous and isotropic.
It may readily be seen that

$$\nabla . \mathbf{J} + \frac{\partial \rho}{\partial t} = \nabla . \left(\nabla \times \mathbf{H} - \epsilon \frac{\partial \mathbf{E}}{\partial t}\right) + \frac{\partial}{\partial t} \epsilon \nabla . \mathbf{E} = 0.$$

This statement is the law of conservation of electric charge. That is, charge is neither created nor destroyed. When **J** and ρ are zero, the system is charge-free.

13. Show that if ψ is a scalar which satisfies

$$\nabla^2 \psi + k^2 \psi = 0$$

then the vector wave equation

$$\nabla^2 \mathbf{Q} + k^2 \mathbf{Q} = 0$$

is satisfied by the three independent solutions

$$\mathbf{L} = \nabla \psi, \qquad T = \nabla \times (\mathbf{u}\psi), \qquad \mathbf{S} = \nabla \times \{\nabla \times (\mathbf{u}\psi)\}$$

where **u** is a constant unit vector, **S** is termed a poloidal vector field, and **T** is termed a toroidal vector field. **L** is irrotational and both **S** and **T** are solenoidal.

$$\begin{aligned} \nabla^2 \mathbf{L} + k^2 \mathbf{L} &= \nabla^2 \nabla \psi + k^2 \nabla \psi \\ &= \nabla(\nabla^2 + k^2)\psi = \mathbf{0} \end{aligned}$$

thereby proving that **L** is a solution.
Similarly, using $\mathbf{T} = -\mathbf{u} \times \nabla \psi = -\mathbf{u} \times \mathbf{L}$

$$\begin{aligned} \nabla^2 \mathbf{T} + k^2 \mathbf{T} &= \nabla^2(-\mathbf{u} \times \mathbf{L}) + k^2(-\mathbf{u} \times \mathbf{L}) \\ &= -\mathbf{u} \times (\nabla^2 + k^2)\mathbf{L} = \mathbf{0} \\ \nabla^2 \mathbf{S} + k^2 \mathbf{S} &= \nabla^2 \nabla \times \mathbf{T} + k^2 \nabla \times \mathbf{T} \\ &= \nabla \times (\nabla^2 + k^2)\mathbf{T} = \mathbf{0} \end{aligned}$$

It may be noted that

$$\mathbf{T}.\mathbf{L} = 0$$
$$\nabla.\mathbf{L} = \nabla^2\psi = -k^2\psi$$
$$\nabla \times \mathbf{S} = \nabla \times (\nabla \times \mathbf{T})$$
$$= \nabla(\nabla.\mathbf{T})$$
$$-\nabla^2\mathbf{T} = -\nabla^2\mathbf{T} = k^2\mathbf{T}.$$

14. V is a function of U only where U is some point function. Prove that if V is harmonic, then

$$\frac{\nabla^2 U}{(\nabla U)^2}$$

is a function of U only.

Let

$$V = f(U)$$
$$\nabla^2 V = 0.$$

That is

$$\nabla.(\nabla U) = 0$$
$$\nabla.\{f'(U)\nabla U\} = 0$$
$$f''(U)(\nabla U)^2 + f(U)\nabla^2 U = 0$$

whence

$$\frac{\nabla^2 U}{(\nabla U)^2} = -\frac{f''(U)}{f'(U)}$$

$$= -\frac{d}{dU}\{\log f'(U)\}.$$

The right-hand side is a function of U only, and this is the required result.

15. If $\mathbf{F} = \nabla\phi$, where $\phi = \mathbf{p}.\nabla(1/R)$, $\mathbf{p}$ being a constant vector, prove that

$$\mathbf{F} = \nabla \times \mathbf{C}$$

where

$$\mathbf{C} = \nabla\left(\frac{1}{R}\right) \times \mathbf{p}$$

these results holding except at $\mathbf{R} = \mathbf{0}$.

$$\nabla \times \mathbf{C} = \nabla \times \left\{\nabla\left(\frac{1}{R}\right) \times \mathbf{p}\right\}$$

$$= -\mathbf{p}\nabla^2\left(\frac{1}{R}\right) + \mathbf{p}.\nabla\left\{\nabla\left(\frac{1}{R}\right)\right\}$$

$$= -\mathbf{p}\nabla^2\left(\frac{1}{R}\right) + \nabla\phi$$

$$= \nabla\phi, \text{ except at } \mathbf{R} = \mathbf{0}.$$

Clearly, $\mathbf{R} = \mathbf{0}$ has to be excluded because R^{-1}, and its derivatives also are singular there and hence cannot obey any differential relation there.

16. (*a*) A necessary and sufficient condition that two functions $U(\mathbf{r})$, $V(\mathbf{r})$ are functionally dependent is given by

$$\nabla U \times \nabla V = \mathbf{0}.$$

If $f(U, V) = 0$,

$$\frac{\partial f}{\partial U}\nabla U + \frac{\partial f}{\partial V}\nabla V = \mathbf{0}.$$

∇U, ∇V are parallel and so $\nabla U \times \nabla V = \mathbf{0}$. Conversely, if $\nabla U \times \nabla V = \mathbf{0}$, either U is a constant, V is a constant or ∇U, ∇V are parallel. If either U or V is a constant, a trivial functional relationship exists. $U = f(V)$ where $f(V)$ is a function which is in fact a constant. If ∇U, ∇V are parallel, the normals to the two surfaces $U =$ constant, $V =$ constant are the same. That is a surface of constant U is a surface of constant V. That is there exists a functional relationship between U and V.

(*b*) A necessary and sufficient condition that three functions $U(\mathbf{r})$, $V(\mathbf{r})$, $W(\mathbf{r})$ are functionally dependent is that

$$[\nabla U \nabla V \nabla W] = 0 \qquad \text{or} \qquad \frac{\partial(U, V, W)}{\partial(x, y, z)} = 0.$$

If U, V, W are functionally independent,

$$W = f(U, V)$$

$$\nabla W = \frac{\partial f}{\partial U}\nabla U + \frac{\partial f}{\partial V}\nabla V$$

Multiplying scalarly by $\nabla U \times \nabla V$

$$(\nabla U \times \nabla V).\nabla W = \frac{\partial f}{\partial U}(\nabla U \times \nabla V).\nabla U + \frac{\partial f}{\partial V}(\nabla U \times \nabla V).\nabla V$$
$$= 0.$$

Thus the condition is necessary. The sufficiency is proved as follows. $(\nabla U \times \nabla V).\nabla W$ is zero if (*i*) $\nabla U \times \nabla V = 0$, (*ii*) $\nabla W = 0$, or (*iii*) $(\nabla U \times \nabla V)$ is perpendicular to ∇W. In case (*i*) there exists a relation between U and V and so functional dependence exists. In case (*ii*) there exists the functional dependence $W = f(U, V)$ where f is a constant. In case (*iii*) $U =$ constant, $V =$ constant defines some curve. ∇U, ∇V are normal to the two surfaces, and therefore to the tangent to this curve. Also ∇U, ∇V, ∇W are always coplanar. Thus ∇W is always normal to the tangent to the curve and so has no component along the curve. Thus W will be constant for U and V constant and so there will be a functional relationship.

17. A rigid body moves so that the velocity at any arbitrary point is $\mathbf{v}$. Prove that the angular velocity is given by $\frac{1}{2}\nabla \times \mathbf{v}$.

The velocity $\mathbf{v}$ at any point may be expressed in the form

$$\mathbf{v} = \mathbf{v}_G + \boldsymbol{\omega} \times (\mathbf{r} - \mathbf{r}_G)$$

where $\mathbf{v}_G$ is the velocity of the centroid, $\boldsymbol{\omega}$ the angular velocity and $\mathbf{r}_G$ the position vector of the centroid

$$\begin{aligned}\nabla \times \mathbf{v} &= \nabla \times \{\mathbf{v}_G + \boldsymbol{\omega} \times (\mathbf{r} - \mathbf{r}_G)\} = \nabla \times (\boldsymbol{\omega} \times \mathbf{r}) \\ &= \boldsymbol{\omega}\nabla . \mathbf{r} - (\boldsymbol{\omega} . \nabla)\mathbf{r} = 3\boldsymbol{\omega} - \boldsymbol{\omega} = 2\boldsymbol{\omega}.\end{aligned}$$

This indicates the origin of the term irrotational for a vector field with zero curl. It will be noted that

$$\nabla . \mathbf{v} = \nabla . \{\mathbf{v}_G + \boldsymbol{\omega} \times (\mathbf{r} - \mathbf{r}_G)\} = \nabla . (\boldsymbol{\omega} \times \mathbf{r}) = -\boldsymbol{\omega} . (\nabla \times \mathbf{r}) = 0.$$

18. In an ideal fluid, the relation between density ρ, acceleration $\boldsymbol{\alpha}$ and pressure p is given by

$$\rho\boldsymbol{\alpha} = \nabla p.$$

Prove that

$$\begin{aligned}\boldsymbol{\alpha} . (\nabla \times \boldsymbol{\alpha}) &= 0 \\ \nabla \times (\rho\boldsymbol{\alpha}) &= \mathbf{0} \\ \rho(\nabla \times \boldsymbol{\alpha}) - \boldsymbol{\alpha} \times (\nabla\rho) &= \mathbf{0}.\end{aligned}$$

$\nabla \times \boldsymbol{\alpha}$ is parallel to $\boldsymbol{\alpha} \times \nabla\rho$ which is perpendicular to $\boldsymbol{\alpha}$. It follows that $\boldsymbol{\alpha} . (\nabla \times \boldsymbol{\alpha}) = 0$.

19. $V(\mathbf{r}) = 0$ is some surface. Find the components of a vector $\mathbf{Q}$ normal and tangential to this surface.

The unit normal to $V(\mathbf{r}) = 0$ is

$$\mathbf{n} = \frac{\nabla V}{|\nabla V|}$$

$$\begin{aligned}\mathbf{Q} &= (\mathbf{n} . \mathbf{Q})\mathbf{n} + \mathbf{Q} - (\mathbf{n} . \mathbf{Q})\mathbf{n} \\ &= (\mathbf{n} . \mathbf{Q})\mathbf{n} + (\mathbf{n} . \mathbf{n})\mathbf{Q} - (\mathbf{n} . \mathbf{Q})\mathbf{n} \\ &= (\mathbf{n} . \mathbf{Q})\mathbf{n} + (\mathbf{n} \times \mathbf{Q}) \times \mathbf{n}.\end{aligned}$$

The first term is normal to the surface and the second is tangential to the surface.

20. A vector $\mathbf{F}$ is both solenoidal and irrotational. Show that it can be written either in the form

$$\mathbf{F} = \nabla V, \text{ where } \nabla^2 V = 0, \qquad \text{or} \qquad \mathbf{F} = \nabla \times \mathbf{A}$$

where $\nabla^2 \mathbf{A} = \mathbf{0}$.

If the vector is solenoidal then it is expressible as a curl and $\mathbf{F} = \nabla \times \mathbf{A}$. Because $\mathbf{F}$ is irrotational $\nabla \times \mathbf{F} = \mathbf{0}$ and so $\nabla \times (\nabla \times \mathbf{A}) = \mathbf{0}$ and

$$-\nabla^2 \mathbf{A} + \nabla(\nabla . \mathbf{A}) = \mathbf{0}.$$

The vector potential **A** is indefinite, however, and by the addition of a suitable gradient it is always possible to make $\nabla.\mathbf{A}=0$. It follows that $\nabla^2\mathbf{A}=\mathbf{0}$.

If the vector is irrotational, it is expressible as a gradient and $\mathbf{F}=\nabla V$. Because **F** is solenoidal $\nabla.\mathbf{F}=0$, and so $\nabla.(\nabla V)=\nabla^2 V=0$.

21. A particle moves on the surface $\psi(\mathbf{r})=0$. Show that its acceleration $\boldsymbol{\alpha}$, velocity $\mathbf{v}$, position vector $\mathbf{r}$ and angular momentum per unit mass $\mathbf{h}$ are related by

$$\mathbf{v}(\mathbf{r}\,.\,\nabla\psi)=\mathbf{h}\times\nabla\psi$$

and

$$\boldsymbol{\alpha}=(\mathbf{v}\,.\,\nabla)\,\mathbf{v}.$$

If the particle moves on $\psi=0$, its velocity is perpendicular to the normal, hence

$$\mathbf{v}\,.\,\nabla\psi=0$$

and

$$\begin{aligned}\mathbf{h}\times\nabla\psi=(\mathbf{r}\times\mathbf{v})\times\nabla\psi&=\mathbf{v}(\mathbf{r}\,.\,\nabla\psi)-\mathbf{r}(\mathbf{v}\,.\,\nabla\psi)\\&=\mathbf{v}(\mathbf{r}\,.\,\nabla\psi).\end{aligned}$$

Now v does not contain t implicitly, and so

$$\boldsymbol{\alpha}=\frac{d\mathbf{v}}{dt}=\frac{\partial\mathbf{v}}{\partial t}+(\mathbf{v}\,.\,\nabla)\,\mathbf{v}=(\mathbf{v}\,.\,\nabla)\,\mathbf{v}.$$

22. Show that for a function of the form

$$f\left(\mathbf{t}-\frac{\mathbf{n}.\mathbf{r}}{c}\right)$$

where t is time, and $\mathbf{n}$ a unit vector

$$\frac{\partial f}{\partial t}+c(\mathbf{n}\,.\,\nabla)\,f=0.$$

$$\frac{\partial f}{\partial t}=f'\left(t-\frac{\mathbf{n}.\mathbf{r}}{c}\right)$$

$$\nabla f=-\frac{\mathbf{n}}{c}f'\left(t-\frac{\mathbf{n}.\mathbf{r}}{c}\right)$$

whence the result follows.

Exercises

1. Prove that the lines drawn from a point on an ellipse, to the foci make equal angles with the normal to the ellipse.

2. Prove that

$$\nabla.(U\nabla V-V\nabla U)=U\nabla^2 V-V\nabla^2 U$$

where U, V are point functions.

3. Prove that

$$\nabla^2 r^m = m(m+1) r^{m-2}$$

(Note that this is true for all r if $m > 2$, but is not true for $r = 0$ when $m < 2$, as then there is a singularity there.)

4. Prove that if $\psi = e^{\pm ikR}/R$, where $R = |\mathbf{r} - \mathbf{r}'|$

$$\nabla^2 \psi + k^2 \psi = 0$$

except at $R = 0$.

5. Evaluate

$$\boldsymbol{\nabla}.\{\mathbf{r}(\mathbf{r}.\mathbf{Q})\}$$

where $\mathbf{Q}$ is a constant vector.

6. Show that, if $\boldsymbol{\nabla}.\mathbf{B} = 0$, $\mathbf{B}$ is of the form

$$\mathbf{B} = \boldsymbol{\nabla} \times \mathbf{S}$$

where

$$\mathbf{S} = \mathbf{k} \times \int_0^z \mathbf{B}(\mathbf{r}^*)\, dz' + \mathbf{f}(\mathbf{r}_\perp) + \boldsymbol{\nabla}\psi$$

where

$$\mathbf{r}_\perp = x\mathbf{i} + y\mathbf{j}$$
$$\mathbf{r}^* = \mathbf{r}_\perp + z'\mathbf{k}$$

and

$$\mathbf{k}.\mathbf{f} = 0, \qquad \mathbf{k}.(\boldsymbol{\nabla} \times \mathbf{f}) = \mathbf{k}.\mathbf{B}(\mathbf{r}_\perp).$$

7. Prove that

$$\boldsymbol{\nabla} \times \mathbf{A} = \frac{\mathbf{r}}{r^3}$$

when

$$\mathbf{A} = \frac{1}{r} \tan\frac{\theta}{2}\, \hat{\boldsymbol{\varphi}}.$$

8. If

$$\mathbf{B} = \frac{2\cos\theta}{r^3}\hat{\mathbf{r}} + \frac{\sin\theta}{r^3}\hat{\boldsymbol{\theta}}$$

then

$$\mathbf{B} = \boldsymbol{\nabla}\psi = \boldsymbol{\nabla} \times \mathbf{A}$$

where

$$\psi = -\frac{\cos\theta}{r^2}, \qquad \mathbf{A} = \frac{\sin\theta}{r^2}\hat{\boldsymbol{\varphi}}.$$

9. Prove that

$$\tfrac{1}{2}\boldsymbol{\nabla}P^2 = (\mathbf{P}.\boldsymbol{\nabla})\mathbf{P} + \mathbf{P} \times (\boldsymbol{\nabla} \times \mathbf{P}).$$

10. Prove that

$$\boldsymbol{\nabla}.(r^2\mathbf{c}) = 2\mathbf{c}.\mathbf{r}$$

and that

$$\nabla \times (r^2 \mathbf{c}) = -2\mathbf{c} \times \mathbf{r}.$$

11. Find $\nabla^2 \psi$ in oblate spheroidal coordinates defined by

$$x = a \cosh \xi \cos \eta \cos \phi$$
$$y = a \cosh \xi \cos \eta \sin \phi$$
$$z = a \sinh \xi \sin \eta.$$

For this system, the circular disc $z = 0$, $0 < \rho < a$ is defined by $\xi = 0$, and $\eta = 0$ defines the z plane excluding the disc.

The range of variables is as follows.

In all cases $-\pi < \phi < \pi$.

If the boundary of the region considered is the disc, then $0 < \xi$, $-\pi/2 < \eta < \pi/2$.

If the boundary of the region considered is the z plane less the disc ξ has all values, but $0 < \eta < \pi/2$.

12. Prove that

$$\nabla . (\mathbf{r} \times \nabla \psi) = 0.$$

13. Prove that

$$\nabla^2 (\mathbf{r} \times \nabla \psi) = \mathbf{r} \times \nabla^2 (\nabla \psi).$$

14. (*a*) Prove that a solution of two of Maxwell's equations (example 12)

$$\nabla \times \mathbf{E} + \mu \frac{\partial \mathbf{H}}{\partial t} = \mathbf{0}, \qquad \nabla . \mathbf{H} = 0$$

is given by

$$\mathbf{E} = -\nabla V - \frac{\partial \mathbf{A}}{\partial t}, \qquad \mathbf{H} = \frac{1}{\mu} \nabla \times \mathbf{A}$$

where $\mathbf{A}$ and V are arbitrary and that the fields $(\mathbf{E}, \mathbf{H})$ are unaffected if $\mathbf{A}$ is replaced by $\mathbf{A} + \nabla \psi$, and V is replaced by $V - (\partial \psi / \partial t)$, where ψ is arbitrary. $\mathbf{A}$ is a vector potential and V is a scalar potential.

(*b*) Show that a possible solution to the other Maxwell equations

$$\nabla \times \mathbf{H} - \epsilon \frac{\partial \mathbf{E}}{\partial t} = \mathbf{J}, \qquad \nabla . \mathbf{E} = \rho / \epsilon$$

is given when $\mathbf{A}$ and V satisfy the equations

$$\nabla^2 \mathbf{A} - \mu \epsilon \frac{\partial^2 \mathbf{A}}{\partial t^2} = -\mu \mathbf{J}$$

$$\nabla^2 V - \mu \epsilon \frac{\partial^2 V}{\partial t^2} = -\rho / \epsilon$$

$$\nabla . \mathbf{A} + \mu \epsilon \frac{\partial V}{\partial t} = 0.$$

This relation between $\mathbf{A}$ and V is termed a gauge condition.

15. A solution in charge free space to Maxwell's equations is given by

$$\mathbf{E} = \mathbf{E}_0 f\left(t - \frac{\mathbf{n}.\mathbf{r}}{c}\right)$$

$$\mathbf{H} = \mathbf{H}_0 f\left(t - \frac{\mathbf{n}.\mathbf{r}}{c}\right)$$

n being a unit vector.

Show that

$$c^2 \mu\epsilon = 1,$$

that

$$\mathbf{E}_0 = \sqrt{\left(\frac{\mu}{\epsilon}\right)}\, \mathbf{n} \times \mathbf{H}_0$$

and

$$\mathbf{E}_0 . \mathbf{n} = 0, \qquad \mathbf{H}_0 . \mathbf{n} = 0.$$

(These equations represent a plane electromagnetic wave. The speed of propagation is c, and **n** is the direction of propagation. The magnetic field, the electric field and the direction of propagation are mutually perpendicular.)

16. If **c** is a constant vector and **F** is a vector which is both solenoidal and irrotational, prove that

$$\nabla(\mathbf{c}.\mathbf{F}) + \nabla \times (\mathbf{c} \times \mathbf{F}) = \mathbf{0}.$$

17. Prove that

$$\nabla\{(\mathbf{a} \times \mathbf{r}).(\mathbf{b} \times \mathbf{r})\} = \mathbf{a} \times (\mathbf{r} \times \mathbf{b}) + \mathbf{b} \times (\mathbf{r} \times \mathbf{a}).$$

18.
$$\mathbf{P} = \gamma(\nabla\alpha) \times (\nabla\beta).$$

Show that the necessary and sufficient condition for **P** to be solenoidal is that γ is a function of α and β.

19. Show that if **P** is a constant vector

$$(\mathbf{P} \times \nabla) \times \mathbf{r} = -2\mathbf{P}.$$

20. Show that, if **P** is of the form $\mathbf{P}(\mathbf{r})\hat{\mathbf{P}}$, where $\hat{\mathbf{P}}$ is fixed,

$$\mathbf{P}.(\nabla \times \mathbf{P}) = 0.$$

4 Line Surface and Volume Integrals

4.0. General

The concept of the integral

$$\int_a^b f(x)\,dx$$

where f is a scalar function of some real variable x, is well known. The question immediately arises as to how the concept of integral can be extended when the integral may, for example, be over a surface or a volume, or when the integrand is a vector. In some cases it is easier to consider these by reduction to ordinary integrals, in other cases it is better to use a definition, analogous to one of the definitions of an ordinary integral

$$\int_a^b f(x)\,dx = \lim_{n\to\infty} \sum_{n=1}^{n} f_p \delta_p \qquad (4.0.1)$$

where the set δ_p spans $a < x < b$ and f_p is a value of $f(x)$ within δ_p. It will be assumed always that the curves, surfaces and volumes considered are well enough behaved for the operations discussed to be carried out, and existence conditions will not be discussed.

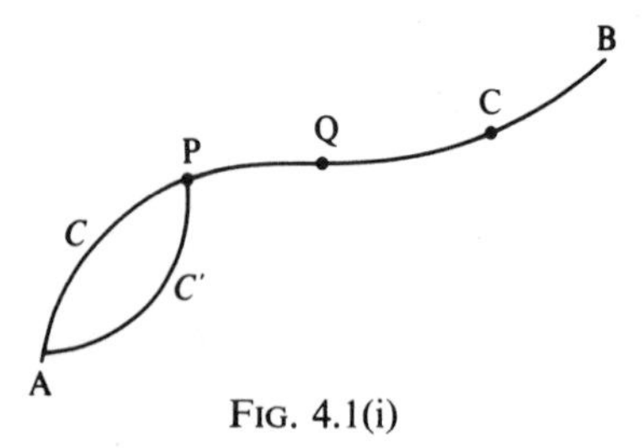

FIG. 4.1(i)

4.1. Integral along a curve

The curve $\mathscr{C}$ in figure 4.1(i) is composed of the total of elementary displacements such as $\overrightarrow{PQ} = \delta\mathbf{r}$.

$\delta\mathbf{r}$ may be considered as equivalent to three cartesian displacements δx, δy, δz. Alternatively, it is possible to consider $\overrightarrow{PQ}$ as associated with an elementary arc distance δs where s is the arc distance along the curve $\mathscr{C}$ measured from some point on it together with an associated unit tangent vector $\hat{\mathbf{t}}$ at P. The curve is always described in a specific sense, and consequently due regard must be paid to signs. Corresponding to the concept of a definite integral is the integral from A to B along the curve. Corresponding to the concept of an indefinite integral is the integral from A to a variable point such as P on the curve. It should be noted that there is no reason *a priori* for an integral from A to B over two different curves $\mathscr{C}$ and $\mathscr{C}'$ to be the same, and in many cases, this will not, in fact be the case.

4.11. *Tangential line integral of a vector*

Let $\mathbf{F}(\mathbf{r})$ be some vector field.

$$\int_{A,\mathscr{C}}^{B} \mathbf{F}.d\mathbf{r},$$

the integral of $\mathbf{F}$ from A to B along $\mathscr{C}$, may be defined as follows. At any point of the curve

$$\mathbf{F}.\delta\mathbf{r} = \mathbf{F}.\hat{\mathbf{t}}\,\delta s = F_t\,\delta s \tag{4.11.1}$$

where F_t is the tangential component of $\mathbf{F}$. Clearly F_t will be a function of s, the arc length along the curve $\mathscr{C}$, and so

$$\int_{A,\mathscr{C}}^{B} \mathbf{F}.d\mathbf{r} = \int_{s_A}^{s_B} F_t(s)\,ds. \tag{4.11.2}$$

Clearly

$$\int_{A,\mathscr{C}}^{B} \mathbf{F}.d\mathbf{r} = -\int_{B,\mathscr{C}}^{A} \mathbf{F}.d\mathbf{r}. \tag{4.11.3}$$

The right-hand side of (4.11.2) is an ordinary definite integral and may be evaluated by the usual methods. An alternative method is to write the quantities in terms of, say, cartesian components,

$$\int_{A,\mathscr{C}}^{B} \mathbf{F}.d\mathbf{r} = \int F_x\,dx + \int F_y\,dy + \int F_z\,dz \tag{4.11.4}$$

where the appropriate limits are understood.

The meaning of $\int F_x dx$ and similar terms is determined as follows. In full $F_x = F_x(x,y,z)$ and the equation of the curve $\mathscr{C}$ may be written as $y = y(x)$, $z = z(x)$. (If these functions are not one valued, the curve must be split up into arcs for which this is the case.) Thus

$$F_x = F_x(x, y(x), z(x)) = F_x^*(x),$$

say, on the curve and so

$$\int_{A,\mathscr{C}}^{B} \mathbf{F}.d\mathbf{r} = \int_{x_A}^{x_B} F_x^*(x)\,dx + \int_{y_A}^{y_B} F_y^*(y) + \int_{z_A}^{z_B} F_z^*(z)\,dz \quad (4.11.5)$$

Alternatively, if the equation of the curve $\mathscr{C}$ is given parametrically through one valued functions $x = x(u)$, $y = y(u)$, $z = z(u)$, then

$$\int_{A,\mathscr{C}}^{B} \mathbf{F}.d\mathbf{r} = \int_{u_A}^{u_B} \left\{ F_x(u)\frac{dx}{du} + F_y(u)\frac{dy}{du} + F_z(u)\frac{dz}{du} \right\} du \quad (4.11.6)$$

If $u = s$, this is equivalent to (4.11.2).

It is easy to see that

$$\int_{A,\mathscr{C}}^{B} \mathbf{F}.d\mathbf{r} = \int_{A,\mathscr{C}}^{C} \mathbf{F}.d\mathbf{r} + \int_{C,\mathscr{C}}^{B} \mathbf{F}.d\mathbf{r}. \quad (4.11.7)$$

Consider now the quantity

$$\int_{A,\mathscr{C}}^{P} \mathbf{F}.d\mathbf{r}$$

which corresponds to the concept of an indefinite integral when P is a variable point. In general it depends both upon P, and upon the path $\mathscr{C}$. For convenience let

$$\int_{A,\mathscr{C}}^{P} \mathbf{F}.d\mathbf{r} = U_{\mathscr{C}}(P). \quad (4.11.8)$$

The case when the integral depends only on P is of great interest. In this case, a scalar field

$$U(P) = \int_{A}^{P} \mathbf{F}.d\mathbf{r} \quad (4.11.9)$$

is defined, the integral being independent of the path. If this be the case,

$$U(Q) - U(P) = \int_A^Q \mathbf{F}.d\mathbf{r} - \int_A^P \mathbf{F}.d\mathbf{r} = \int_P^Q \mathbf{F}.d\mathbf{r}$$

and if Q is infinitesimally near P

$$\delta U = \mathbf{F}.\delta\mathbf{r} \tag{4.11.10}$$

and by what has been done previously

$$\mathbf{F} = \nabla U. \tag{4.11.11}$$

Conversely, working backwards, if $\mathbf{F} = \nabla U$ where U is a one valued scalar function of position,

$$\int_A^P \mathbf{F}.d\mathbf{r} = U(P) - U(A) \tag{4.11.12}$$

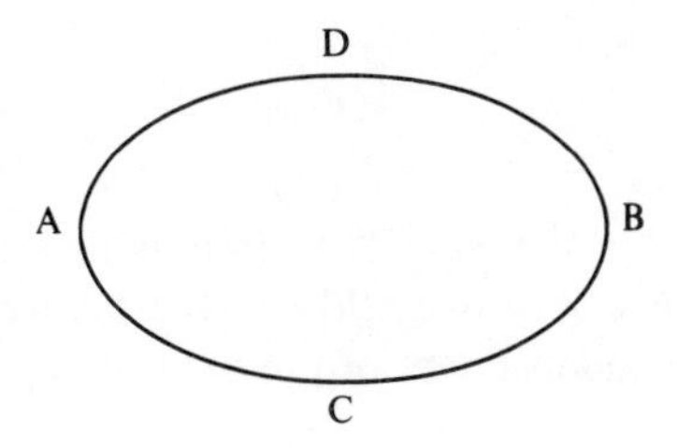

FIG. 4.11(i)

is a function of P only and independent of the path from A to P. Thus the necessary and sufficient condition that

$$\int_A^P \mathbf{F}.d\mathbf{r}$$

should be independent of the path between A and P is that F should be a gradient, that is

$$\nabla \times \mathbf{F} = \mathbf{0}. \tag{4.11.13}$$

Fields of this character are said to be conservative.

If $\mathbf{F}$ is a conservative field and $\mathscr{C}$ is a closed curve, then

$$\int_{\mathscr{C}} \mathbf{F}.d\mathbf{r} = 0. \tag{4.11.14}$$

The proof is as follows.

With an obvious notation

$$\int_{ACB} \mathbf{F}.d\mathbf{r} = \int_{ADB} \mathbf{F}.d\mathbf{r} = \int_{A}^{B} \mathbf{F}.d\mathbf{r}$$

$$= -\int_{BDA} \mathbf{F}.d\mathbf{r}$$

whence

$$\int_{ACB} \mathbf{F}.d\mathbf{r} + \int_{BDA} \mathbf{F}.d\mathbf{r} = 0$$

that is

$$\int_{ACBDA} \mathbf{F}.d\mathbf{r} = 0$$

which proves the theorem. The proof of the converse is left to the reader.

$$\int_{A,\mathscr{C}}^{B} \mathbf{F}.d\mathbf{r}$$

may be interpreted as the work done in moving a particle from A to B under the influence of a force field $\mathbf{F}$. When $\mathbf{F}$ is a conservative force, and the integral is independent of the path, it is possible to define a potential energy

$$V(P) = \int_{P}^{A} \mathbf{F}.d\mathbf{r}, \qquad \mathbf{F} = -\nabla V. \tag{4.11.15}$$

Clearly a particle moving on a closed curve and returning to its initial point does not change its potential energy.

4.12. *Vector integrals along a curve*

Consider the quantity defined by

$$\int_{A,\mathscr{C}}^{B} f(\mathbf{r})\,d\mathbf{r} = \lim_{n\to\infty} \sum_{p=1}^{n} f_p\,\boldsymbol{\epsilon}_p \tag{4.12.1}$$

where f is a scalar function of position and $\mathscr{C}$ is some curve. The $\boldsymbol{\epsilon}_p$ are a set of vector displacements along chords of $\mathscr{C}$ such that

$$\sum_{p=1}^{n} \boldsymbol{\epsilon}_p$$

is equivalent to the net displacement from one end A of the curve to the other, end B.

If $(\lambda_p, \mu_p, \nu_p)$ are the direction cosines of $\boldsymbol{\epsilon}_p$ and $\delta_p = |\boldsymbol{\epsilon}_p|$,

$$\boldsymbol{\epsilon}_p = (\lambda_p, \mu_p, \nu_p)\,\delta_p. \tag{4.12.2}$$

f_p is the value of f at the initial point of $\boldsymbol{\epsilon}_p$.

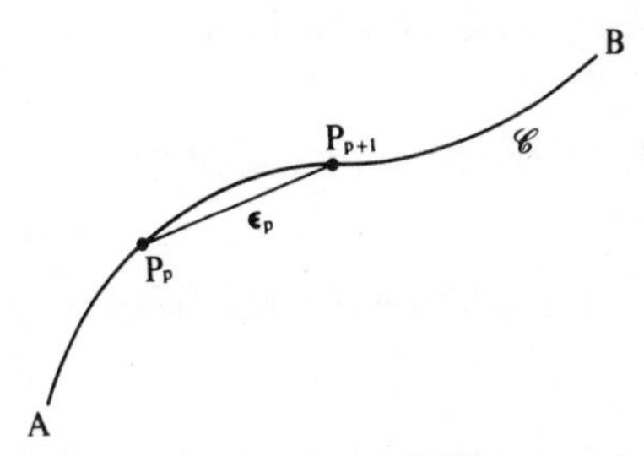

FIG. 4.12(i)

Suppose that the arc length along the curve $\mathscr{C}$ from some fixed point on it is s, and the arc length at P_p is s_p. On the curve, it is possible to write $f(\mathbf{r}) = f(s)$ and $f_p = f(s_p)$.

Similarly, the unit tangent at any point on the curve is

$$(\lambda(s), \mu(s), \nu(s)).$$

Thus

$$\int_{A,\mathscr{C}}^{B} f(r)\,d\mathbf{r} = \lim_{n\to\infty} \sum_{p=1}^{n} f_p(\lambda_p \mathbf{i} + \mu_p \mathbf{j} + \nu_p \mathbf{k})\,\delta_p$$

$$= \mathbf{i} \int_{S_A}^{S_B} f(s)\lambda(s)\,ds + \mathbf{j} \int_{S_A}^{S_B} f(s)\mu(s)\,ds + \mathbf{k} \int_{S_A}^{S_B} f(s)\nu(s)\,ds \tag{4.12.3}$$

and the integrand is reduced to the evaluation of three definite integrals. The extension to integration in terms of a parameter u along the curve is obvious.

Similarly, it is possible to define a vector $\mathbf{F}_{\mathscr{C}}(\mathbf{r})$ by

$$\mathbf{F}_{\mathscr{C}}(\mathbf{r}) = \int_{A,\mathscr{C}}^{P} f\,d\mathbf{r} = \mathbf{i} \int_{0}^{s} f(s')\lambda(s')\,ds'$$

$$+ \mathbf{j} \int_{0}^{s} f(s')\mu(s')\,ds' + \mathbf{k} \int_{0}^{s} f(s')\nu(s')\,ds'. \tag{4.12.4}$$

It will be shown in section 5.1. that this quantity depends in general upon the path $\mathscr{C}$ from A to P.

Similarly,

$$\mathbf{E} = \int_{A,\mathscr{C}}^{B} \mathbf{R} \times d\mathbf{r}$$

where $\mathbf{R}$ is some vector field, may be defined as

$$\left.\begin{aligned} &\mathbf{i}\int_{s_A}^{s_B} \{R_y(s)\,\nu(s) - R_z(s)\,\mu(s)\}\,ds \\ &\quad + \mathbf{j}\int_{s_A}^{s_B} \{R_z(s)\,\lambda(s) - R_x(s)\,\nu(s)\}\,ds \\ &\quad + \mathbf{k}\int_{s_A}^{s_B} \{R_x(s)\,\mu(s) - R_y(s)\,\lambda(s)\}\,ds \end{aligned}\right\} \tag{4.12.5}$$

with an analogous definition for

$$\mathbf{E}_{\mathscr{C}}(\mathbf{r}) = \int_{A,\mathscr{C}}^{P} \mathbf{R} \times d\mathbf{r}. \tag{4.12.6}$$

It will be shown later in section 5.1 that this quantity depends in general upon the path from A to P.

Integrals around closed curves can often be evaluated by using surface integrals related to them. This involves Stokes's theorem, which is discussed in section 5.1.

4.2. Surface integrals

Before defining a surface integral, it will be helpful to consider certain properties of surfaces. In this book, only surfaces which have two sides will be considered. Surfaces such as the Mobius strip and Klein bottle are excluded. If a surface is closed, the positive normal is conventionally supposed to be in the outward direction. If a surface S is open, it will be bounded by a rim $\mathscr{C}$, and the relation between the direction of traversing the rim and the positive normal is conventionally as shown in figure 4.2(i). (For example, the hemisphere $x^2 + y^2 + z^2 = a^2$, $z > 0$ has a rim defined by $x = a\cos\phi$, $y = a\sin\phi$, $z = 0$. If the rim is transversed in the direction of ϕ increasing, the positive direction of the normal is the one with $\mathbf{n}.\mathbf{k} > 0$, that is the one with an upwards vertical component.) If S is closed, $\mathscr{C}$, of course, does not exist.

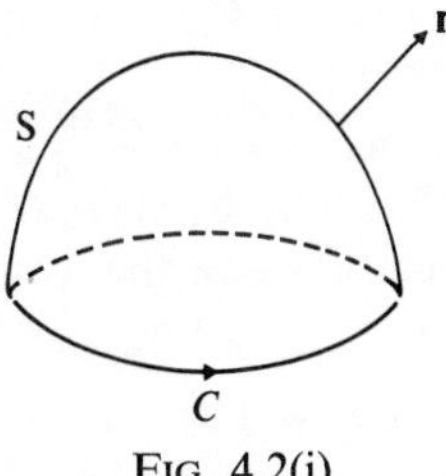

FIG. 4.2(i)

4.21. *Integral over a plane domain*

Before going into the details of surface integrals, it will be helpful to consider the simplest possible case, the integral of a function over a plane domain.

If D is a domain in the xy plane, surrounded by a curve $\mathscr{C}$

$$\iint_D f(x,y)\,dx\,dy = \lim_{m\to\infty}\lim_{n\to\infty}\sum_{i=1}^{m}\sum_{j=1}^{n} f(x_i,y_j)\,\delta A_{ij} \qquad (4.21.1)$$

this being an extension of the usual method of defining integrals by the limit of a sum. This integral may be interpreted as the volume of

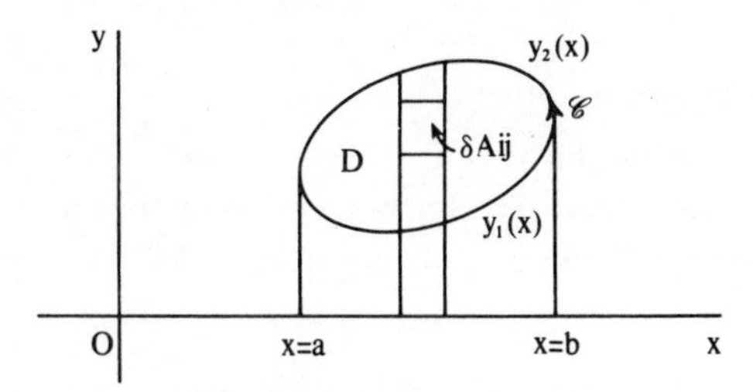

FIG. 4.21(i)

the solid cut off by the cylinder parallel to the z axis which has $\mathscr{C}$ as its cross-section in the xy plane, the plane $z=0$ and the surface $z=f(x,y)$. It may be evaluated as follows.

Suppose that D is dissected relative to the y axis by dividing it by lines parallel to the y axis into a number of strips, each strip being bounded below and above by curves $y=y_1(x)$ and $y=y_2(x)$ respectively, $y_1(x)$ and $y_2(x)$ being single valued continuous functions of x such that $y_1(a)=y_2(a)$; $y_1(b)=y_2(b)$. (The extension to the case where D is bounded by either $x=a$, or $x=b$ is obvious.) If this is not possible, D may be split up into sub-domains for which it is possible. Then

$$\iint_D f(x,y)\,dx\,dy = \int_a^b dx \int_{y_1(x)}^{y_2(x)} f(x,y)\,dy. \tag{4.21.2a}$$

Alternatively, the dissection may be carried out with respect to the x axis in which case the relation is of the form

$$\iint_D f(x,y)\,dx\,dy = \int_\alpha^\beta dy \int_{x_1(y)}^{x_2(y)} f(x,y)\,dx. \tag{4.21.2b}$$

If now, instead of expressing everything in terms of x and y, everything be expressed in terms of orthogonal coordinates u_1, u_2 the element of area is of the form $h_1 h_2 \delta u_1 \delta u_2$, and

$$f(x,y) = f^*(u_1, u_2) \tag{4.21.3}$$

and

$$\iint_D f(x,y)\,dx\,dy = \iint_{D^*} f^*(u_1,u_2)\,h_1 h_2\,du_1\,du_2 \tag{4.21.4}$$

D^* being the image domain in the u_1, u_2 plane of D in the x, y plane. The integration over D^* may be carried through by the methods of (4.21.2).

4.22. *General surface integrals*

The definitions of the various types of surface integral follow on without any difficulty. An element of surface has associated with it an elementary area and an outward drawn unit normal.

$$\delta\boldsymbol{\mathscr{A}} = \mathbf{n}\,\delta\mathscr{A}. \tag{4.22.1}$$

The fundamental surface integral is defined by

$$\int_S \mathbf{F}.d\boldsymbol{\mathscr{A}} = \lim_{n\to\infty} \sum_{\alpha=1}^{n} \mathbf{F}_\alpha.\delta\boldsymbol{\mathscr{A}}_\alpha \tag{4.22.2}$$

where the set $\delta\boldsymbol{\mathscr{A}}_\alpha$ cover between them the whole of S.

A particular case of this is

$$\iint_D f(x,y)\,dx\,dy,$$

where $\mathbf{F} = f\mathbf{k}$ and $d\boldsymbol{\mathscr{A}} = \mathbf{k}\,dx\,dy$.

Many other types of surface integral may be expressed in terms of integrals of the form (4.22.2). It is customary to use a single integral sign only for an integral over a surface. It must be remembered, however, that two variables are involved and when these appear explicitly, two integral signs are used.

$$\int_S \mathbf{G} \times d\mathscr{A} = \mathbf{i}\int_S (\mathbf{i} \times \mathbf{G}).d\mathscr{A} + \mathbf{j}\int_S (\mathbf{j} \times \mathbf{G}).d\mathscr{A} + \mathbf{k}\int_S (\mathbf{k} \times \mathbf{G}).d\mathscr{A} \tag{4.22.3}$$

has been expressed as three integrals of this form.

Similarly

$$\int_S U\,d\mathscr{A} = \int_S U\left(\frac{\nabla\psi}{|\nabla\psi|}.\mathbf{n}\right)d\mathscr{A}$$

where $\psi(\mathbf{r}) = 0$ is the equation of the surface S, giving an expression for $\mathbf{n}$ the outward unit normal. $\mathbf{n} = \nabla\psi/|\nabla\psi|$. Thus

$$\int_S U\,d\mathscr{A} = \int \mathbf{F}.d\mathscr{A} \tag{4.22.4}$$

where $\mathbf{F} = (U\nabla\psi)/|\nabla\psi|$.

More generally

$$\int_S \mathbf{F}\,d\mathscr{A} = \mathbf{i}\int_S F_x\,d\mathscr{A} + \mathbf{j}\int_S F_y\,d\mathscr{A} + \mathbf{k}\int_S F_z\,d\mathscr{A} \tag{4.22.5}$$

which can be expressed as three integrals of the type (4.22.4).

There are two possible ways of evaluating surface integrals. If the surface is closed, it is possible to use the divergence theorem which will be discussed in section 5.2. This turns the surface integral into a volume integral. The following method is, however, applicable for open surfaces, and hence for closed surfaces which are capable of being split up into a number of open surfaces.

It is possible to express a surface integral as a double integral taken over the projected area of the surface S on one of the coordinate planes, provided that any line perpendicular to the coordinate plane chosen meets the surface in no more than one point. Any surface, such as a closed surface or re-entrant surface which does not satisfy this condition, can be split up into portions which do satisfy the condition.

$$\iint_S \mathbf{F}.d\mathscr{A}$$

is the sum of three quantities such as

$$\iint_S F_z(x, y, z)\mathbf{k}.d\mathscr{A}$$

and if S_0 is the projection of S upon the xy plane, and $z = z(x,y)$ is the equation of the surface S, the integral may be written as

$$\iint_{S_0} F_z(x, y, z(x, y))\, dx\, dy. \tag{4.22.6}$$

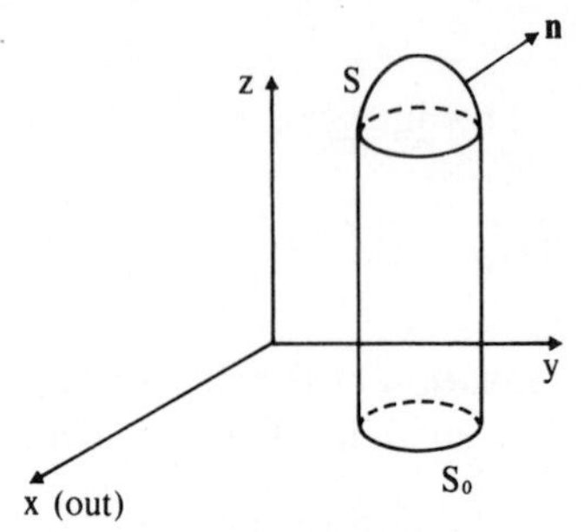

FIG. 4.22(i)

Similarly

$$\iint_S \psi\, d\mathscr{A}$$

has three components such as

$$\iint_S \psi\mathbf{k}\,.\,d\mathscr{A} = \iint_{S_0} \psi(x, y, z(x, y))\, dx\, dy. \tag{4.22.7}$$

Integrals such as (4.22.6) and (4.22.7) may be evaluated by the methods outlined in section 4.21.

4.23. *Solid angle*

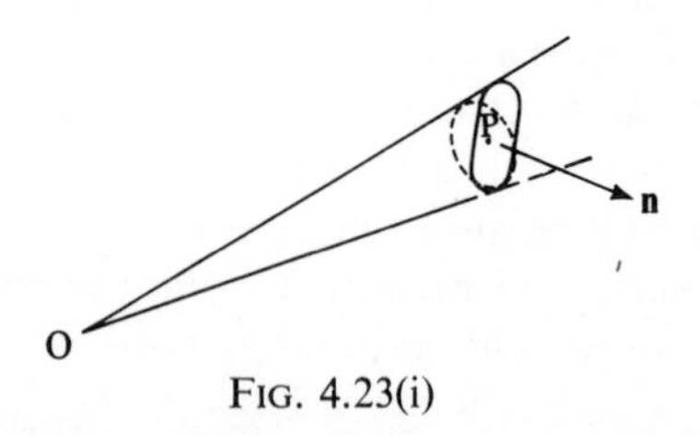

FIG. 4.23(i)

Suppose that, at a point P, there is an element of surface $\delta\mathscr{A}$, then the elementary solid angle subtended at O by $\delta\mathscr{A}$ is defined to be

$$\delta\Omega = \frac{\delta\mathscr{A}\,.\,\hat{\mathbf{r}}}{r^2}. \tag{4.23.1}$$

This quantity is dimensionless. It is easy to see that the area cut off by an elementary cone with vertex O and base the rim of $\delta\mathscr{A}$ on the sphere of radius r is given by

$$\delta\Sigma = \delta\mathscr{A}\,.\,\hat{\mathbf{r}} = r^2\,\delta\Omega \qquad (4.23.2)$$

and if the sphere is of radius r', $r'^2\,\delta\Omega$.

If the elementary cone cuts the surface an even number of times as in figure 4.23(ii), it is easy to see that the two elementary solid angles cancel one another out. It follows from this that, if O is within a surface S, each elementary cone cuts the surface an odd number of

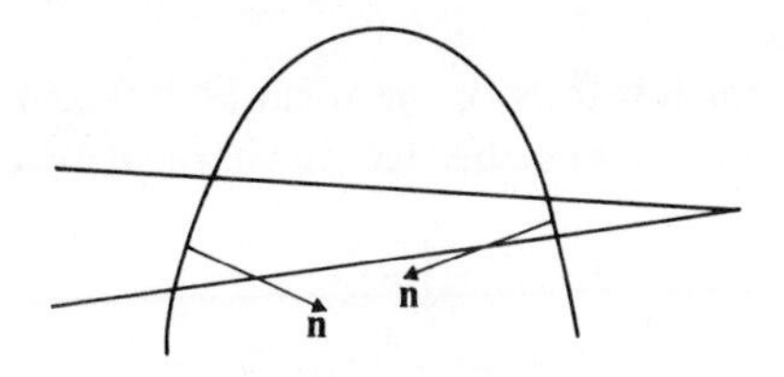

FIG. 4.23(ii)

times and so cuts off a solid angle $\delta\Omega$ on a sphere of unit radius, centre O. Hence

$$\int_S \frac{d\mathscr{A}\,.\,\hat{\mathbf{r}}}{r^2} = 4\pi, \qquad (4.23.3a)$$

the area of the unit sphere.

If, however, O is outside S, the elementary cones cut the surface an even number of times giving a zero solid angle, and so

$$\int_S \frac{d\mathscr{A}\,.\,\hat{\mathbf{r}}}{r^2} = 0. \qquad (4.23.3b)$$

4.3. **Volume integral**

The integral over a volume of a point function $f(\mathbf{r})$ is defined by

$$\int_V f(\mathbf{r})\,d\tau = \lim_{n\to\infty} \sum_{\alpha=1}^{n} f_\alpha \tau_\alpha \qquad (4.3.1)$$

where τ denotes volume. The set τ_α covers the volume V and f_α is the value of $f(\mathbf{r})$ at some point within τ_α.

It is customary to use only one integral sign for the integral over a volume although in fact three variables occur. When this is explicit, three integral signs are used and the following are alternative expressions for the same quantity

$$\int_V f(\mathbf{r})\,d\tau = \iiint_V f(x, y, z)\,dx\,dy\,dz. \tag{4.3.2}$$

The extension to the integral of a vector field over a volume follows immediately

$$\int_V F(\mathbf{r})\,d\tau = \mathbf{i} \int F_x\,d\tau + \mathbf{j} \int F_y\,d\tau + \mathbf{k} \int F_z\,d\tau. \tag{4.3.3}$$

Interpretations for integrals of the form (4.3.2) and (4.3.3) come to mind immediately. If $\rho(\mathbf{r})$ is the electric charge density,

$$\int_V \rho\,d\tau$$

is the total charge contained within V, and if $\mathbf{F}(\mathbf{r})$ is the force density per unit volume acting upon some medium $\int \mathbf{F}\,d\tau$ is the total force

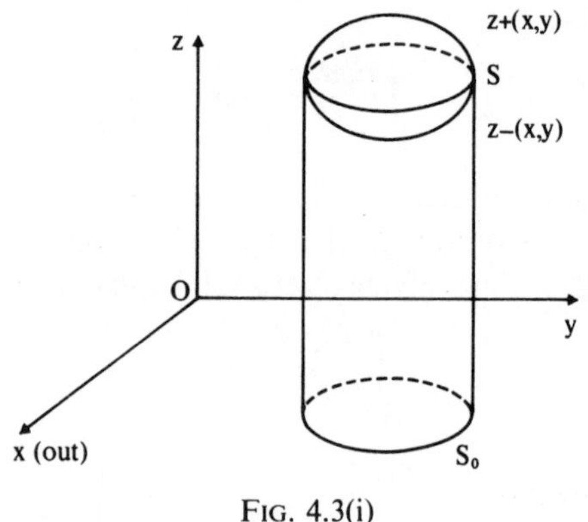

FIG. 4.3(i)

acting on the volume of interest and $\int \mathbf{r} \times \mathbf{F}\,d\tau$ is the moment of this force distribution about the origin.

In order to evaluate a volume integral, it is necessary to express it in terms of some coordinate system. In cartesian coordinates, it is possible to proceed as follows.

Suppose that the volume can be dissected parallel to the z axis so that lines parallel to the z axis cut the boundary of the volume in two points only. (If this is not the case the volume can always be divided into portions for which it is true.) The boundary surface can be split into an upper part defined by $z = z_+(x, y)$ and a lower part defined by $z = z_-(x, y)$. These may or may not have a common rim, e.g., for the cylinder $x^2 + y^2 < a^2$, $z^2 < c^2 z_+(x, y) = c$, $z_-(x, y) = -c$, and there is no common rim, whereas for the sphere $x^2 + y^2 + z^2 < a^2$ the surfaces are defined by

$$z_+(x, y) = \sqrt{(a^2 - x^2 - y^2)} \quad \text{and} \quad z_-(x, y) = -\sqrt{(a^2 - x^2 - y^2)},$$

both having a common rim $x^2 + y^2 = a^2$, $z = 0$. The projection of the volume boundary encloses a domain S_0 in the xy plane. Consider now the integral

$$\int_V f(\mathbf{r})\, d\tau.$$

The contribution to this integral from an element of volume is

$$f(\mathbf{r})\, \delta\tau = f(x, y, z)\, \delta x\, \delta y\, \delta z$$

and that due to the elementary cylinder with cross-sectional area indicated in figure 4.3(i) is given by

$$\left\{ \int_{z_-(x,y)}^{z_+(x,y)} f(x, y, z)\, dz \right\} \delta x\, \delta y \tag{4.3.5}$$

giving the value of the integral as

$$\iint_D dx\, dy \left\{ \int_{z_-}^{z_+} f(x, y, z)\, dz \right\}. \tag{4.3.6}$$

In a general orthogonal coordinate system

$$\delta\tau = h_1 h_2 h_3\, \delta u_1\, \delta u_2\, \delta u_3$$

and

$$\int_V f(\mathbf{r})\, d\tau = \iiint_U F(u_1, u_2, u_3)\, du_1\, du_2\, du_3 \tag{4.3.7}$$

where U is the set of values of (u_1, u_2, u_3) corresponding to V and

$$F(u_1, u_2, u_3) = f(\mathbf{r})\, h_1 h_2 h_3. \tag{4.3.8}$$

Worked examples

1. Let $\mathbf{F} = ax^2\mathbf{i} + y^2z\mathbf{j} + by^2\mathbf{k}$ and let $\mathscr{C}_1$ be the straight line jointing O to the point (c, c, c) and let $\mathscr{C}_2$ be the curve composed of the three segments of straight lines

$$\begin{array}{lll} x = 0, & y = 0, & 0 < z < c \\ x = 0, & 0 < y < c, & z = c \\ 0 < x < c, & y = c, & z = c. \end{array}$$

Find $\int \mathbf{F}.d\mathbf{r}$ over both curves and explain why, despite the fact that the beginning and end points of the two curves are the same, the two integrals are not equal.

Over $\mathscr{C}_1$, the relation between s, x, y, z is given by

$$\frac{s}{\sqrt{3}} = x = y = z,$$

and so

$$\int_{\mathscr{C}_1} \mathbf{F}.d\mathbf{r} = \int_0^{\sqrt{(3)}c} \left(a\frac{s^2}{3} + \frac{s^3}{3\sqrt{3}} + b\frac{s^2}{3}\right)\frac{ds}{\sqrt{3}}$$

$$= \frac{ac^3}{3} + \frac{c^4}{4} + \frac{bc^3}{3}.$$

$$\int_{\mathscr{C}_2} \mathbf{F}.d\mathbf{r} = \int_{\substack{0 \\ (x=0,\, y=0)}}^{c} by^2\,dz + \int_{\substack{0 \\ (x=0,\, z=c)}}^{c} y^2x\,dy + \int_{\substack{0 \\ y=c,\, z=c}}^{c} ax^2\,dx$$

$$= \frac{c^4}{4} + \frac{ac^3}{3} \neq \int_{\mathscr{C}_1} \mathbf{F}.d\mathbf{r}.$$

The reason that the two integrals are not the same is that

$$\nabla \times \mathbf{F} = 2by\mathbf{i} \neq \mathbf{0}.$$

Thus $\mathbf{F}$ is not of the form $\nabla\phi$.

2. Prove that, if $\mathbf{B}$ is a constant vector

$$\int_{\mathscr{C}} \mathbf{B}.d\mathbf{r} = 0$$

if $\mathscr{C}$ is any closed curve.

$\mathbf{B} = \nabla(\mathbf{B}.\mathbf{r})$, and so the result follows immediately.

3.

$$\mathbf{P} = \frac{-y\mathbf{i} + x\mathbf{j}}{x^2 + y^2}$$

Calculate $\nabla \times \mathbf{P}$, and

$$\int_{\mathscr{C}} \mathbf{P}.d\mathbf{r}$$

where $\mathscr{C}$ is the circle $x^2 + y^2 = a^2$, $z = 0$. Comment.

$$\mathbf{P} = \frac{\mathbf{k} \times \mathbf{r}_\perp}{r_\perp^2} \qquad (\text{where } \mathbf{r}_\perp = x\mathbf{i} + y\mathbf{j})$$

$$= \mathbf{k} \times \nabla_\perp \log r_\perp$$

$$\nabla \times \mathbf{P} = \nabla_\perp \times \mathbf{P} = \nabla_\perp \times [\mathbf{k} \times \nabla_\perp \log r_\perp]$$

$$= \mathbf{k}\nabla_\perp^2 \log r_\perp$$

$$= \mathbf{0} \qquad (\text{except at } r_\perp = 0).$$

Let $x = a\cos\phi$, $y = a\sin\phi$. Then

$$\int_{\mathscr{C}} \mathbf{P}.d\mathbf{r} = \int_{\mathscr{C}} \frac{-y\,dx + x\,dy}{x^2 + y^2}$$

$$= \int_{-\pi}^{\pi} \frac{-a\sin\phi(-a\sin\phi\,d\phi) + a\cos\phi(a\cos\phi\,d\phi)}{a^2}$$

$$= 2\pi.$$

$\nabla \times \mathbf{P}$ is zero except along $x = 0$, $y = 0$. This means that it is not possible to write $\mathbf{P} = \nabla V$ over the whole of space, and so an integral round a closed curve is not necessarily zero. It is possible to write, however,

$$\mathbf{P} = \nabla \tan^{-1}\left(\frac{y}{x}\right).$$

The quantity $\tan^{-1}(y/x)$ is not, however, one-valued, because if a curve which surrounds the z axis is traced out, there is a change in $\tan^{-1}(y/x)$ of 2π.

4. $\mathbf{Q} = yz^2\mathbf{i} + xz^2\mathbf{j} + 2xyz\mathbf{k}$. Find

$$\int_{\mathscr{C}} \mathbf{Q}.d\mathbf{r}$$

where $\mathscr{C}$ is defined by the two equations $x^2 + y^2 + z^2 = a^2$, $x = y$, being taken from the point

$$\left(\frac{a}{\sqrt{2}}, \frac{a}{\sqrt{2}}, 0\right)$$

to the point

$$\left(\frac{a\cos\alpha}{\sqrt{2}}, \frac{a\cos\alpha}{\sqrt{2}}, a\sin\alpha\right).$$

$\nabla \times \mathbf{Q} = \mathbf{0}$ everywhere, and so it follows that $\mathbf{Q}$ is of the form ∇U. U is easily seen to be xyz^2. The integral is independent of the path and depends only on the end points. Thus

$$\begin{aligned}\int_{\mathscr{C}} \mathbf{Q}.d\mathbf{r} &= U\left(\frac{a\cos\alpha}{\sqrt{2}}, \frac{a\cos\alpha}{\sqrt{2}}, a\sin\alpha\right) - U\left(\frac{a}{\sqrt{2}}, \frac{a}{\sqrt{2}}, 0\right) \\ &= \tfrac{1}{2}a^4\cos^2\alpha\sin^2\alpha.\end{aligned}$$

5. Prove that, if $\mathbf{Q}$ is a constant vector

$$\mathbf{Q}.\int_S \mathbf{F} \times d\mathscr{A} = \int_S (\mathbf{Q} \times \mathbf{F}).d\mathscr{A}$$

$$\begin{aligned}\mathbf{Q}.\int_S \mathbf{F} \times d\mathscr{A} &= \mathbf{Q}.\left|\lim_{\alpha\to\infty} \mathbf{F}_\alpha \times \delta\mathscr{A}_\alpha\right| \\ &= \lim_{\alpha\to\infty} |\mathbf{Q}.\{\mathbf{F}_\alpha \times \delta\mathscr{A}_\alpha\}| \\ &= \lim_{\alpha\to\infty} |(\mathbf{Q} \times \mathbf{F}_\alpha).\delta\mathscr{A}_\alpha| \\ &= \int_S (\mathbf{Q} \times \mathbf{F}).d\mathscr{A}.\end{aligned}$$

Many other similar formulae can be proved, involving integrals over curves surfaces or volumes.

6. Prove that if $\mathbf{B}$ is irrotational and $\nabla^2\mathbf{B} + k^2\mathbf{B} = \mathbf{0}$, then

$$\nabla.\mathbf{B} + \int_{P_0}^{P} \mathbf{B}.d\mathbf{r} = 0.$$

where P_0 is arbitrary.

$$\begin{aligned}\mathbf{0} = \nabla^2\mathbf{B} + k^2\mathbf{B} &= \nabla(\nabla.\mathbf{B}) - \nabla \times (\nabla \times \mathbf{B}) + k^2\mathbf{B} \\ &= \nabla(\nabla.\mathbf{B}) + k^2\mathbf{B}.\end{aligned}$$

Thus if $\nabla.\mathbf{B} = U$, $k^2\mathbf{B} = -\nabla U$.

It follows that

$$U = -k^2 \int_{P_0}^{P} \mathbf{B}.d\mathbf{r}$$

whence the result follows.

7. If $\psi(\mathbf{r}) = 0$ is the equation of a surface S and p is the perpendicular from the origin to the tangent plane at any point on the surface, prove that

$$\int_S \frac{d\mathscr{A}}{p} = \int_S \mathbf{F}.d\mathscr{A}$$

where

$$\mathbf{F} = \frac{\nabla\psi}{\mathbf{r}.\nabla\psi}$$

$$\int_S \frac{d\mathscr{A}}{p} = \int_S \frac{\mathbf{n}.d\mathscr{A}}{\mathbf{n}.\mathbf{r}}$$

where $\mathbf{n}$ is the unit normal

$$\mathbf{n} = \frac{\nabla\psi}{|\nabla\psi|}$$

and the result follows.

8. $\mathscr{C}$ is the curve $(x-a)^2 + (y-2a)^2 = a^2$, $z = 0$ and $\mathbf{F}$ is the vector $(x-a,\ y,\ z+a)$. Calculate

$$\int_{\mathscr{C}} \mathbf{F} \times d\mathbf{r} \qquad \text{and} \qquad \int_{\mathscr{C}} \mathbf{F}.d\mathbf{r}$$

where the integral is taken round the whole curve in the anticlockwise direction. Parametrically, $\mathscr{C}$ is given by

$$x = a + a\cos\phi, \qquad y = 2a + a\sin\phi, \qquad z = 0$$

and $\mathbf{F}$, on the curve, is therefore

$$(a\cos\phi, 2a + a\sin\phi, 0)$$

$$\delta\mathbf{r} = (-a\sin\phi, a\cos\phi, 0)\,\delta\phi$$

$$\mathbf{F} \times \delta\mathbf{r} = \mathbf{i}(-a^2\cos\phi)\,\delta\phi + \mathbf{j}(-a^2\sin\phi)\,\delta\phi + \mathbf{k}(a^2 + 2a^2\sin\phi)\,\delta\phi$$

$$\int_{\mathscr{C}} \mathbf{F} \times d\mathbf{r} = a^2 \int_{-\pi}^{\pi} [-\cos\phi\mathbf{i} - \sin\phi\mathbf{j} + (1 + 2\sin\phi)\mathbf{k}]\,d\phi$$

$$= 2\pi a^2\mathbf{k}$$

$$\nabla \times \mathbf{F} = \mathbf{0}\left[\mathbf{F} = \nabla\left\{\frac{(x-a)^2 + y^2 + (z+a)^2}{2}\right\}\right]$$

and so

$$\int_{\mathscr{C}} \mathbf{F}.d\mathbf{r} = 0.$$

This can be verified by direct integration.

9. Calculate

$$\int_S \mathbf{F}.d\mathscr{A}$$

where $\mathbf{F} = x\mathbf{i} + y^2\mathbf{j} + z\mathbf{k}$ and S is the cap $0 < \theta < \alpha$ of the sphere $r = a$,

$$\int_S \mathbf{F}.d\mathscr{A} = \int_S (\mathbf{F}.\mathbf{n})\,d\mathscr{A} = \int_S (\mathbf{F}.\hat{\mathbf{r}})\,d\mathscr{A}$$

$$= \int_S (x\mathbf{i}.\hat{\mathbf{r}} + y^2\mathbf{j}.\hat{\mathbf{r}} + z\mathbf{k}.\hat{\mathbf{r}})\,d\mathscr{A}$$

$$= \int_S (x\sin\theta\cos\phi + y^2\sin\theta\sin\phi + z\cos\theta)\,d\mathscr{A}$$

$$= \int_{-\pi}^{\pi} d\phi \int_0^{\alpha} a^2\sin\theta\,d\theta(a\sin^2\theta\cos^2\phi + a^2\sin^3\theta\sin^3\phi + a\cos^2\theta)$$

$$= a^3 \int_{-\pi}^{\pi} d\phi \int_0^{\alpha} \sin\theta(\sin^2\theta\cos^2\phi + \cos^2\theta)\,d\theta$$

upon using a symmetry argument

$$= \pi a^3 \int_0^{\pi} (\sin^3\theta + 2\cos^3\theta\sin\theta)\,d\theta$$

$$= \pi a^3 \left\{ -\left(\cos\theta + \frac{\cos^3\theta}{3}\right)\right\}_0^{\alpha}$$

$$= a^3\left(\frac{4 - 3\cos\alpha - \cos^3\alpha}{3}\right)$$

10. Calculate

$$\int_V z^2\,d\tau$$

where V is the volume in the first octant ($x > 0, y > 0, z > 0$, bounded by the plane $x + y + z = a$).

$$\int_{\mathscr{V}} z^2\,d\tau = \iiint z^2\,dx\,dy\,dz$$

$$= \iint_D dx\,dy \int_0^{a-z-y} z^2\,dz$$

where D is the projection of the xy plane of that portion of the plane which is in the first quadrant. This is $x > 0$, $y > 0$, $x + y < a$, and so

$$\int_V z^2\,d\tau = \int_0^a dx \int_0^{a-x} dy \int^{a-x-y} z^2\,dz$$

$$= \int_0^a dx \int_0^{a-x} dy\,\tfrac{1}{3}\{(a-x-y)^3\}$$

$$= \int_0^a dx\,\Big[-\tfrac{1}{12}(z-x-y)^4\Big]_0^{a-x}$$

$$= \frac{1}{12}\int_0^a dx(a-x)^4 = \frac{1}{60}a^5.$$

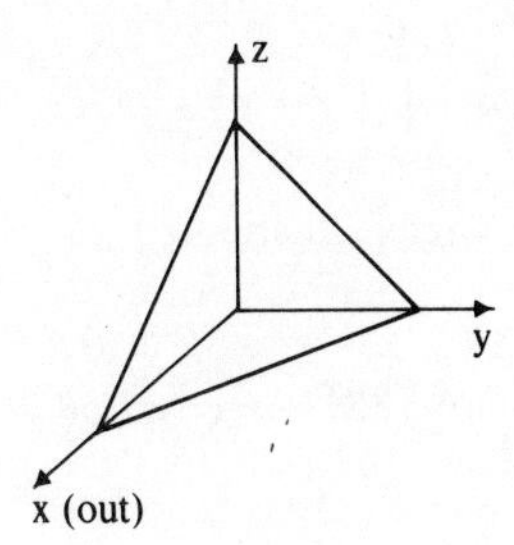

FIG. 4A

11. Calculate

$$\int_V z^2\,d\tau$$

where V is the volume of the ellipsoid

$$\frac{x^2}{a^2}+\frac{y^2}{b^2}+\frac{z^2}{c^2}<1.$$

Let $a = a\xi$, $y = b\eta$, $z = c\zeta$. Then the required integral is equal to

$$abc^3\int_{V*}\zeta^2\,d\tau*$$

where V^* is the volume of the unit sphere $\xi^2+\eta^2+\zeta^2<1$, and $\delta\tau^*$ is an element of volume in the (ξ,η,ζ) space. By symmetry

$$\int_{V^*} \zeta^2\,d\tau^* = \frac{1}{3}\int_{V^*} (\xi^2+\eta^2+\zeta^2)\,d\tau^*$$

$$= \frac{1}{3}\int_0^1 r^2.4\pi r^2\,d\tau$$

$$= \frac{4\pi}{15}.$$

Examples 10 and 11 are not strictly examples involving vectors. They are however included as examples of how volume integrals, which occur if the divergence theorem (section 5.2) is used to evaluate surface integrals, may be evaluated.

12. Evaluate

$$\int_{A,\mathscr{C}}^{B} (x^2+y^2+z^2)\,ds$$

where A is (0,0,0), B is $(a\cos\alpha, a\sin\alpha, c\alpha)$ and $\mathscr{C}$ is the helix $x=a\cos\theta$, $y=a\sin\theta$, $z=c\theta$

$$x^2+y^2+z^2 = a^2+c^2\theta^2$$
$$s^2 = a^2\sin^2\theta\,\delta\theta^2 + a^2\cos^2\theta\,\delta\theta^2 + c^2\,\delta\theta^2$$
$$= (a^2+c^2)\,\delta\theta^2.$$

Thus the required integral is

$$(a^2+c^2)^{1/2}\int_0^\alpha (a^2+c^2\theta^2)\,d\theta = (a^2+c^2)^{1/2}\left(a^2\alpha+\frac{c^2\alpha^3}{3}\right).$$

13. Evaluate

$$\int_{A,\mathscr{C}}^{B} (a^2+y^2+z^2)\,d\mathbf{r}$$

where A, B, $\mathscr{C}$ are the same as in example 12. The integral is given by

$$-\mathbf{i}\int_0^\alpha (a^2+c^2\theta^2)\,a\sin\theta\,d\theta + \mathbf{j}\int_0^\alpha (a^2+c^2\theta^2)\,a\cos\theta\,d\theta + \mathbf{k}\int_0^\alpha (a^2+c^2\theta^2)\,c\,d\theta$$

$$= \mathbf{i}\{a^3(\cos\alpha-1) + ac^2(\alpha^2\cos\alpha - 2\alpha\sin\alpha + 2\cos\alpha - 2)\} + \mathbf{j}\{a^3\sin\alpha$$
$$+ ac^2(\alpha^2\sin\alpha + 2\alpha\cos\alpha - 2\sin\alpha)\} + \mathbf{k}\left\{a^2c\alpha + \frac{c^3\alpha^3}{3}\right\}.$$

14. S comprises the curved surface of part $-\alpha < \theta < \alpha$ of the torus

$$x = (a + b\cos\theta)\cos\phi$$
$$y = (a + b\cos\theta)\sin\phi$$
$$z = b\sin\theta.$$

Calculate

$$\int_S d\mathscr{A}$$

Suppose that S^* comprises the plane ends of the torus.

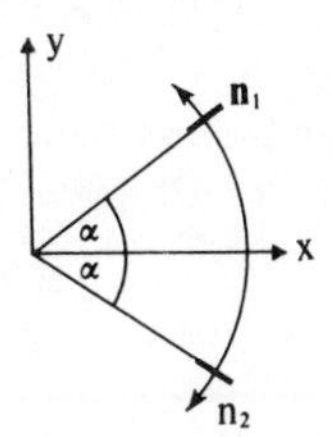

FIG. 4B

Then

$$\int_{S+S^*} d\mathscr{A} = \mathbf{0}$$

as $S + S^*$ is a closed surface.

Whence

$$\int_S d\mathscr{A} = -\int_{S^*} d\mathscr{A}$$

$= -\pi b^2\{\mathbf{n}_1 + \mathbf{n}_2\}$ (where $\mathbf{n}_1$ and $\mathbf{n}_2$ are unit normals as indicated)

$= 2\pi b^2 \sin\alpha\mathbf{i}.$

15. Calculate

$$\iint_D \mathbf{r}\,dx\,dy$$

where D is the domain $x > 0$, $x^{2/3} + y^{2/3} < a^{2/3}$

$$\iint_D \mathbf{r}\,dx\,dy = \mathbf{i}\iint_D x\,dx\,dy.$$

By symmetry

$$= 2\mathbf{i}\int_0^a dy \int_0^{(a^{2/3}-y^{2/3})^{3/2}} x\,dx,$$

on dissecting parallel to the x axis

$$= \mathbf{i} \int_0^a (a^{2/3} - y^{2/3})^3 \, dy$$

$$= \frac{97}{140} a^3 \mathbf{i}.$$

16. $\mathbf{F} = x^2\mathbf{i} + y^2\mathbf{j} + z^2\mathbf{k}$, O is the origin and A the point (a, a, a). Calculate

$$\int_O^A \mathbf{F} \times d\mathbf{r}$$

along the straight line OA, and along the contour defined by $O < x < a$, $y = 0$, $z = 0$; $x = a$, $O < y < a$, $z = 0$; $x = a$, $y = a$, $O < z < a$. Along OA,

$$x = y = z = \frac{s}{\sqrt{3}}$$

and

$$\mathbf{r} = x\mathbf{i} + y\mathbf{j} + z\mathbf{k}$$

whence

$$\delta\mathbf{r} = \frac{\delta s}{\sqrt{3}}(\mathbf{i} + \mathbf{j} + \mathbf{k}).$$

$$\int_O^A \mathbf{F} \times d\mathbf{r} = \int_0^{\sqrt{(3a)}} \left[\frac{s^2}{3}(\mathbf{i} + \mathbf{j} + \mathbf{k}) \times \frac{ds}{\sqrt{3}}(\mathbf{i} + \mathbf{j} + \mathbf{k})\right]$$

$$= \mathbf{0}.$$

For integrating along the second contour, it is necessary to split the integral up into three sections. The values of $\mathbf{F}$ and $\delta\mathbf{r}$ over these are respectively

(i) $\mathbf{F} = x^2\mathbf{i}$ $\quad \delta\mathbf{r} = \delta x\mathbf{i}$

(ii) $\mathbf{F} = a^2\mathbf{i} + y^2\mathbf{j}$ $\quad \delta\mathbf{r} = \delta y\mathbf{j}$

(iii) $\mathbf{F} = a^2\mathbf{i} + a^2\mathbf{j} + z^2\mathbf{k}$ $\quad \delta\mathbf{r} = \delta z\mathbf{k}$

The integral is thus given by

$$\int_0^a (x^2\mathbf{i}) \times dx\mathbf{i} + \int_0^a (a^2\mathbf{i} + y^2\mathbf{j}) \times dy\mathbf{j} + \int_0^a (a^2\mathbf{i} + a^2\mathbf{j} + z^2\mathbf{k}) \times dz\mathbf{k}$$

$$= \int_0^a a^2 dy(\mathbf{i} \times \mathbf{j}) + \int_0^a a^2\{(\mathbf{i} \times \mathbf{k}) + (\mathbf{j} \times \mathbf{k})\} \, dz = a^3(\mathbf{i} - \mathbf{j} + \mathbf{k}).$$

Thus the value of

$$\int_O^A \mathbf{F} \times d\mathbf{r}$$

depends upon the path of integration.

17. Calculate

$$\int_S x\, d\mathscr{A} \qquad \text{and} \qquad \int_S x\, d\mathscr{A}$$

when S is the octant of the sphere $r = a$ in $0 < x,\ y,\ z < a$. In spherical coordinates, the octant is $0 < \theta < \pi/2$, $0 < \phi < \pi/2$.

$$\int_S x\, d\mathscr{A} = a^2 \int_O^{\pi/2} d\theta \sin\theta \int_O^{\pi/2} d\phi\, a \sin\theta \cos\phi$$

$$= \frac{\pi a^3}{4}$$

$$\int_S x\, d\mathscr{A} = \int_S x\hat{\mathbf{r}}\, d\mathscr{A}$$

$$= \int_S x(\mathbf{i} \sin\theta \cos\phi + \mathbf{j} \cos\theta \sin\phi + \mathbf{k} \cos\theta)\, d\mathscr{A}$$

$$= a^3 \int_O^{\pi/2} d\theta \int_O^{\pi/2} d\phi \sin^2\theta \cos\phi (\mathbf{i} \sin\theta \cos\phi + \mathbf{j} \sin\theta \sin\phi + \mathbf{k} \cos\theta)$$

$$= a^3 \left(\mathbf{i}\frac{\pi}{6} + \mathbf{j}\frac{1}{3} + \mathbf{k}\frac{1}{3} \right).$$

18. Calculate

$$\int_S \mathbf{F} . d\mathscr{A}$$

where $\mathbf{F} = x^2\mathbf{i} + y^2\mathbf{j} + z^2\mathbf{k}$ and S is that portion of the plane $x + y + z = a$ which lies in the first octant. It is assumed that the positive normal to the plane is that with an upward component.

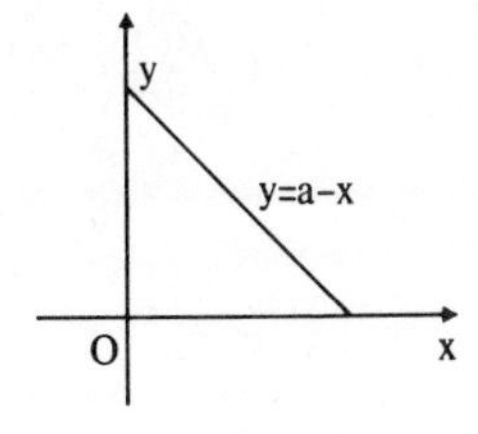

FIG. 4C.

It is convenient to use the result of exercise 4 in this chapter. Because the integral will be evaluated over a domain in the xy plane it is convenient to eliminate z. This is done as follows.

On

$$x + y + z - a = 0,$$

the values of

$$\mathbf{F} = x^2\mathbf{i} + y^2\mathbf{j} + z^2\mathbf{k}$$

and

$$\mathbf{F}^* = x^2\mathbf{i} + y^2\mathbf{j} + (a - x - y)^2\mathbf{k}$$

are the same.

It follows that

$$\int_S \mathbf{F} . d\mathscr{A} = \int_S \mathbf{F}^* . d\mathscr{A}$$

and consequently, attention may be concentrated upon $\mathbf{F}^*$ instead of $\mathbf{F}$. Thus

$$\int_S \mathbf{F}^* . d\mathscr{A} = \iint_{S_0} (\mathbf{F}^* . \nabla\psi) \frac{dx\,dy}{\nabla\psi . \mathbf{k}}$$

S_0 being the projection of S on the xy plane and $\psi(\mathbf{r}) = 0$ being the equation of S. Clearly S_0 is the domain $x > 0$, $y > 0$, $x + y < a$.

$$\nabla\psi = \mathbf{i} + \mathbf{j} + \mathbf{k}.$$

Then

$$\mathbf{F}^* . \nabla\psi = x^2 + y^2 + (a - x - y)^2$$

and

$$\mathbf{k} . \nabla\psi = 1.$$

Thus

$$\int_S \mathbf{F} . d\mathscr{A} = \int_0^a dx \int_0^{a-z} dy\{x^2 + y^2 + (a - x - y)^2\}$$

$$= \int_0^a dx \left[x^2 y + \frac{y^3}{3} - \frac{(a - x - y)^3}{3} \right]_0^{a-x}$$

$$= \int_0^a \left\{ x^2(a - x) + \frac{2(a - x)^3}{3} \right\} dx = \tfrac{1}{4}a^4.$$

19. Calculate

$$\int_S \mathbf{F} . d\mathscr{A}$$

where $\mathbf{F} = xz^2\mathbf{i} + yz^2\mathbf{j} + xyz\mathbf{k}$ and S is the surface of the cylinder

$$x^2 = y^2 < a^2, \qquad |z| < c.$$

This can clearly be written in the form

$$\int_{S_1} (xz^2\mathbf{i} + yz^2\mathbf{j}) . d\mathscr{A} + \int_{S_2} xyz\mathbf{k} . d\mathscr{A}$$

where S_1 is the curved part of S and S_2 is the plane part of S. Putting everything into cylindrical polar coordinates, this becomes

$$\int_{-\pi}^{\pi} d\phi \int_{-c}^{c} dz(a^2\cos^2\phi + a^2\sin^2\phi)z^2 + 2c \int_{-\pi}^{\pi} d\phi \int_0^a \rho\, d\rho \,.\, \rho^2 \sin\phi\cos\phi$$

$$= \frac{4\pi}{3} a^2 c^3.$$

20. Calculate

$$\int_S \mathbf{F}\,.\,d\mathscr{A}$$

where $\mathbf{F}$ is the same as in example 19, and S is the outside of the open cone

$$\frac{x^2+y^2}{a^2} = \frac{z^2}{c^2} \qquad 0 < z < c.$$

In cylindrical coordinates

$$\mathbf{F} = \rho z^2 \cos\phi \mathbf{i} + \rho z^2 \sin\phi \mathbf{j} + \rho^2 \sin\phi\cos\phi z\mathbf{k}.$$

The unit normal to the cone pointing out is

$$\begin{aligned} \mathbf{n} &= \hat{\boldsymbol{\rho}}\cos\gamma - \mathbf{k}\sin\gamma, \tan\gamma = (a/c) \\ \mathbf{F}\,.\,\mathbf{n} &= (\rho z^2\cos^2\phi + \rho z^2\sin^2\phi)\cos\gamma - \rho^2 z\cos\phi\sin\phi\sin\gamma \\ &= \rho z^2\cos\gamma - \rho^2 z\cos\phi\sin\phi\sin\gamma \\ \delta\mathscr{A} &= \rho\,\delta z\,\delta\phi\sec\gamma \end{aligned}$$

and so

$$\int_S \mathbf{F}\,.\,d\mathscr{A} = \int_{-\pi}^{\pi} d\phi \int_0^c dz \rho\sec\gamma(\rho z^2\cos\gamma - \rho^2 z\sin\phi\cos\phi\sin\gamma)$$

$$= 2\pi \int_0^c \rho^2 z^2\, dz.$$

$\rho = z\tan\gamma$ and so the result is

$$\frac{2\pi}{5} c^5 \tan^2\gamma = \frac{2\pi}{5} a^2 c^3.$$

Exercises

1. $\mathbf{F}$ is a constant vector. Prove that

$$\int_{\mathscr{C}} \mathbf{F} \times d\mathbf{r} = \mathbf{0}$$

if $\mathscr{C}$ is any closed curve.

2. Prove that, if S is a closed surface

$$\int_S d\mathscr{A} = \mathbf{0}.$$

3. Show that

$$\int_{\mathscr{C}} \hat{\mathbf{t}}.d\mathbf{r}$$

where $\hat{\mathbf{t}}$ is the unit tangent vector, is equal to the length of $\mathscr{C}$.

4. An open surface S has projection S_0 on the xy plane. Prove that

$$\int_S \mathbf{F}.d\mathscr{A} = \iint_{S_0} \mathbf{F}.\mathbf{n}\frac{dx\,dy}{\mathbf{k}.\mathbf{n}}.$$

5. The electric displacement $\mathbf{D}$ due to an electric charge Q situated at the origin is $(O/4\pi r^2)\hat{\mathbf{r}}$, prove that if S is a closed surface

$$\int_S \mathbf{D}.d\mathscr{A} = \text{the net charge within } S$$

(Gauss's electric flux theorem).

6. Prove that the solid angle subtended by a circular disc at a point on its axis is $2\pi(1 - \cos\alpha)$ where α is the semi-vertical angle of the cone with the point as vertex and the circular disc as base.

7. Evaluate

$$\int_V (x^2 + y^2 + z^2)^{3/2}\, d\tau$$

over the volume $x^2 + y^2 + z^2 \leqslant a^2$.

8. If the radius vector from O traces out a curve $\mathscr{C}$, with end-points A and B, prove that the net vector area of the surface defined by OA, OB and $\mathscr{C}$ between A and B is

$$\frac{1}{2}\int_{A,\mathscr{C}}^{B} \mathbf{r} \times d\mathbf{r}.$$

9. Prove that

$$\int_{\mathscr{C}} \mathbf{F}.d\mathbf{r}$$

where $\mathscr{C}$ is a closed contour, is zero when

$$\mathbf{F} = 3x^2y^2 \sin z\mathbf{i} + 2x^3y \sin z\mathbf{j} + x^3y^2 \cos z\mathbf{k}.$$

10. A surface S has the equation $F(\mathbf{r}) = 0$. Show that, if D is the projection of S upon a plane with unit normal $\mathbf{n}$, the surface area of S is

$$\int_D \frac{|\nabla F|}{\mathbf{n}.\nabla F} d\Sigma$$

where $\delta\Sigma$ is an element of area in the plane. It is assumed that a line in the direction of $\mathbf{n}$ cuts S once only.

11. Evaluate

$$\int_S \mathbf{F}.d\mathscr{A}$$

where $\mathbf{F} = r^2\mathbf{r}$ and S is the curved surface of the cylinder $x^2 + y^2 = a^2$ which lies in the first octant between $z = 0$ and $z = c$.

12. Evaluate

$$\int_S d\mathscr{A}.(\nabla \times \mathbf{F})$$

where S is the hemisphere $x^2 + y^2 + z^2 = a^2$, $z < 0$ and $\mathbf{F} = y^2\mathbf{i} + x^2\mathbf{j} + xy\mathbf{k}$.

13. If a path $\mathscr{C}$ encloses a wire carrying a steady current I, then the following relation exists

$$\int_{\mathscr{C}} \mathbf{H}.d\mathbf{r} = I$$

where $\mathbf{H}$ is the magnetic field strength. If $\mathbf{J}$ is the electric current density vector, S is the open surface with edge $\mathscr{C}$, prove Ampere's circuital law

$$\int_{\mathscr{C}} \mathbf{H}.d\mathbf{r} = \int_S \mathbf{J}.d\mathscr{A}$$

(The electric current density vector is defined by the relation that $\mathbf{J}.\mathbf{n}$ is the current flow per unit area across a surface with unit normal n.)

14. Calculate

$$\int_S P d\mathscr{A}$$

where $P = xyz$ and S is that part of the plane

$$\alpha x + \beta y + \gamma z = q$$

which lies in the first octant.

15. Evaluate

$$\int_S \mathbf{F} \times d\mathscr{A}$$

when $\mathbf{F} = x^2\mathbf{i} + y^2\mathbf{j} + xy\mathbf{k}$ and S is the outside of the hemisphere

$$x^2 + y^2 + z^2 = a^2,\ z > 0.$$

16. Calculate

$$\int_{A,\mathscr{C}}^{B} \mathbf{F} \times d\mathbf{r}$$

when $\mathbf{F} = ax\mathbf{i} + by\mathbf{j} + cz\mathbf{k}$, $\mathscr{C}$ is the curve $x = \cos\phi$, $y = \sin\phi$, $z = \phi$, and A, B are the points defined by $\phi = \alpha$ and $\phi = \beta$. Does this differ from the result when $\mathscr{C}$ is the straight line joining the two points A and B?

17. S is a closed surface. Prove that

$$\frac{1}{3}\int_S \mathbf{r}\,.\,d\mathscr{A}$$

is the volume enclosed by the surface.

18. $\mathbf{F} = xy(\mathbf{i} + \mathbf{j}) + z^2\mathbf{k}$. Calculate

$$\int_{\mathscr{C}} \mathbf{F}\,.\,d\mathbf{r}$$

when $\mathscr{C}$ is the straight line from the origin to the point with spherical coordinates (r_0, θ_0, ϕ_0) and when $\mathscr{C}$ comprises the three arcs

(*i*) $0 < r < r_0$, $\theta = 0$, $\phi = 0$
(*ii*) $r = r_0$, $0 < \theta < \theta_0$, $\phi = 0$
(*iii*) $r = r_0$, $\theta = \theta_0$, $0 < \phi < \phi_0$

taken in that order.

19. Calculate

$$\int_{\mathscr{C}} xyz\,d\mathbf{r}$$

when $\mathscr{C}$ is that portion of the helix $x = a\cos\phi$, $y = a\sin\phi$, $z = c\phi$ which lines in $2n\pi < \phi < 2(n+1)\pi$.

20. Calculate

$$\int_S \mathbf{F}\,.\,d\mathscr{A}$$

when S is the surface of the cube defined by $|x| < a$, $|y| < a$, $|z| < a$, and $\mathbf{F} = \alpha x\mathbf{i} + \beta y\mathbf{j} + xy\mathbf{k}$.

21. S is a surface and $\mathbf{F}$ is some vector field such that $\mathbf{F}\,.\,\mathbf{n}$ is non-negative over S. ϕ and ψ are scalar fields such that $\phi > \psi$ over S. Prove that

$$\int_S (\phi - \psi)\mathbf{F}\,.\,d\mathscr{A} > 0.$$

5 Stokes's and Gauss's Theorems and their Applications

5.1. Stokes's theorem

Stokes's theorem states that, if $\mathbf{F}$ is a vector field and S is an open orientable surface with a rim $\mathscr{C}$ bounding it

$$\int_S d\mathscr{A}.(\nabla \times \mathbf{F}) = \int_{\mathscr{C}} \mathbf{F}.d\mathbf{r}. \tag{5.1.1}$$

If S is a closed surface, the rim does not exist and the right-hand side of (5.1.1) is replaced by zero. If $\mathbf{F}$ is a gradient, the theorem is trivially true as both sides vanish. For a counter example which shows that the theorem does not hold when the surface is not orientable, see exercise 20 in this chapter.

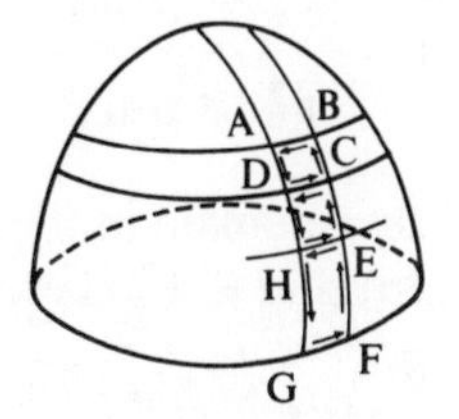

Fig. 5.1(i)

Suppose that the surface is divided by two systems of curves into elements of surface S_i such as $ABCD$.

If $\mathscr{C}_i$ is the rim $ADCBA$ which surrounds S_i

$$\int_{\mathscr{C}} \mathbf{F}.d\mathbf{r} = \sum_i \int_{\mathscr{C}_i} \mathbf{F}.d\mathbf{r}. \tag{5.1.2}$$

This follows because for two adjacent S_i there is a common boundary and the line integrals along this will cancel one another out. In the end it is only integrals such as

$$\int_{GF} \mathbf{F}.d\mathbf{r}$$

which remain, as they are only traversed once. Summing over all arc portions such as GF, (5.1.2) follows.

Consider now the left-hand side of (5.1.1)

$$\int_S d\mathscr{A}.(\nabla \times \mathbf{F}) = \sum_i \int_{S_i} d\mathscr{A}.(\nabla \times \mathbf{F}). \tag{5.1.3}$$

It will be possible to prove Stokes's theorem if the relation

$$\int_{S_i} d\mathscr{A}.(\nabla \times \mathbf{F}) = \int_{\mathscr{C}_i} \mathbf{F}.d\mathbf{r} \tag{5.1.4}$$

holds when S_i is infinitesimal, because if this is so, the result follows by summation. Now, if S_i is infinitesimal, it may be treated as plane. Thus, it remains to prove that (5.1.4) is true for a plane element of surface. As the result is independent of the frame of reference, it will be convenient to take the element as in the xy plane.

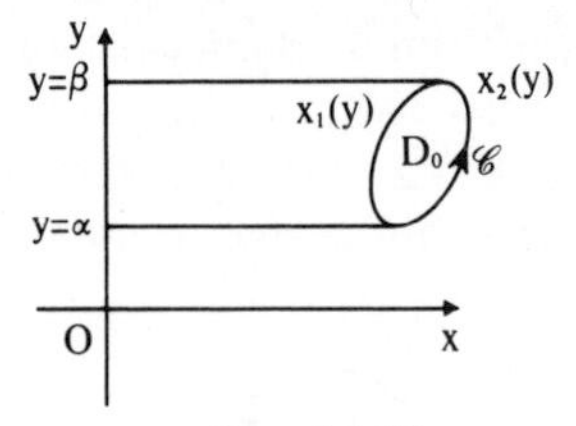

FIG. 5.1(ii)

Let $\mathbf{F} = (P, Q, R)$ and D_0 a domain in the xy plane surrounded by a boundary $\mathscr{C}_0 . d\boldsymbol{\sigma} = \mathbf{k}\, dx\, dy$ and (5.1.4) becomes

$$\iint_{D_0} \left(\frac{\partial Q}{\partial x} - \frac{\partial P}{\partial y}\right) dx\, dy = \int_{\mathscr{C}_0} (P\, dx + Q\, dy). \tag{5.1.5}$$

Dissecting parallel to the x axis

$$\iint_{D_0} \frac{\partial Q(x, y)}{\partial x}\, dx\, dy = \int_\alpha^\beta [Q(x, y)]_{x=x_1(y)}^{x=x_2(y)}\, dy$$

$$= \int_\alpha^\beta [Q(x_2(y), y) - Q(x_1(y), y)]\, dy = \int_{\mathscr{C}_0} Q\, dy. \tag{5.1.6a}$$

Similarly, on dissecting parallel to the x axis

$$\iint_D -\frac{\partial P}{\partial y}\, dx\, dy = \int_{\mathscr{C}_0} P\, dx. \tag{5.1.6b}$$

Thus (5.1.5) is proved. (This proof in fact holds for arbitrary domains in the xy plane, not only elementary ones.) It follows that Stokes's theorem has been shown to be true for plane surfaces (in this form it is sometimes termed Green's theorem in two dimensions). By what has gone before, the proof of Stokes's theorem for a plane surface implies the proof for an arbitrary surface.

The area of a domain D_0 in the xy plane is given by

$$A = \iint_{D_0} dx\, dy.$$

By putting successively $P = 0$, $Q = x$; $P = -y$, $Q = 0$; $P = -y/2$, $Q = x/2$, three different expressions follow

$$A = \int_{\mathscr{C}_0} x\, dy = \int_{\mathscr{C}_0} -y\, dx = \frac{1}{2} \int_{\mathscr{C}_0} (x\, dy - y\, dx). \qquad (5.1.7)$$

Integrals such as, for example,

$$\int_{\mathscr{C}_0} Q(x, y)\, dx,$$

are best evaluated by the relation

$$\int_{\mathscr{C}_0} Q(x, y)\, dx = \int_{\mathscr{C}_0} Q[x(s), y(s)] \frac{dx}{ds}\, ds$$

where s is arc length measured along $\mathscr{C}_0$.

5.11. *Extensions of Stokes's theorem*

Let $\mathbf{F} = \mathbf{A}V$ where $\mathbf{A}$ is a constant vector. Then

$$\int_S d\mathscr{A}.\{\nabla \times (\mathbf{A}V)\} = \int_{\mathscr{C}} (\mathbf{A}V).d\mathbf{r}$$

$$\int_S d\mathscr{A}.\{(\nabla V) \times \mathbf{A}\} = \left\{\int_S d\mathscr{A} \times \nabla V\right\}.\mathbf{A} = \mathbf{A}.\int_{\mathscr{C}} d\mathbf{r}\, V.$$

$\mathbf{A}$ is arbitrary and so

$$\int_S d\mathscr{A} \times \nabla V = \int_{\mathscr{C}} d\mathbf{r}\, V. \qquad (5.11.1)$$

Let $\mathbf{F} = \mathbf{G} \times \mathbf{A}$ where $\mathbf{A}$ is a constant vector

$$\int_S d\mathscr{A}.\{\nabla \times (\mathbf{G} \times \mathbf{A})\} = \int_{\mathscr{C}} d\mathbf{r}.(\mathbf{G} \times \mathbf{A}).$$

$$\left\{\int_S (d\mathscr{A} \times \nabla) \times \mathbf{G}\right\}.\mathbf{A} = \mathbf{A}.\left\{\int_{\mathscr{C}} d\mathbf{r} \times \mathbf{G}\right\}.$$

As $\mathbf{A}$ is arbitrary, it follows that

$$\int_S (d\mathscr{A} \times \nabla) \times \mathbf{G} = \int_{\mathscr{C}} d\mathbf{r} \times \mathbf{G} \tag{5.11.2}$$

Now (5.1.1) can be rewritten as

$$\int_S (d\mathscr{A} \times \nabla).\mathbf{F} = \int_{\mathscr{C}} d\mathbf{r}.\mathbf{F} \tag{5.11.3}$$

Comparing (5.11.1), (5.11.2) and (5.11.3), it follows that they may all be expressed formally by

$$\int_S d\mathscr{A} \times \nabla\{\ \ \} = \int_{\mathscr{C}} d\mathbf{r}\{\ \ \} \tag{5.11.4}$$

Stokes's theorem and its extensions may be used to evaluate line integrals around closed contours through using surface integrals and vice versa. It is now apparent why, in many cases, line integrals between two points depend upon the curve joining the points, this ambiguity being equivalent to the non-vanishing of line integrals around closed contours.

5.2. The Gauss divergence theorem

Let $\mathbf{F}$ be a vector function defined within a volume V bounded by a surface S. Then

$$\int_S \mathbf{F}.d\mathscr{A} = \int_V \nabla.\mathbf{F}\,d\tau. \tag{5.2.1}$$

This may be written as

$$\int_S (F_x\mathbf{i} + F_y\mathbf{j} + F_z\mathbf{k}).d\mathscr{A} = \int_V \left(\frac{\partial F_x}{\partial x} + \frac{\partial F_y}{\partial y} + \frac{\partial F_z}{\partial z}\right) d\tau. \tag{5.2.2}$$

It will clearly be sufficient to prove

$$\int_S F_z\mathbf{k}.d\mathscr{A} = \iiint_V \frac{\partial F_z}{\partial z}\,dx\,dy\,dz. \tag{5.2.3}$$

Suppose that the boundary surface can be split up into two portions S_+, S_- whose equations are respectively $z = z_+(x,y)$ and $z = z_-(x,y)$ $z_+ > z_-$. On $S_+ \mathbf{n}.\mathbf{k} > 0$, on $S_- \mathbf{n}.\mathbf{k} < 0$. If this is not the case the volume may be split up into portions for which this does hold. There will not be any contribution from the common surface between two portions because the elements of surface are equal and opposite there

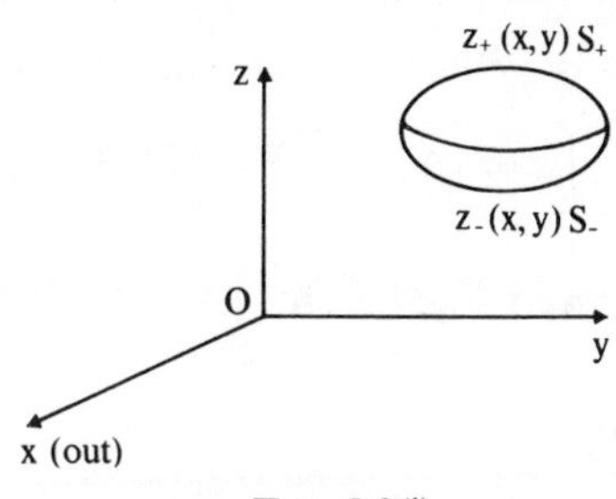

FIG. 5.2(i)

and the surface contribution is exactly that from the original boundary surface.

$$\int_S F_z \mathbf{k}.d\mathscr{A} = \int_{S_+} F_z \mathbf{k}.d\mathscr{A} + \int_{S_-} F_z \mathbf{k}.d\mathscr{A} \tag{5.2.4}$$

$$= \iint_{D_0} F_z(x,y,z_+)\,dx\,dy + \iint_{D_0} F_z(x,y,z_-)(-dx\,dy)$$

where D_0 is the domain in the xy plane enclosed by the projection of the rim or rims of S_+ and S_-

$$= \iint_{D_0} \{F_z(x,y,z_+) - F_z(x,y,z_-)\}\,dx\,dy$$

$$= \iint_{D_0} dx\,dy \int_{z_-}^{z_+} \frac{\partial F_z}{\partial z}\,dz = \iiint_V \frac{\partial F_z}{\partial z}\,d\tau. \tag{5.2.5}$$

Similar arguments hold for the formulae involving F_z and F_y. Thus

$$\int_S \mathbf{F}.d\mathscr{A} = \int_V \nabla.\mathbf{F}\,d\tau. \tag{5.2.6}$$

It will be seen that if $\mathbf{F} = \nabla \times \mathbf{G}$, $\nabla.\mathbf{F}$ vanishes and hence

$$\int_S \mathbf{F}.d\mathscr{A} = 0.$$

The same result would follow from Stokes's theorem as

$$\int_S d\mathscr{A}.(\nabla \times \mathbf{G}) = 0$$

because S is closed.

5.21. *Extensions of the divergence theorem*

Let $\mathbf{F} = \mathbf{A}U$ where $\mathbf{A}$ is a constant vector. Then

$$\int_S d\mathscr{A}.\mathbf{A}U = \int_V \nabla.(\mathbf{A}U)\, d\tau$$

$$\mathbf{A}.\int_S d\mathscr{A}U = \mathbf{A}.\int_V \nabla U\, d\tau.$$

$\mathbf{A}$ is arbitrary and so

$$\int_S d\mathscr{A}U = \int_V d\tau\, \nabla U \tag{5.21.1}$$

Let $\mathbf{F} = \mathbf{G} \times \mathbf{A}$

$$\int_S d\mathscr{A}.(\mathbf{G} \times \mathbf{A}) = \int_V \nabla.(\mathbf{G} \times \mathbf{A})\, d\tau.$$

Using the scalar triple product formula

$$\mathbf{A}.\int_S d\mathscr{A} \times \mathbf{G} = \mathbf{A}.\int_V d\tau(\nabla \times \mathbf{G})$$

As $\mathbf{A}$ is arbitrary, it follows that

$$\int_S d\mathscr{A} \times \mathbf{G} = \int_V d\tau(\nabla \times \mathbf{G}). \tag{5.21.2}$$

Now (5.2.1) may be rewritten in the form

$$\int_S d\mathscr{A}.\mathbf{F} = \int_V d\tau\, \nabla.\mathbf{F}. \tag{5.21.3}$$

Comparing (5.21.1), (5.21.2) and (5.21.3), it follows that they may all be expressed formally by

$$\int_S d\mathscr{A}\{\quad\} = \int_V d\tau\, \nabla\{\quad\} \tag{5.21.4}$$

The divergence theorem and its extensions may be used to evaluate surface integrals in terms of volume integrals and vice versa. The relation (5.21.4) may be used to provide an alternative definition of the operator ∇. If the volume V is very small, (5.21.4) may be written as

$$\int_S d\mathscr{A}\{\quad\} = V\nabla\{\quad\}$$

and so there follows a definition

$$\nabla\{\quad\} = \lim_{V\to 0} \frac{1}{V} \int_S d\mathscr{A}\{\quad\}. \tag{5.21.5}$$

This is sometimes used as a definition of ∇. It is unsatisfactory, however, in that there is no reason *a priori* for the limit to be independent of the shape of S and the proof that it is, is lengthy.

5.22. *Green's theorems*

Let ϕ, ψ be scalar point functions

$$\nabla.(\phi\nabla\psi) = \nabla\phi.\nabla\psi + \phi\nabla^2\psi. \tag{5.22.1}$$

Integrating over a volume V, this becomes

$$\int_V \nabla.(\phi\nabla\psi)\,d\tau = \int_V \{(\nabla\phi).(\nabla\psi) + \phi\nabla^2\psi\}\,d\tau$$

and on using the divergence theorem

$$\int_S \phi(\nabla\psi).d\mathscr{A} = \int_V \{(\nabla\phi).(\nabla\psi) + \phi\nabla^2\psi\}\,d\tau. \tag{5.22.2}$$

This is known as Green's first theorem.

If $\psi = \phi$

$$\int_S \phi(\nabla\phi).d\mathscr{A} = \int_V \{(\nabla\phi)^2 + \phi\nabla^2\phi\}\,d\tau. \tag{5.22.3}$$

Exchanging ϕ and ψ in (5.22.2)

$$\int_S \psi(\nabla\phi).d\mathscr{A} = \int_V \{(\nabla\phi).(\nabla\psi) + \psi\nabla^2\phi\}\,d\tau. \tag{5.22.4}$$

Subtracting (5.22.4) from (5.22.2)

$$\int_S (\phi\nabla\psi - \psi\nabla\phi).d\mathscr{A} = \int_V (\phi\nabla^2\psi - \psi\nabla^2\phi)\,d\tau. \tag{5.22.5}$$

This is known as Green's second theorem. Often a quantity such as

$$\int_S \phi(\nabla\psi).d\mathscr{A}$$

is written

$$\int_S \phi \frac{\partial \psi}{\partial n} d\mathscr{A},$$

$\partial\psi/\partial n$ being the rate of change of ψ in the direction of the outward normal.

A number of consequences of Green's theorems are of importance.

(*i*) If ϕ and ψ are harmonic, that is $\nabla^2\phi = 0$, $\nabla^2\psi = 0$

$$\int_S \phi \frac{\partial \psi}{\partial n} d\mathscr{A} = \int_S \psi \frac{\partial \phi}{\partial n} d\mathscr{A} \tag{5.22.6}$$

(*ii*) If ϕ and ψ obey the same boundary condition $\alpha\psi + \beta(\partial\psi/\partial n) = 0$, $\alpha\phi + \beta(\partial\phi/\partial n) = 0$, α, β arbitrary functions including the special cases $\partial\psi/\partial n = 0$, $\partial\phi/\partial n = 0$ (the Neumann condition) and $\phi = 0$, $\psi = 0$ (the Dirichlet condition).

$$\int_S \left(\phi \frac{\partial \psi}{\partial n} - \psi \frac{\partial \phi}{\partial n}\right) d\mathscr{A} = 0$$

and so

$$\int_V \phi \nabla^2 \psi \, d\tau = \int_V \psi \nabla^2 \phi \, d\tau. \tag{5.22.7}$$

(*iii*) If ϕ is harmonic, it follows from (5.22.3) that

$$\int_S \phi \frac{\partial \phi}{\partial n} d\mathscr{A} = \int_V (\nabla\phi)^2 \, d\tau. \tag{5.22.8}$$

If ϕ is constant on S – say, $\phi = \phi_0$

$$\int_V (\nabla\phi)^2 \, d\tau = \phi_0 \int_S (\nabla\phi) . d\mathscr{A} = \phi_0 \int_V \nabla^2 \phi \, d\tau = 0.$$

Now $\int (\nabla\phi)^2 d\tau$ can only be zero if $\nabla\phi = \mathbf{0}$, that is, if ϕ is constant throughout the region. This constant value is the value on the boundary surface ϕ_0. Thus if a function is harmonic within a region and has a constant value on the boundary surface, it is constant throughout the region. In particular, if a harmonic function vanishes on the boundary of its region of definition, it vanishes everywhere there. Similarly, if $\partial\phi/\partial n$ vanishes everywhere over the boundary, $\int (\nabla\phi)^2 d\tau$ is zero and must be constant everywhere.

There are vectorial analogues of Green's theorems which arise as follows.

$$\nabla . \{\mathbf{P} \times (\nabla \times \mathbf{Q})\} = (\nabla \times \mathbf{P}) . (\nabla \times \mathbf{Q}) - \mathbf{P} . \{\nabla \times (\nabla \times \mathbf{Q})\}. \tag{5.22.9}$$

Integrating over a volume V and using the divergence theorem, an analogue of Green's first theorem follows.

$$\int_S d\mathscr{A}.\{\mathbf{P}\times(\nabla\times\mathbf{Q})\} = \int_V [(\nabla\times\mathbf{P}).(\nabla\times\mathbf{Q}) - \mathbf{P}.\{\nabla\times(\nabla\times\mathbf{Q})\}]\,d\tau. \tag{5.22.10}$$

If $\mathbf{P} = \mathbf{Q}$, this takes the form

$$\int_S d\mathscr{A}.\{\mathbf{P}\times(\nabla\times\mathbf{P})\} = \int_V [(\nabla\times\mathbf{P})^2 - \mathbf{P}.\{\nabla\times(\nabla\times\mathbf{P})\}]\,d\tau. \tag{5.22.11}$$

Exchanging $\mathbf{P}$ and $\mathbf{Q}$ in (5.22.10)

$$\int_S d\mathscr{A}.\{\mathbf{Q}\times(\nabla\times\mathbf{P})\} = \int_V [(\nabla\times\mathbf{P}).(\nabla\times\mathbf{Q}) - \mathbf{Q}.\{\nabla\times(\nabla\times\mathbf{P})\}]\,d\tau \tag{5.22.12}$$

and subtracting (5.22.12) from (5.22.10)

$$\int_S d\mathscr{A}.[\mathbf{P}\times(\nabla\times\mathbf{Q}) - \mathbf{Q}\times(\nabla\times\mathbf{P})]$$

$$= \int_V [\mathbf{Q}.(\nabla\times(\nabla\times\mathbf{P})) - \mathbf{P}.(\nabla\times(\nabla\times\mathbf{Q}))]\,d\tau, \tag{5.22.13}$$

which is the vector analogue of Green's second theorem.

If $\nabla\times(\nabla\times\mathbf{P}) = \mathbf{0}$, it follows from (5.22.11) that

$$\int_V (\nabla\times\mathbf{P})^2\,d\tau = \int_S \{d.\mathscr{A}\times\mathbf{P}\}.(\nabla\times\mathbf{P}) = \int_S \mathbf{P}.\{d\mathscr{A}\times(\nabla\times\mathbf{P})\}$$

and so $\nabla\times\mathbf{P} = \mathbf{0}$ everywhere within V if any of the following conditions hold. (*i*) $\mathbf{P} = \mathbf{0}$ on S, (*ii*) $\nabla\times\mathbf{P} = \mathbf{0}$ on S, (*iii*) $\mathbf{n}\times\mathbf{P} = \mathbf{0}$ on S, (*iv*) $\mathbf{n}\times(\nabla\times\mathbf{P}) = \mathbf{0}$ on S, (*v*) $\mathbf{P}\times(\nabla\times\mathbf{P}) = \mathbf{0}$ on S.

5.23. *Green's formula*

Suppose that ψ is a scalar function defined over a volume V enclosed in a surface S and suppose that P' is some arbitrary point. If P' lies within V, enclose it in a sphere of radius δ centre P'. The volume of this sphere is V_0 and its boundary surface is S_0. The normal out of V over S_0 is inwards towards P'. If P' is outside S, V_0 and S_0 do not exist.

$$\nabla^2\left(\frac{1}{R}\right) = 0$$

over $V - V_0$.

From Green's second theorem, it follows, on putting $\phi = 1/R$, that

$$\int_{S+S_0} \left\{\psi \frac{\partial}{\partial n}\left(\frac{1}{R}\right) - \frac{1}{R}\frac{\partial \psi}{\partial n}\right\} d\mathscr{A} + \int_{V-V_0} \frac{\nabla^2 \psi}{R}\, d\tau = 0. \qquad (5.23.1)$$

This may be rewritten as

$$\int_{S} \left\{\psi \frac{\partial \psi}{\partial n}\left(\frac{1}{R}\right) - \frac{1}{R}\frac{\partial \psi}{\partial n}\right\} d\mathscr{A} + \int_{V} \frac{\nabla^2 \psi}{R}\, d\tau$$

$$= \int_{S_0} \left\{\frac{1}{R}\frac{\partial \psi}{\partial n} - \psi \frac{\partial}{\partial n}\left(\frac{1}{R}\right)\right\} d\mathscr{A} + \int_{V_0} \frac{\nabla^2 \psi}{R}\, d\tau. \qquad (5.23.2)$$

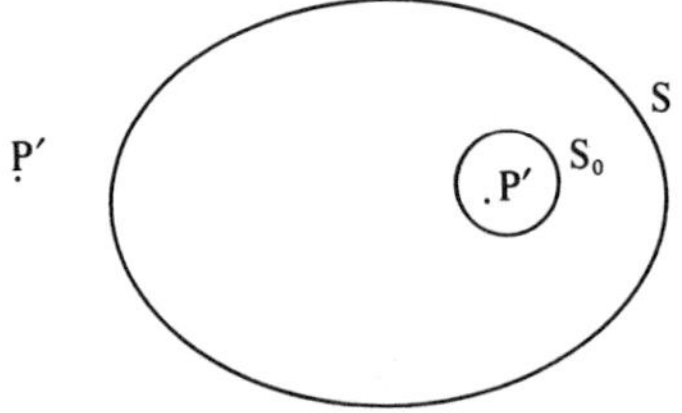

FIG. 5.23(i)

If P' is outside S, S_0 and V_0 do not exist and the right-hand side of (5.23.2) is zero. Consider now what happens as δ tends to zero. $\delta\tau = R^2\, \delta\Omega\, \delta R$ where Ω is solid angle and

$$\int_{V_0} \frac{\nabla^2 \psi\, d\tau}{R} = \int_0^{\delta} R\, dR \int_{S_0} d\Omega \nabla^2 \psi.$$

If $\nabla^2 \psi$ is not infinite at $R = 0$ this tends to zero as δ tends to zero. On S_0

$$\mathbf{n} = -\hat{\mathbf{R}}, \qquad \delta = R^2\, \delta\Omega, \qquad R = \delta$$

and

$$\int_{S_0} \frac{1}{R}\frac{\partial \psi}{\partial n}\, d\mathscr{A} = \frac{1}{\delta}\int_{S_0} \delta^2\, d\Omega \frac{\partial \psi}{\partial n} = \delta \int_{S_0} \frac{\partial \psi}{\partial n}\, d\Omega$$

which also tends to zero as δ tends to zero.

$$\int_{S_0} \psi \frac{\partial}{\partial n}\left(\frac{1}{R}\right) d\mathscr{A} = \int_{S_0} d\Omega \psi. \qquad (5.23.3)$$

This is independent of δ and making δ tend to zero, it becomes $4\pi\psi(\mathbf{r}')$. Thus

$$\int_S \left\{\psi \frac{\partial}{\partial n}\left(\frac{1}{R}\right) - \frac{1}{R}\frac{\partial \psi}{\partial n}\right\} d\mathscr{A} + \int_V \frac{\nabla^2 \psi}{R}\, d\tau$$

$$= 0 \text{ if } P' \text{ is outside } S \qquad (5.23.4a)$$

$$= -4\pi\psi(\mathbf{r}') \text{ if } P' \text{ is inside } S. \qquad (5.23.4b)$$

This is known as Green's formula.

Suppose that V becomes the whole of space, S receding to infinity (S is the sphere $r = a$, where a tends to infinity, the so-called 'large sphere at infinity').

If ψ vanishes sufficiently rapidly at infinity, the surface integral vanishes and

$$\int \frac{\nabla^2 \psi}{R}\, d\tau = -4\pi\psi(\mathbf{r}') \qquad (5.23.5)$$

everywhere.

Suppose that the solution is required of Poisson's partial differential equation

$$\nabla^2 \psi = f(\mathbf{r}) \qquad (5.23.6)$$

and ψ vanishes sufficiently rapidly at infinity for (5.23.5) to be valid. It follows that from (5.23.5) that

$$\psi(\mathbf{r}) = -\frac{1}{4\pi}\int \frac{f(\mathbf{r}')}{R}\, d\tau \qquad (5.23.7)$$

the integration being over the whole of space.

Thus (5.23.8) is the solution of the differential equation (5.23.6) subject to ψ behaving suitably at infinity.

It will be seen that

$$\int_S \psi \frac{\partial}{\partial n}\left(\frac{1}{R}\right) d\mathscr{A}$$

and

$$\int_S \frac{\partial \psi}{\partial n}\frac{1}{R}\, d\mathscr{A}$$

will certainly vanish on the 'large sphere at infinity' if

$$\lim_{r\to\infty} \psi = 0 \qquad \text{and} \qquad \lim_{r\to\infty} r\frac{\partial \psi}{\partial r} = 0.$$

These are thus suitable boundary conditions at infinity.

5.3. Green's function for Laplace's equation

Consider the function G_0 defined by the differential equation

$$\nabla^2 G_0 = \delta(\mathbf{r}) \tag{5.3.1}$$

and the boundary condition $G_0 \to 0$ as $r \to \infty$ $\delta(\mathbf{r})$ is the three-dimensional delta function with the singularity at the origin. G_0 will depend only on r and

$$\nabla^2 G_0 = \frac{1}{r^2}\frac{d}{dr}\left(r^2 \frac{dG}{dr}\right) = 0 \qquad (r > 0). \tag{5.3.2}$$

Thus for $r > 0$,

$$G_0 = A + \frac{B}{r}.$$

Because G_0 vanishes at infinity, A is zero and

$$G_0 = \frac{B}{r} \qquad (r > 0). \tag{5.3.3}$$

Integrate equation (5.3.1) over a sphere of radius a, centre the origin.

$$\int \nabla^2 G_0 \, d\tau = \int \delta(\mathbf{r}) \, d\tau. \tag{5.3.4}$$

Using the divergence theorem

$$\int_{r=a} d\mathscr{A} . \boldsymbol{\nabla} G_0 = 1. \tag{5.3.5}$$

This gives

$$a^2 \int d\Omega \hat{\mathbf{r}} . -\frac{B}{r^2} \qquad (\hat{\mathbf{r}} = 1)$$

whence

$$B = -\frac{1}{4\pi}.$$

Thus

$$G_0 = -\frac{1}{4\pi r} \tag{5.3.6}$$

Similarly if

$$\nabla^2 G(\mathbf{r}, \mathbf{r}') = \delta(\mathbf{r} - \mathbf{r}') \tag{5.3.7}$$

and

$$\lim_{r \to \infty} G(\mathbf{r}, \mathbf{r}') = 0$$

$$G(\mathbf{r}, \mathbf{r}') = -\frac{1}{4\pi R} \tag{5.3.8}$$

This follows by a simple change of origin.

It will be observed that $G(\mathbf{r},\mathbf{r}') = G(\mathbf{r}',\mathbf{r})$. Thus, (5.23.8) may be rewritten

$$\psi(\mathbf{r}) = \int G(\mathbf{r},\mathbf{r}')\,f(\mathbf{r}')\,d\tau' \tag{5.3.9}$$

the integral being over the whole of space and the solution of (5.23.6) with its associated boundary condition is thus equivalent to finding the solution of (5.3.1) with its associated boundary condition. $-\dfrac{1}{4\pi R}$ is termed the Green's function of Laplace's equation which vanishes at infinity.

Suppose now that ψ is defined within a volume V by Poisson's equation

$$\nabla^2\psi = f$$

with the boundary condition

$$\alpha\psi + \beta\frac{\partial\psi}{\partial n} = 0 \tag{5.3.10}$$

on the boundary surface S. This includes $\psi = 0$ and $\partial\psi/\partial n = 0$ as special cases. Let $G(\mathbf{r},\mathbf{r}')$ be defined by

$$\nabla^2 G = \delta(\mathbf{r}-\mathbf{r}') \text{ over } V$$

and

$$\alpha G + \beta\frac{\partial G}{\partial n} = 0 \text{ over } S. \tag{5.3.11}$$

Then

$$\psi(\mathbf{r}) = \int_V G(\mathbf{r},\mathbf{r}')\,f(\mathbf{r}')\,d\tau'. \tag{5.3.12}$$

Before proceeding to the proof, it is convenient to prove the reciprocity relation

$$G(\mathbf{r},\mathbf{r}') = G(\mathbf{r}',\mathbf{r}). \tag{5.3.13}$$

Using Green's theorem

$$\int_S \left\{G(\mathbf{r},\mathbf{r}'')\frac{\partial G}{\partial n}(\mathbf{r},\mathbf{r}') - G(\mathbf{r},\mathbf{r}')\frac{\partial G}{\partial n}(\mathbf{r},\mathbf{r}'')\right\} d\mathscr{A}$$

$$= \int_V \{G(\mathbf{r},\mathbf{r}'')\,\nabla^2 G(\mathbf{r},\mathbf{r}') - G(\mathbf{r},\mathbf{r}')\,\nabla^2 G(\mathbf{r},\mathbf{r}'')\}\,d\tau. \tag{5.3.14}$$

The left-hand side of (5.3.14) vanishes by virtue of the boundary conditions and using the differential equation for Green's function

$$0 = \int_V \{G(\mathbf{r},\mathbf{r}'')\,\delta(\mathbf{r}-\mathbf{r}') - G(\mathbf{r},\mathbf{r}')\,\delta(\mathbf{r}-\mathbf{r}'')\}\,d\tau$$

whence

$$0 = G(\mathbf{r}', \mathbf{r}'') - G(\mathbf{r}'', \mathbf{r}')$$

showing that the Green's function is symmetrical. This is a general property of Green's function which holds for many other differential equations. Loosely, if f be regarded as a cause, and ψ as an effect, the symmetry of the Green's function is equivalent to saying that the effect at P' due to a cause at P, is the same as the effect at P due to the same cause at P'.

Now, consider the application of Green's theorem to ψ and G.

$$\int_S \left(\psi \frac{\partial G}{\partial n} - G \frac{\partial \psi}{\partial n}\right) d\mathscr{A} = \int_V \{\psi \nabla^2 G - G \nabla^2 \psi\} \, d\tau. \qquad (5.3.15)$$

Under the specified boundary conditions on ψ and G, the left-hand side vanishes, and (5.3.15) becomes, writing the right-hand side in full

$$0 = \int \{\psi(\mathbf{r}) \, \delta(\mathbf{r} - \mathbf{r}') - G(\mathbf{r}, \mathbf{r}') \, \nabla^2 \psi(\mathbf{r})\} \, d\tau$$

and

$$\psi(\mathbf{r}') = \int_V G(\mathbf{r}', \mathbf{r}) \, f(\mathbf{r}) \, d\tau$$

by the symmetry property and the definition of the delta function. It should be pointed out, however, that this solution of Poisson's equation is not necessarily unique. The concept of uniqueness will be discussed in the next section.

It is possible to write

$$G(\mathbf{r}, \mathbf{r}') = -\frac{1}{4\pi R} + H \qquad (5.3.16)$$

where H is harmonic.

For

$$\nabla^2 G = -\nabla^2 \left(\frac{1}{4\pi R}\right) + \nabla^2 H = \delta(\mathbf{r} - \mathbf{r}')$$

H obeys the boundary condition on S

$$\alpha H + \beta \frac{\partial H}{\partial n} = -\alpha \left(\frac{1}{R}\right) - \beta \frac{\partial}{\partial n} \left(\frac{1}{R}\right). \qquad (5.3.17)$$

The solution of the differential equation is equivalent to finding the Green's function. Once the Green's function has been found, the equation can be solved for arbitrary $f(\mathbf{r})$. It should be mentioned, however, that, in many cases, it is no easier to find the Green's function than it is to solve the equation directly.

5.31. *Uniqueness theorem*

If $\nabla^2\psi = 0$ over a volume V surrounded by a surface S, it has been proved earlier (5.22.8) that

$$0 = \int_V (\nabla\psi)^2\, d\tau - \int_S \psi \frac{\partial\psi}{\partial n}\, d\mathscr{A}. \tag{5.31.1}$$

It has been shown that, if ψ vanishes on S, ψ vanishes everywhere and that if the normal derivative of ψ vanishes over S, it is constant over V. If the more general boundary condition

$$\alpha\psi + \beta\frac{\partial\psi}{\partial n} = 0 \tag{5.31.2}$$

holds over S, matters are more complicated. Suppose that S' is the portion of S for which β is zero, that is $\psi = 0$ (S' may not, of course, exist)

$$\int_V (\nabla\psi)^2\, d\tau + \int_{S-S'} \frac{\alpha}{\beta}\psi^2\, d\mathscr{A} = 0. \tag{5.31.3}$$

If $(\alpha/\beta) > 0$, the left-hand side of (5.31.3) is intrinsically positive and the equation can hold only if ψ is zero.

If, however, (α/β) is not always positive, the left-hand side of (5.31.3) may be zero for some non-zero ψ. Such solutions are termed eigen solutions of the system. These considerations may be used to determine the uniqueness of solutions of Poisson's equation under given boundary conditions.

Let ψ_1, ψ_2 be two solutions of $\nabla^2\psi = f$ which obey the boundary condition

$$\alpha\psi + \beta\frac{\partial\psi}{\partial n} = 0.$$

Then $\Psi = \psi_1 - \psi_2$ obeys Laplace's equation

$$\nabla^2\Psi = 0 \tag{5.31.4}$$

and the boundary condition

$$\alpha\Psi + \beta\frac{\partial\Psi}{\partial n} = 0. \tag{5.31.5}$$

From what has gone before, the following results hold:

(*i*) If $\partial\psi/\partial n$ is zero everywhere on S, $\psi_1 - \psi_2$ differ by only a constant.

(*ii*) If either $\psi = 0$, or $\alpha\psi + \beta(\partial\psi/\partial n) = 0$ $(\alpha/\beta) > 0$ on S, then $\psi_1 - \psi_2$ is zero and ψ is unique. The condition $(\alpha/\beta) > 0$ may be replaced by $(\alpha/\beta) \geqslant 0$, provided that α is non-zero on some part of S.

(*iii*) If $(\alpha/\beta) \nrightarrow 0$ everywhere on S, eigen solutions for the system may exist, and any solution obtained can differ from another solution by the eigen solutions. Thus the solution is not unique.

5.32. *Helmholtz's theorem*

Helmholtz's theorem states that, if $\mathbf{F}$ is an arbitrary vector field, it is possible to express $\mathbf{F}$ in the form

$$\mathbf{F} = \nabla \times \mathbf{A} + \nabla V \tag{5.32.1}$$

where $\mathbf{A}$ is the vector potential and V the scalar potential. For two special cases the method is straightforward.

If $\nabla . \mathbf{F} = 0$, V is zero and $\mathbf{A}$ may be found by the methods of exercise 3.6 or example 3.9. If $\nabla \times \mathbf{F} = \mathbf{0}$

$$V(\mathbf{r}) = \int_{P_0}^{P} \mathbf{F}(\mathbf{r}').d\mathbf{r}'.$$

For the general case, however, the solution is somewhat more complicated and it is necessary to assume that $\mathbf{F}$ has a certain behaviour at infinity. The actual details will not be given here. To determine V

$$\begin{aligned} \nabla . \mathbf{F} &= \nabla .(\nabla \times \mathbf{A}) + \nabla . \nabla V \\ &= \nabla^2 V \end{aligned} \tag{5.32.2}$$

whence

$$V(\mathbf{r}) = -\frac{1}{4\pi} \int \frac{\nabla' . \mathbf{F}(\mathbf{r}')\, d\tau'}{R} \tag{5.32.3}$$

the integral being over the whole of space.

The solution for $\mathbf{A}$ follows similar lines

$$\nabla \times \mathbf{F} = \nabla \times (\nabla \times \mathbf{A}) + \nabla \times (\nabla V) \tag{5.32.4}$$

whence

$$\nabla^2 \mathbf{A} = \nabla(\nabla . \mathbf{A}) - \nabla \times \mathbf{F}.$$

$\mathbf{A}$, as has been pointed out, is indefinite by a gradient. Suppose that $\nabla . \mathbf{A}$ vanishes. Then

$$\nabla^2 \mathbf{A} = -\nabla \times \mathbf{F} \tag{5.32.5}$$

and if the solution actually found is solenoidal, it is the required solution. The solution of (5.32.5) is

$$\mathbf{A}(\mathbf{r}) = \frac{1}{4\pi} \int \frac{\nabla' \times \mathbf{F}(\mathbf{r}')\, d\tau'}{R} \tag{5.32.6}$$

the integral being over the whole of space

$$\nabla . \mathbf{A} = \frac{1}{4\pi} \int \nabla . \left(\frac{\nabla' \times \mathbf{F}(\mathbf{r}')}{R} \right) d\tau'$$

$$= \frac{1}{4\pi} \int \{\nabla' \times \mathbf{F}(\mathbf{r}')\} . \nabla \left(\frac{1}{R} \right) d\tau'$$

$$= -\frac{1}{4\pi} \int \{\nabla' \times \mathbf{F}(\mathbf{r}')\} . \nabla' \left(\frac{1}{R} \right) d\tau'$$

$$= -\frac{1}{4\pi} \int \nabla' . \left\{ \frac{\nabla' \times \mathbf{F}(\mathbf{r}')}{R} \right\} d\tau' + \frac{1}{4\pi} \int \frac{\nabla' . \{\nabla' \times \mathbf{F}(\mathbf{r}')\}}{R} d\tau'$$

$$= -\frac{1}{4\pi} \int d\mathscr{A}' . \left\{ \frac{\nabla' \times \mathbf{F}(\mathbf{r}')}{R} \right\}$$

where the surface integral is over the large sphere at infinity. **F** was assumed to behave suitably at infinity and so

$$\nabla . \mathbf{A} = 0 \tag{5.32.7}$$

Thus the set of values given for **A** by (5.32.6) satisfy the required solenoidal condition, and so

$$\mathbf{F} = \frac{1}{4\pi} \nabla \times \int \frac{\nabla' \times \mathbf{F}(\mathbf{r}')}{R} d\tau' - \frac{1}{4\pi} \nabla \int \frac{\nabla' . \mathbf{F}(\mathbf{r}')}{R} d\tau'. \tag{5.32.8}$$

5.4. Bounds for the Dirichlet integral

Many physical problems can be reduced to the determination of the so-called Dirichlet integral

$$\int_V (\nabla U)^2 \, d\tau \tag{5.4.1}$$

where $\nabla^2 U = 0$ within V and $U = g_0(\mathbf{r})$ on the boundary surface S. In many cases, an exact solution for U is impossible and recourse must be had to approximate methods. It is possible in fact to obtain lower and upper bounds for the expression (5.4.1). These are based on the fact that U, which can easily be shown to be unique by the methods of section 5.31, is the only function which is common to the set of functions which are harmonic and the set of functions satisfying the boundary condition. An upper bound is determined by the Ritz method, where the approximating function obeys the boundary condition and a lower bound by the Trefftz method where the approximating function obeys the differential equation.

For the Ritz method consider an approximating function Y where $Y = g_0(\mathbf{r})$ on S

$$\int_V (\nabla Y)^2\, d\tau = \int_V \{\nabla(U + Y - U)\}^2\, d\tau$$

$$= \int_V (\nabla U)^2\, d\tau + \int_V \{\nabla(Y - U)\}^2\, d\tau + 2\int_V \nabla U . \nabla(Y - U)\, d\tau \tag{5.4.2}$$

Thus,

$$\int_V (\nabla Y)^2\, d\tau > \int_V (\nabla U)^2\, d\tau \tag{5.4.3}$$

if the last integral on the right-hand side of (5.4.2) vanishes. This is so because

$$\int_V \nabla U . \nabla(Y - U)\, d\tau = \int_S (Y - U)\frac{\partial U}{\partial n}\, d\mathscr{A} - \int_V (Y - U)\nabla^2 U\, d\tau$$

$Y = U$ on S and $\nabla^2 U = 0$ within V, and so the result is proved.

For the Trefftz method, consider a harmonic function W

$$\int_V (\nabla U)^2\, d\tau = \int \{\nabla(W + U - W)\}^2\, d\tau$$

$$= \int_V (\nabla W)^2\, d\tau + \int_V \{\nabla(U - W)\}^2\, d\tau$$

$$+ 2\int_V \nabla W . \nabla(U - W)\, d\tau. \tag{5.4.4}$$

Thus

$$\int_V (\nabla U)^2\, d\tau > \int_V (\nabla W)^2\, d\tau \tag{5.4.5}$$

if the third integral on the right-hand side of (5.4.4) vanishes. This can be made to happen as follows

$$\int_V \nabla W . \nabla(U - W) = \int_S (W - U)\frac{\partial W}{\partial n}\, d\mathscr{A} - \int_V (W - U)\nabla^2 W\, d\tau.$$

$\nabla^2 W = 0$ within V, and $U = g_0(\mathbf{r})$ on S, and so the integral becomes

$$\int_S [W - g_0(\mathbf{r})]\frac{\partial W}{\partial n}\, d\mathscr{A}.$$

If this does not vanish, it may be made to do so by taking $W' = \lambda W$, λ being a constant such that

$$\lambda \int_S W \frac{\partial W}{\partial n} d\mathscr{A} = \int_S g_0(\mathbf{r}) \frac{\partial W}{\partial n} d\mathscr{A}. \tag{5.4.6}$$

Thus

$$\int_V (\nabla Y)^2 d\tau > \int_V (\nabla U)^2 d\tau > \int_V (\nabla W)^2 d\tau. \tag{5.4.7}$$

Thus upper and lower bounds have been found for the Dirichlet integral. Clearly it is desirable to make the upper bound as small as possible and the lower bound as great as possible. The details of this may be seen in any book on the calculus of variations (e.g. S. G. Mikhlin, *Variational Methods in Mathematical Physics*, Pergamon, Oxford (1964)).

Worked examples

1. Use Stokes's theorem to prove that $\nabla \times \nabla V = \mathbf{0}$, V being any scalar point function.

If S is any open surface and $\mathscr{C}$ the rim

$$\int_S d\mathscr{A} . (\nabla \times \mathbf{F}) = \int_{\mathscr{C}} \mathbf{F} . d\mathbf{r}.$$

If $\mathbf{F}$ is a gradient, the right-hand side vanishes. Thus

$$\int_S d\mathscr{A} . (\nabla \times \mathbf{F}) = 0 \text{ for all } S.$$

S is arbitrary and so it follows that the integrand $\nabla \times \mathbf{F}$ vanishes which is the required result.

2. Verify Stokes's theorem when

$$\mathbf{F} = \frac{1}{r} \tan \frac{\theta}{2} \hat{\boldsymbol{\varphi}}$$

and S is the spherical cap $r = a$, $0 < \theta < \alpha$, spherical coordinates being used. It may be shown that

$$\nabla \times \mathbf{F} = \frac{1}{r^3} \mathbf{r}$$

$$\delta \mathscr{A} = \hat{\mathbf{r}} a^2 \sin \theta \, \delta\theta \, \delta\phi$$

$$\int_S (\nabla \times \mathbf{F}) . d\mathscr{A} = \int_{-\pi}^{\pi} d\phi \int_0^{\alpha} a^2 \sin \theta \, d\theta \frac{1}{a^3} \hat{\mathbf{r}} . \mathbf{r}$$

$$= 2\pi(1 - \cos \alpha).$$

On $\mathscr{C}$, $\delta\mathbf{r} = a \sin\alpha\, \delta\phi \boldsymbol{\hat\phi}$, and

$$\int_{\mathscr{C}} \mathbf{F}.d\mathbf{r} = \int_{-\pi}^{\pi} d\phi\, a \sin\alpha \boldsymbol{\hat\phi}.\left(\frac{1}{a}\tan\frac{\alpha}{2}\hat{\phi}\right)$$

$$= 2\pi \sin\alpha \tan\frac{\alpha}{2} = 2\pi(1 - \cos\alpha).$$

3. Prove that Green's theorem in the plane is true for a multiply connected domain.

The proof given in section 5.1 is only applicable for simply connected domains. It may be extended to multiply connected domains as follows.

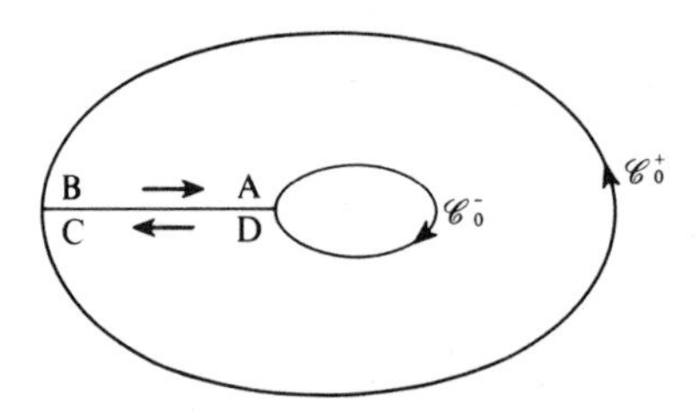

FIG. 5A.

Suppose that $\mathscr{C}$, the curve surrounding the domain D_0 has two portions $\mathscr{C}_0^+, \mathscr{C}_0^-$ indicated. If there are more portions, the extension is obvious.

Cut the domain D_0 as indicated. Then the domain with boundary $\mathscr{C}_0^+ + BA + \mathscr{C}_0^- + DC$ is simply connected. When integrating over D_0 it does not make any difference whether the cutting line is included or not. And so

$$\iint_{D_0} \left(\frac{\partial Q}{\partial x} - \frac{\partial P}{\partial y}\right) dx\, dy = \int_{(\mathscr{C}_0^+ + BA + \mathscr{C}_0^- + DC)} (P\, dx + Q\, dy)$$

$$= \int_{\mathscr{C}_0^+ + \mathscr{C}_0^-} (P\, dx + Q\, dy) + \int_{BA+DC} (P\, dx + Q\, dy)$$

$$= \int_{\mathscr{C}_0^+ + \mathscr{C}_0^-} (P\, dx + Q\, dy) = \int_{\mathscr{C}_0} (P\, dx + Q\, dy)$$

because the integrals along BA and DC are equal and opposite.

4. The equation of a surface is $\psi(\mathbf{r}) = 0$. Find a volume integral giving the surface area of the surface.

If $\delta\mathscr{A}$ is an element of surface, the elementary area is

$$\mathbf{n}\,.\,\delta\mathscr{A} = \frac{\nabla\psi}{|\nabla\psi|}\,.\,\delta\mathscr{A}$$

Thus the surface area is given by

$$\int\limits_S \frac{\nabla\psi}{|\nabla\psi|}\,.\,d\mathscr{A} = \int\limits_V \nabla\,.\left\{\frac{\nabla\psi}{|\nabla\psi|}\right\}d\tau$$

$$= \int\limits_V \left\{\frac{\nabla^2\psi}{|\nabla\psi|} - \frac{\nabla\psi\,.\,\nabla\{|\nabla\psi|\}}{|\nabla\psi|^2}\right\}d\tau.$$

5. Prove that

$$\int\limits_S (W\phi\nabla\psi)\,.\,d\mathscr{A} = \int\limits_V \phi\nabla\,.\,(W\nabla\psi)\,d\tau + \int\limits_V W(\nabla\phi)\,.\,(\nabla\psi)\,d\tau$$

$$\int\limits_S (W\phi\nabla\psi)\,.\,d\mathscr{A} = \int\limits_V \nabla\,.\{W\phi\nabla\psi\}\,d\tau$$

$$= \int\limits_V \phi\nabla\,.\{W\nabla\psi\}\,d\tau + \int\limits_V W(\nabla\phi)\,.\,(\nabla\psi)\,d\tau$$

If $\phi = \psi$

$$\int\limits_V W(\nabla\psi)^2\,d\tau = \int\limits_S W\psi\frac{\partial\psi}{\partial n}\,d\mathscr{A} - \int\limits_V \psi\nabla\,.\,(W\nabla\psi)\,d\tau.$$

6. $\mathbf{F}$ and ψ are vector and scalar fields respectively which vanish suitably at infinity.

Show that

$$\int \nabla^2\psi\mathbf{F}\,d\tau = \int (\nabla^2\mathbf{F})\psi\,d\tau$$

where the volume integrals are over the whole of space.

Let χ be any one of the rectangular components of $\mathbf{F}$. Then for a volume V surrounded by a surface S

$$\int\limits_V (\chi\nabla^2\psi - \psi\nabla^2\chi)\,d\tau = \int\limits_S \left(\chi\frac{\partial\psi}{\partial n} - \psi\frac{\partial\chi}{\partial n}\right)d\mathscr{A}$$

If V is the whole of space, S is the 'surface at infinity' and the integral over it vanishes. It follows that

$$\int \chi\nabla^2\psi\,d\tau = \int \psi\nabla^2\chi\,d\tau$$

where the volume integration is over the whole of space.

The proof of the result follows immediately.

7. Prove that, with the usual notation

$$\mathbf{F}(\mathbf{r}) = -\frac{1}{4\pi}\int \frac{\nabla'^2 \mathbf{F}(\mathbf{r}')}{R}\, d\tau'$$

the integration being over the whole of space and **F** vanishing suitably at infinity.

Using example 6, the result is equivalent to the truth of

$$\mathbf{F}(\mathbf{r}) = \int_V \nabla'^2 \left(-\frac{1}{4\pi R}\right) \mathbf{F}(\mathbf{r}')\, d\tau'$$

$$= \int_V \delta(\mathbf{r} - \mathbf{r}')\, \mathbf{F}(\mathbf{r}')\, d\tau' = \mathbf{F}(\mathbf{r}).$$

The result may be used to prove Helmholtz's theorem for

$$\int \frac{\nabla'^2 \mathbf{F}(\mathbf{r}')}{R}\, d\tau' = \int \frac{[\nabla'.\{\nabla'.\mathbf{F}(\mathbf{r}')\} - \nabla' \times \{\nabla' \times \mathbf{F}(\mathbf{r}')\}]}{R}\, d\tau'$$

$$= \int \left[\nabla' \left\{ \frac{\nabla'.\mathbf{F}(\mathbf{r}')}{R} \right\} - \nabla' \times \left\{ \frac{\nabla' \times \mathbf{F}(\mathbf{r}')}{R} \right\} \right.$$

$$\left. - \nabla'\left(\frac{1}{R}\right) \nabla'.\mathbf{F}(\mathbf{r}') + \nabla'\left(\frac{1}{R}\right) \times \{\nabla' \times \mathbf{F}(\mathbf{r}')\} \right] d\tau'$$

$$= \int d\mathscr{A} \left\{ \frac{\nabla'.\mathbf{F}(\mathbf{r}')}{R} \right\} - \int d\mathscr{A} \times \left\{ \frac{\nabla' \times \mathbf{F}(\mathbf{r}')}{R} \right\}$$

$$+ \int \left[\nabla\left(\frac{1}{R}\right) \nabla'.\mathbf{F}(\mathbf{r}') - \nabla\left(\frac{1}{R}\right) \times \{\nabla' \times \mathbf{F}(\mathbf{r}')\} \right] d\tau'.$$

The surface integrals vanish because of the conditions imposed upon F at infinity and the last two terms become

$$\nabla \left\{ \int \frac{\nabla'.\mathbf{F}(\mathbf{r}')}{R}\, d\tau' \right\} - \nabla \times \left\{ \int \frac{\nabla' \times \mathbf{F}(\mathbf{r}')}{R}\, d\tau' \right\}.$$

8. Prove that

$$-\frac{e^{-ikR}}{4\pi R}$$

is the solution of $\nabla^2 G + k^2 G = \delta(\mathbf{r} - \mathbf{r}')$ which obeys the condition

$$\lim_{r\to\infty} r\left(\frac{\partial G}{\partial r} + ikG\right) = 0$$

(Sommerfeld's radiation condition).

Consider first the solution of $\nabla^2 G_0 + k^2 G_0 = \delta(\mathbf{r})$. It will be observed that this is a generalization of Laplace's equation and that the solution required should reduce to that associated with Laplace's equation when $k = 0$. The first point to note is that G_0 is spherically symmetrical and so

$$\frac{1}{r^2}\frac{d}{dr}\left(r^2\frac{dG_0}{dr}\right) + k^2 G_0 = \delta(\mathbf{r})$$

Now except at the origin

$$\frac{1}{r^2}\frac{d}{dr}\left(r^2\frac{dG_0}{dr}\right) + k^2 G_0 = 0$$

and it can be verified that solutions of this are

$$\frac{Ae^{-ikr}}{r} \quad \text{and} \quad \frac{Be^{ikr}}{r}.$$

It is the first of these which satisfies the Sommerfeld radiation condition.

Now integrate both sides of the differential equation for G_0 over the volume of a sphere of radius a, centre O.

$$\int_V \nabla^2 G_0\, d\tau + k^2 \int_V G_0\, d\tau = \int_V \delta(\mathbf{r})\, d\tau.$$

This may be rewritten as

$$\int_{r=a} d\mathscr{A}\,.\,\nabla G_0 + k^2 \int_V G_0\, d\tau = 1$$

or

$$4\pi a^2 \left\{\frac{d}{dr}\left(\frac{Ae^{-ikr}}{r}\right)\right\}_{r=a} + k^2 \int_0^a 4\pi r^2 \frac{Ae^{-ikr}}{r}\, dr = 1.$$

Simplifying

$$-4\pi A[1 + ika]e^{-ika} + 4\pi A k^2 \int_0^a r e^{-ikr}\, dr = 1$$

whence

$$-4\pi A = 1.$$

Thus

$$G_0 = -\frac{1}{4\pi r} e^{-ikr}$$

and

$$G(\mathbf{r}, \mathbf{r}') = -\frac{1}{4\pi R} e^{-ikR}$$

on making a change of origin.

As

$$\lim_{r\to\infty} \frac{R}{r} = 1,$$

it can easily be seen that the Sommerfeld radiation condition is satisfied.

This is the Green's function for the wave equation in free space when there is an $\exp\{i\omega t\}$ time variation. The Sommerfeld radiation condition is the condition that the field is outgoing at infinity.

9. If **D** is the electric displacement and ρ is the change density, show that $\nabla . \mathbf{D} = \rho$.

From exercise 5, chapter 4

$$\int_S \mathbf{D} . d\mathscr{A} = \text{total charge within } S$$

$$= \int_V \rho \, d\tau.$$

Also, by the divergence theorem

$$\int_S \mathbf{D} . d\mathscr{A} = \int_V \nabla . \mathbf{D} \, d\tau.$$

Thus

$$\int_V \nabla . \mathbf{D} \, d\tau = \int_V \rho \, d\tau.$$

This is true for all volumes, and so $\nabla . \mathbf{D} = \rho$.

10. Prove that, for a steady magnetic field $\nabla \times \mathbf{H} = \mathbf{J}$, where **J** is the current density vector.

From exercise 13, chapter 4

$$\int_{\mathscr{C}} \mathbf{H} . d\mathbf{r} = \int_S \mathbf{J} . d\mathscr{A}$$

Using Stokes's theorem, the left-hand side becomes

$$\int_S (\nabla \times \mathbf{H}) . d\mathscr{A} = \int_S \mathbf{J} . d\mathscr{A}.$$

This holds for all S, and so $\nabla \times \mathbf{H} = \mathbf{J}$.

11. ψ is a point function which obeys Poisson's equation $\nabla^2 \psi = f$ within a volume V and has the values $\psi_0(\mathbf{r})$ on the surrounding surface S. Derive an explicit expression for ψ.

Let G be a solution of $\nabla^2 G = \delta(\mathbf{r} - \mathbf{r}')$.

For the present, the boundary condition on G is not specified. By Green's theorem

$$\int_V (\psi\nabla^2 G - G\nabla^2\psi)\,d\tau = \int_S \left(\psi\frac{\partial G}{\partial n} - G\frac{\partial\psi}{\partial n}\right) d\mathscr{A}$$

$$\int_V \{\psi(\mathbf{r})\,\delta(\mathbf{r}-\mathbf{r}') - G(\mathbf{r},\mathbf{r}')\,f(\mathbf{r})\}\,d\tau = \int_S \left\{\psi_0(\mathbf{r})\frac{\partial G}{\partial n} - G\frac{\partial\psi}{\partial n}\right\} d\mathscr{A}.$$

If now G vanishes on the surface S, integration gives

$$\psi(\mathbf{r}') = \int_V G(\mathbf{r},\mathbf{r}')\,f(\mathbf{r})\,d\tau + \int_S \psi_0(\mathbf{r})\frac{\partial G}{\partial n}\,d\mathscr{A}$$

which is the required result.

12. If $\nabla^2\phi = f$ within V, and either $\phi = 0$, or $\partial\phi/\partial n = 0$ over the surface S surrounding V, show that

$$\int_V (\nabla\phi)^2\,d\tau = -\int_V \phi f\,d\tau$$

$$\int_V (\nabla\phi)^2\,d\tau + \int_V \phi f\,d\tau = \int_V \{(\nabla\phi)^2 + \phi\nabla^2\phi\}\,d\tau = \int_V \nabla.(\phi\nabla\phi)\,d\tau$$

$$= \int_S \phi\frac{\partial\phi}{\partial n}\,d\mathscr{A} = 0.$$

13. Show that if ϕ is harmonic within a volume V and $\partial\phi/\partial n = f$ on the boundary surface, then

$$\int_S f\,d\mathscr{A} = 0$$

$$0 = \int_V \nabla^2\phi\,d\tau = \int_V \nabla.(\nabla\phi)\,d\tau = \int_S \nabla\phi.d\mathscr{A} = \int_S f\,d\mathscr{A}$$

If f does not satisfy this relation there is no solution to the differential equation under the given boundary condition. ϕ is in fact overdetermined.

14. A fluid has density ρ and moves at any point with velocity $\mathbf{v}$. If there are no sources of matter present, show that

$$\nabla.(\rho\mathbf{v}) + \frac{\partial\rho}{\partial t} = 0.$$

If an amount of matter q is created per unit volume per unit time, show that the right-hand side is replaced by q.

Consider the fluid under consideration as enclosed in a volume V surrounded by a fixed surface S. The total amount of matter within the volume is

$$\int_V \rho \, d\tau.$$

The quantity $\mathbf{J} = \rho\mathbf{v}$ may be termed the flux vector.

Matter crosses a surface element $\delta\mathscr{A}$ at a rate $\mathbf{J}.\delta\mathscr{A} = \rho v_n \delta\mathscr{A}$. The rate of increase of matter within a volume is equal to the rate at which matter is being created within the volume less the rate at which matter is leaving the volume.

Thus

$$\frac{\partial}{\partial t}\int_V \rho \, d\tau = \int_V q \, d\tau - \int_S \mathbf{J}.d\mathscr{A}$$

$$= \int_V q \, d\tau - \int_V \nabla.\mathbf{J} \, d\tau.$$

The volume is arbitrary, so it follows that

$$\nabla.\mathbf{J} + \frac{\partial \rho}{\partial t} = q$$

or

$$\nabla.(\rho\mathbf{v}) + \frac{\partial \rho}{\partial t} = q.$$

This equation, with q zero, is termed a conservation equation. The equation can also be written

$$\frac{\partial \rho}{\partial t} + (\mathbf{v}.\nabla)\rho + \rho\nabla.\mathbf{v} = q$$

or

$$\frac{d\rho}{dt} + \rho\nabla.\mathbf{v} = q$$

where $d\rho/dt$ is differentiation following the fluid.

If the fluid is incompressible, the density of a portion of the fluid remains constant as it moves about and $d\rho/dt$ is zero. Thus, if there is no creation of matter, it follows that, for an incompressible fluid, $\nabla.\mathbf{v} = 0$.

If a rate m_0 of matter is created at the point $\mathbf{r}'$, there is said to be a source and $q = m_0\,\delta(\mathbf{r} - \mathbf{r}')$. If m_0 is negative and matter is removed, a sink is said to exist.

This theory is applicable equally in electromagnetic theory where $\mathbf{J}$ is the current density vector and ρ the charge density.

15. The electrostatic potential, the electric field and the electric charge density distribution $\rho(\mathbf{r})$ are respectively V, $\mathbf{E}$ and $\mathbf{D}$. Show that the increase in energy of the system if ρ change to $\rho + \delta\rho$ is $\int \mathbf{E}.\delta\mathbf{D}\,d\tau$, the integral being taken over the whole of space.

The energy required to increase ρ to $\rho + \delta\rho$ over an element $\delta\tau$ is $V(\rho\,\delta\tau)$, $\rho\,\delta\tau$ being the amount of charge brought up from infinity. Thus over the whole of space the energy increase is

$$\begin{aligned}\int V\,\delta\rho\,d\tau &= \int V(\nabla.\delta\mathbf{D})\,d\tau\\ &= \int \nabla.(V\delta\mathbf{D}) - \int (\nabla V).\delta\mathbf{D}\,d\tau\\ &= \int_S d\mathscr{A}.(V\delta\mathbf{D}) + \int (\mathbf{E}.\delta\mathbf{D})\,d\tau\end{aligned}$$

The surface integral vanishes as the surface is at infinity and the result follows.

16. The external force per unit mass on a fluid is $\mathbf{F}$, and the pressure at any point in the fluid is p. If v is the fluid velocity at any point, show that

$$\mathbf{F} - \frac{1}{\rho}\nabla p - \tfrac{1}{2}\nabla v^2 - \frac{\partial \mathbf{v}}{\partial t} + \mathbf{v} \times (\nabla \times \mathbf{v}) = \mathbf{0}.$$

Consider a surface S moving with the fluid and the volume V inside it. The force inwards across an element of surface is $-p\,\delta\mathscr{A}$ and the external force on an element of fluid is $\rho\mathbf{F}\,\delta\tau$. Thus the total force on the volume of fluid is

$$\int_V \rho\mathbf{F}\,d\tau - \int_S p\,d\mathscr{A}$$

The rate of increase of momentum with the volume V is

$$\int_V \frac{d}{dt}(\rho\mathbf{v})\,d\tau.$$

Thus

$$\begin{aligned}\int_V \frac{d}{dt}(\rho\mathbf{v})\,d\tau &= \int_V \rho\mathbf{F}\,d\tau - \int_S p\,d\mathscr{A}\\ &= \int_V (\rho\mathbf{F} - \nabla p)\,d\tau.\end{aligned}$$

whence

$$\frac{d}{dt}(\rho\mathbf{v}) = \rho\mathbf{F} - \nabla p.$$

$$\mathbf{F} - \frac{1}{\rho}\nabla p - \frac{d}{dt}(\mathbf{v}) = \mathbf{0}.$$

Now

$$\frac{d\mathbf{v}}{dt} = \frac{\partial\mathbf{v}}{\partial t} + (\mathbf{v}\,.\,\nabla)\,\mathbf{v}$$

and

$$\tfrac{1}{2}\nabla v^2 = (\mathbf{v}\,.\,\nabla)\,\mathbf{v} + \mathbf{v} \times (\nabla \times \mathbf{v})$$

whence the result follows.

17. If $\boldsymbol{\omega} = \frac{1}{2}\nabla \times \mathbf{v}$ be the vorticity of a fluid then the line integral of fluid velocity along a closed curve is equal to twice the normal surface integral of the vorticity over any open surface bounded by that curve.

$$\int_{\mathscr{C}} \mathbf{v}\,.\,d\mathbf{r} = \int (d\mathscr{A} \times \nabla)\,.\,\mathbf{v} = \int d\mathscr{A}\,.\,2\boldsymbol{\omega}.$$

The quantity

$$\int_{\mathscr{C}} \mathbf{v}\,.\,d\mathbf{r}$$

is termed the circulation of $\mathbf{v}$ along $\mathscr{C}$. If $\boldsymbol{\omega}$ is zero the fluid is irrotational.

18. Show that, if $\mathbf{E}$ be the electric field, $\mathbf{D}$ the electric displacement, $\mathbf{J}$ the current density vector, $\mathbf{H}$ the magnetic field, and B the magnetic induction,

$$\int_V \left(\mathbf{E}\,.\,\frac{\partial\mathbf{D}}{\partial t} + \mathbf{H}\,.\,\frac{\partial\mathbf{B}}{\partial t}\right) d\tau + \int_V \mathbf{J}\,.\,\mathbf{E}\,d\tau + \int_S (\mathbf{E} \times \mathbf{H})\,.\,d\mathscr{A} = 0.$$

(Poynting's theorem).

Show that $\mathbf{H}\,.\,(\partial\mathbf{B}/\partial t)$ may be replaced by

$$\frac{\partial\mathbf{A}}{\partial t}\,.\,(\nabla \times \mathbf{H})$$

and $\mathbf{E} \times \mathbf{H}$ by

$$\mathbf{E} \times \mathbf{H} + \frac{\partial\mathbf{A}}{\partial t} \times \mathbf{H}$$

where $\mathbf{A}$ is the vector potential. (Chapter 3, exercise 14.) Interpret the result.

Using Maxwell's equations

$$\nabla\,.\,(\mathbf{E} \times \mathbf{H}) = \mathbf{H}\,.\,(\nabla \times \mathbf{E}) - \mathbf{E}\,.\,(\nabla \times \mathbf{H})$$

$$= -\mathbf{H}\,.\,\frac{\partial\mathbf{B}}{\partial t} - \mathbf{J}\,.\,\mathbf{E} - \mathbf{E}\,.\,\frac{\partial\mathbf{D}}{\partial t}$$

whence

$$\mathbf{H}.\frac{\partial \mathbf{B}}{\partial t}+\mathbf{E}.\frac{\partial \mathbf{D}}{\partial t}+\mathbf{J}.\mathbf{E}+\nabla.(\mathbf{E}\times\mathbf{H})=0.$$

Integrating over a volume it follows that

$$\int_V \mathbf{H}.\frac{\partial \mathbf{B}}{\partial t}+\mathbf{E}.\frac{\partial \mathbf{D}}{\partial t}\,d\tau+\int_V \mathbf{J}.\mathbf{E}\,d\tau=\int_S (\mathbf{E}\times\mathbf{H}).d\mathscr{A}=0$$

the divergence theorem being used.

$$\int_V \mathbf{H}.\frac{\partial \mathbf{B}}{\partial t}\,d\tau$$

is the rate of increase of magnetic energy within the volume.

$$\int_V \mathbf{E}.\frac{\partial \mathbf{D}}{\partial t}\,d\tau$$

is the rate of increase of electric energy within the volume.

$$\int_V \mathbf{J}.\mathbf{E}\,d\tau$$

is the rate of dissipation of energy within the volume.

If the principle of conservation of energy holds, the interpretation of the surface integral is the rate of streaming of energy out of the volume across its boundary surface.

The vector $\mathbf{E}\times\mathbf{H}$ is known as the Poynting vector and the surface integral as the Poynting flux. That is the equations are consistent with a flux of energy $(\mathbf{E}\times\mathbf{H}).\delta\mathscr{A}$ across an elementary surface $\delta\mathscr{A}$. It is not possible to say that this is the actual energy flux distribution, because if $\mathbf{G}$ is any vector

$$\int(\mathbf{E}\times\mathbf{H}+\nabla\times\mathbf{G}).d\mathscr{A}=\int\nabla.[\mathbf{E}\times\mathbf{H}+\nabla\times\mathbf{G}]\,d\tau=\int\nabla.(\mathbf{E}\times\mathbf{H})\,d\tau.$$

All that is possible to say is that the result is consistent with it being the energy flow distribution.

Now

$$\nabla.\left(\frac{\partial \mathbf{A}}{\partial t}\times\mathbf{H}\right)=\mathbf{H}.\left(\nabla\times\frac{\partial \mathbf{A}}{\partial t}\right)-\frac{\partial \mathbf{A}}{\partial t}.(\nabla\times\mathbf{H})$$

$$=\mathbf{H}.\frac{\partial \mathbf{B}}{\partial t}-\frac{\partial \mathbf{A}}{\partial t}.(\nabla\times\mathbf{H})$$

and

$$\int_S d\mathscr{A}.\left(\frac{\partial \mathbf{A}}{\partial t}\times\mathbf{H}\right)=\int_V \mathbf{H}.\frac{\partial \mathbf{B}}{\partial t}\,d\tau-\int_V \frac{\partial \mathbf{A}}{\partial t}.(\nabla\times\mathbf{H})\,d\tau.$$

Thus, it would be possible to interpret the rate of increase of magnetic energy by

$$\int_V \frac{\partial \mathbf{A}}{\partial t}.(\nabla \times \mathbf{H})\,d\tau$$

and the energy flux by

$$\mathbf{E} \times \mathbf{H} + \frac{\partial \mathbf{A}}{\partial t} \times \mathbf{H}.$$

If **D** depends linearly on **E** and **B** depends linearly on **H**

$$\mathbf{E}.\frac{\partial \mathbf{D}}{\partial t} = \frac{\partial}{\partial t}(\tfrac{1}{2}\mathbf{E}.\mathbf{D})$$

and

$$\mathbf{H}.\frac{\partial \mathbf{B}}{\partial t} = \frac{\partial}{\partial t}(\tfrac{1}{2}\mathbf{B}.\mathbf{H}).$$

19. Show that the solution of

$$\nabla^2 \phi - \frac{1}{c^2}\frac{\partial^2 \phi}{\partial t^2} = f(\mathbf{r}, t)$$

which vanishes at infinity is

$$-\frac{1}{4\pi}\int \frac{f(\mathbf{r}', t^*)\,d\tau'}{R}$$

where $t^* = t - (R/c)$, $\mathbf{R} = \mathbf{r} - \mathbf{r}'$.

Consider first the solution of

$$\nabla^2 G_0 - \frac{1}{c^2}\frac{\partial^2 G_0}{\partial t^2} = \delta(\mathbf{r})\,\psi(t).$$

Except at $\mathbf{r} = 0$, G_0 satisfies

$$\nabla^2 G_0 - \frac{1}{c^2}\frac{\partial^2 G_0}{\partial t^2} = 0$$

G_0 depends only on r and t and so

$$\frac{1}{r^2}\frac{\partial}{\partial r}\left(r^2\frac{\partial G_0}{\partial r}\right) - \frac{1}{c^2}\frac{\partial^2 G}{\partial t^2} = 0 \qquad (r > 0)$$

and it may be seen that the solutions of this are

$$\frac{g\{t - (r/c)\}}{r} \quad \text{and} \quad \frac{h\{t + (r/c)\}}{r}$$

where g and h are arbitrary functions. The first function represents a wave going out from the origin, the second a wave coming into the origin. This second wave will be excluded, as a disturbance travels out from an excitation. Thus

$$G_0 = \frac{g\{t-(r/c)\}}{r}$$

and if V is a sphere of radius r_0, centre the origin

$$\int_V \left\{\nabla^2 G_0 - \frac{1}{c^2}\frac{\partial^2 G_0}{\partial t^2}\right\} d\tau = \int_V \delta(\mathbf{r})\,\psi(t)\,d\tau$$

$$\frac{dG_0}{dr} = -\frac{1}{r^2}g\left(t-\frac{r}{c}\right) + \frac{1}{cr}g'\left(t-\frac{r}{c}\right).$$

Now

$$\int_V \nabla^2 G_0\,d\tau = \int_S d\mathscr{A}.\nabla G_0 = \int_S d\mathscr{A}\frac{d}{dr}G_0 = -4\pi g\left(t-\frac{r_0}{c}\right) + 4\pi\frac{r_0}{c}g'\left(t-\frac{r_0}{c}\right)$$

and

$$\frac{1}{c^2}\int_V \frac{\partial^2 G}{\partial t^2}d\tau = \frac{1}{c^2}\int_0^{r_0} g''\left(t-\frac{r}{c}\right)4\pi r\,dr$$

$$= -4\pi\frac{r_0}{c}g'\left(t-\frac{r_0}{c}\right) + 4\pi g\left(t-\frac{r_0}{c}\right) - 4\pi g(t).$$

It follows that

$$\int_V \left(\nabla^2 G_0 - \frac{1}{c^2}\frac{\partial^2 G_0}{\partial t^2}\right) d\tau = -4\pi g(t).$$

Also

$$\int_V \delta(\mathbf{r})\,f(t)\,d\tau = \psi(t).$$

Thus

$$g(t) = -\frac{1}{4\pi}\psi(t).$$

It follows that

$$G_0 = -\frac{1}{4\pi r}\psi\left(t-\frac{r}{c}\right)$$

and

$$G(r,r') = -\frac{1}{4\pi R}\psi\left(t-\frac{R}{c}\right)$$

with an obvious notation, and the result follows on integration.

20. ϕ_1 and ϕ_2 satisfy the relations

$$\nabla^2\phi_1 + k_1^2\phi_1 = 0$$

$$\nabla^2\phi_2 + k_2^2\phi_2 = 0 \qquad (k_1 \neq k_2)$$

within V, and

$$\phi_1 = \phi_2 = 0$$

on S, the surface surrounding V.

Prove that

$$\int_V (\nabla\phi_1).(\nabla\phi_2)\,d\tau = 0 \qquad \text{and} \qquad \int_V \phi_1\phi_2\,d\tau = 0.$$

$$\int_V \phi_1\nabla^2\phi_2\,d\tau = -\int_V (\nabla\phi_1).(\nabla\phi_2)\,d\tau + \int_V \nabla.(\phi_1\nabla\phi_2)\,d\tau$$

$$-k_2^2\int_V \phi_1\phi_2\,d\tau = -\int_V (\nabla\phi_1).(\nabla\phi_2)\,d\tau + \int_S \phi_1\frac{\partial\phi_2}{\partial n}\,d\mathscr{A} = -\int_V (\nabla\phi_1).(\nabla\phi_2)\,d\tau.$$

Similarly

$$-k_1^2\int_V \phi_1\phi_2\,d\tau = -\int_V (\nabla\phi_1).(\nabla\phi_2)\,d\tau.$$

These two last equations are inconsistent unless the required results hold. The results would also hold for the boundary condition $\partial\phi/\partial n = 0$ on S.

21. Find the Green's function for the two-dimensional Laplace equation. This is given by the solution of

$$\nabla^2 G_0 = \delta(\mathbf{r}_\perp) \qquad (\mathbf{r}_\perp = x\mathbf{i} + y\mathbf{j})$$

$$= \frac{1}{2\pi\rho}\,\delta(\rho) \qquad (\rho = (x^2 + y^2)^{1/2}).$$

G will be a function of ρ only and

$$\nabla^2 G_0 = \frac{1}{\rho}\frac{d}{d\rho}\left(\rho\frac{dG}{d\rho}\right)$$

$\nabla^2 G_0 = 0$ for $\rho > 0$, and so in this region

$$\frac{d}{d\rho}\left(\rho\frac{dG_0}{d\rho}\right) = 0$$

the solution of which is $G_0 = A + B\log\rho$.

Thus G_0 cannot be made to vanish at infinity and, dropping the arbitrary constant A, $G_0 = B\log\rho$.

Let D be the domain $0 \leqslant \rho \leqslant \rho_0$

$$\iint_D \nabla^2 G_0 \, dx \, dy = \int_{\mathscr{C}} \frac{\partial G_0}{\partial n} ds = \int_0^{\rho_0} \frac{1}{2\pi\rho} \delta(\rho) \, 2\pi\rho \, d\rho$$

$$\frac{\partial G_0}{\partial n} = \frac{\partial G_0}{\partial \rho} = \frac{B}{\rho}$$

and

$$\int_{\mathscr{C}} \frac{\partial G_0}{\partial n} ds = \frac{B}{\rho_0} . 2\pi\rho_0 = 1$$

whence

$$B = \frac{1}{2\pi} \cdot$$

Thus

$$G_0 = \frac{1}{2\pi} \log \rho$$

and

$$G(\mathbf{r}_\perp, \mathbf{r}'_\perp) = \frac{1}{2\pi} \log |\mathbf{r}_\perp - \mathbf{r}'_\perp|$$

$$= \frac{1}{4\pi} \log \{\rho^2 - 2\rho\rho' \cos(\phi - \phi') + \rho'^2\}$$

$$= \frac{1}{4\pi} \log \{(x - x')^2 + (y - y')^2\}.$$

Exercises

1. Prove that

$$\tfrac{1}{2} \int_{\mathscr{C}} \mathbf{r} \times d\mathbf{r}$$

represents the vector area of an open surface bounded by the closed curve $\mathscr{C}$, and show that this is independent of the origin.

2. Prove that, with the usual notation

$$\int_S [(\nabla V) \times \mathbf{F}] . d\mathscr{A} = \int_{\mathscr{C}} V\mathbf{F} . d\mathbf{r} - \int_S V(\nabla \times \mathbf{F}) . d\mathscr{A} .$$

3. Evaluate

$$\int_S \frac{d\mathscr{A}}{p}$$

where S is the ellipsoid

$$\frac{x^2}{a^2}+\frac{y^2}{b^2}+\frac{z^2}{c^2}=1$$

and p is the perpendicular from the origin to the tangent plane.

4. Use exercise 1 to determine the area of the ellipse $x=a\cos\phi$, $y=b\sin\phi$. $(-\pi<\phi<\pi.)$

5. Show that the volume contained within a surface S is

$$\frac{1}{3}\int_S \mathbf{r}.d\mathscr{A}=\frac{1}{6}\int_S \nabla r^2.d\mathscr{A}.$$

6. ϕ is harmonic within a volume V. Over part of the boundary surface ϕ is zero, and over the remainder $\partial\phi/\partial n$ is zero. Prove that ϕ is zero everywhere.

7. ϕ is harmonic within a sphere. Prove that the mean value of ϕ over the surface of the sphere is equal to its value at the centre of the sphere and hence that a harmonic function cannot have either a maximum or a minimum at any point.

8. Verify Stokes's theorem when $\mathbf{F}=\mathbf{r}\times\mathbf{k}$ and S is the hemisphere $x^2+y^2+z^2=a^2$, $z<0$.

9. D is a domain in the xy plane, and $\mathscr{C}$ is the curve surrounding it. Using the divergence theorem, show that

$$\iint_D \nabla_\perp.\mathbf{P}_\perp\,dx\,dy=\int_{\mathscr{C}}\mathbf{k}.(d\mathbf{r}\times\mathbf{P}_\perp)$$

and hence prove Green's theorem in two dimensions.

10. A scalar function ψ obeys the differential equation $\nabla.\{W\nabla\psi\}=f$ over a volume $\mathscr{V}$. Over the bounding surface S, ψ obeys the boundary condition $\alpha\psi+\beta(\partial\psi/\partial n)=0$. Show that if $W>0$, $(\alpha/\beta)>0$, ψ is unique.

11. Prove that, with the usual notation

$$\int_{\mathscr{C}} f(r)\,d\mathbf{r}=\int_S d\mathscr{A}\times\left\{\frac{\mathbf{r}f'(r)}{r}\right\}.$$

12. $\nabla^2\phi=f$ within a volume V and $\partial\phi/\partial n$ vanishes over the bounding surface S. Prove that for arbitrary ψ

$$-2\int_V f\psi\,d\tau-\int_V(\nabla\psi)^2\,d\tau<\int_V(\nabla\phi)^2\,d\tau.$$

13. If S is an open surface with rim $\mathscr{C}$, prove that

$$\int_{\mathscr{C}} u\,dv = \int_S (\nabla u \times \nabla v) . d\mathscr{A}$$

where u, v are scalar fields.

14. Show that the position vector of the centroid of the volume V bounded by s is given by

$$\frac{1}{2V}\int_S r^2\,d\mathscr{A}.$$

15. **F** is solenoidal and **G** is irrotational. Both may be assumed to be sufficiently small at infinity. Prove that

$$\int \mathbf{F}.\mathbf{G}\,d\tau = 0$$

where the integral is over the whole of space.

16. A distribution charge $\rho(\mathbf{r})$ exists in a deformable medium. The medium is subjected to an infinitesimal deformation in which a particle at a general point $\mathbf{r}$ is displaced to $\mathbf{r}+\delta\mathbf{r}$. Let the particles originally filling any volume V be so displaced that they fill the volume V' and let ρ vary in such a way that its integral over V before the deformation equals its integral over V' after deformation. Show that at a point fixed in space ρ is changed by $\delta\rho = -\nabla.(\rho\,\delta\mathbf{r})$.

A system of steady currents $\mathbf{J}(\mathbf{r})$ flows in a medium subjected to a like deformation. The particles covering any surface S (not necessarily closed) are so displaced as to cover a surface S' and $\mathbf{J}$ varies in such a way that its original flux across S is equal to its final flux across S'. Show that at a fixed point $\delta\mathbf{J} = -\nabla \times (\mathbf{J} \times \delta\mathbf{r})$.

$\mathbf{E}(\mathbf{r})$ and $\mathbf{H}(\mathbf{r})$ are respectively the electric field in the first case and the magnetic field in the second case, and $\rho(\mathbf{r})$, $\mathbf{J}(\mathbf{r})$ are everywhere continuous and vanish for sufficiently large $|\mathbf{r}|$. Consider the integrals taken over all space of $\frac{1}{2}\mathbf{E}.\mathbf{D}$ and of $\frac{1}{2}\mathbf{B}.\mathbf{H}$ and show that in the deformation they change by

$$-\int_V \rho\mathbf{E}.\delta\mathbf{r}\,d\tau \quad \text{and} \quad +\int_V (\mathbf{J}\times\mathbf{H}).\delta\mathbf{r}\,d\tau, \text{ respectively.}$$

Can either or both of these results be interpreted as equations of energy balance? Give reasons for your answers. (V. C. A. Ferraro.)

17. Prove that, in magnetostatics, where the following relations hold, $\mathbf{B} = \nabla \times \mathbf{A}$, $\nabla \times \mathbf{B} = \mu \mathbf{J}$.

$$\mathbf{A}(\mathbf{r}) = \frac{\mu}{4\pi} \int_V \frac{\mathbf{J}(\mathbf{r}')\, d\tau'}{R} - \frac{1}{4\pi} \int_S \frac{d\mathscr{A}' \times \mathbf{B}(\mathbf{r}')}{R}$$

$$- \frac{1}{4\pi} \int_S [d\mathscr{A}' \times \mathbf{A}(\mathbf{r}')] \times \nabla' \left(\frac{1}{R}\right)$$

$$- \frac{1}{4\pi} \int_S [d\mathscr{A}' . \mathbf{A}(\mathbf{r}')] \nabla' \left(\frac{1}{R}\right).$$

18. If a mass of incompressible liquid is set in irrotational motion with prescribed velocities on its bounding surface, the actual kinetic energy is less than the kinetic energy of any other motion consistent with the same boundary conditions.

19. S is a sphere of radius a and centre the origin. C is the point $c\mathbf{k}$ and F is the inverse point of C with respect to S, F thus has position vector $f\mathbf{k} = (a^2/c)\mathbf{k}$ $(c > a)$. Let

$$U(\mathbf{r}) = \frac{1}{|\mathbf{r} - c\mathbf{k}|} - \frac{a}{c} \frac{1}{|\mathbf{r} - f\mathbf{k}|}.$$

Prove that $\nabla^2 U = -4\pi\delta(\mathbf{r} - c\mathbf{k})$, in $r > a$ and that on $r = a$, $U = 0$ and

$$\frac{\partial U}{\partial r} = -\frac{(a^2 - c^2)}{a|\mathbf{r} - c\mathbf{k}|^3}.$$

20. Let S be the surface (the Mobius strip) described by

$$x = (a + ct \sin \tfrac{1}{2}\phi) \cos \phi, \qquad y = (a + ct \sin \tfrac{1}{2}\phi) \sin \phi, \qquad z = ct \cos \tfrac{1}{2}\phi.$$
$$-1 \leqslant t \leqslant 1, \qquad -\pi < \phi \leqslant \pi, \qquad 0 < c < a.$$

Show that S is bounded by the simple closed curve $\mathscr{C}$ defined by

$$x = (a + c \sin \psi) \cos 2\psi, \qquad y = (a + c \sin \psi) \sin 2\psi, \qquad z = c \cos \psi.$$
$$-\pi < \psi \leqslant \pi.$$

Let

$$\mathbf{A} = \left(\frac{-y}{x^2 + y^2}, \frac{x}{x^2 + y^2}, 0\right).$$

Show that

$$\int_S (\nabla \times \mathbf{A}) . d\mathscr{A} = 0$$

and that

$$\int \mathbf{A} . d\mathbf{r} \neq 0.$$

and explain why this does not contradict Stokes's theorem. (S. Brenner.)

6 Miscellaneous Topics

6.1. Linear vector function and dyadics

Suppose that **f** and **F** are two vectors such that **f** depends on **F**. It is possible to write

$$\mathbf{f} = \mathbf{f}(\mathbf{F}). \tag{6.1.1}$$

f is a vector function of **F**. Up till now, such functions have been met with mainly in the particular form of vector fields where a vector is a function of the position vector.

A vector function is said to be linear if

$$\mathbf{f}(\mathbf{F}_1 + \mathbf{F}_2) = \mathbf{f}(\mathbf{F}_1) + \mathbf{f}(\mathbf{F}_2) \tag{6.1.2}$$

and

$$\mathbf{f}(\lambda \mathbf{F}) = \lambda \mathbf{f}(\mathbf{F}) \tag{6.1.3}$$

λ being a scalar.

If now $\mathbf{c}_1$, $\mathbf{c}_2$, $\mathbf{c}_3$ are three non-coplanar vectors

$$\mathbf{F} = F_1\,\mathbf{c}_1 + F_2\,\mathbf{c}_2 + F_3\,\mathbf{c}_3 \tag{6.1.4}$$

and

$$\mathbf{f}(\mathbf{F}) = F_1\,\mathbf{f}(c_1) + F_2\,\mathbf{f}(c_2) + F_3\,\mathbf{f}(c_3) \tag{6.1.5}$$

on using (6.1.2) and (6.1.3).

By (2.32.6), $F_1 = \mathbf{F}.\mathbf{a}_1$, $F_2 = \mathbf{F}.\mathbf{a}_2$, $F_3 = \mathbf{F}.\mathbf{a}_3$, where $\mathbf{a}_1$, $\mathbf{a}_2$, $\mathbf{a}_3$ are the set of vectors reciprocal to $\mathbf{c}_1$, $\mathbf{c}_2$, $\mathbf{c}_3$. Thus

$$\mathbf{f} = (\mathbf{F}.\mathbf{a}_1)\,\mathbf{f}(\mathbf{c}_1) + (\mathbf{F}.\mathbf{a}_2)\,\mathbf{f}(\mathbf{c}_2) + (\mathbf{F}.\mathbf{a}_3)\,\mathbf{f}(\mathbf{c}_3). \tag{6.1.6}$$

Let $\mathbf{f}(\mathbf{c}_r) = \mathbf{b}_r$,

$$\begin{aligned}\mathbf{f} &= (\mathbf{F}.\mathbf{a}_1)\,\mathbf{b}_1 + (\mathbf{F}.\mathbf{a}_2)\,\mathbf{b}_2 + (\mathbf{F}.\mathbf{a}_3)\,\mathbf{b}_3 \\ &= \mathbf{b}_1(\mathbf{a}_1.\mathbf{F}) + \mathbf{b}_2(\mathbf{a}_2.\mathbf{F}) + \mathbf{b}_3(\mathbf{a}_3.\mathbf{F}).\end{aligned} \tag{6.1.7}$$

Consider now the dyad, or indefinite product of two vectors **a** and **b**, **ab**. **a** is the antecedent, **b** is the consequent. This has the properties that

$$(\mathbf{ab}).\mathbf{F} = \mathbf{a}(\mathbf{b}.\mathbf{F}) \tag{6.1.8a}$$

and

$$\mathbf{F}.(\mathbf{ab}) = (\mathbf{F}.\mathbf{a})\,\mathbf{b} \tag{6.1.8b}$$

It is not difficult to see that this obeys the vector multiplication laws (2.0.1) and (2.0.2). For

$$(x\mathbf{a}y\mathbf{b}).\mathbf{F} = x\mathbf{a}(y\mathbf{b}.\mathbf{F}) = x\mathbf{a}y\mathbf{b}.\mathbf{F}$$
$$= xy\mathbf{a}(\mathbf{b}.\mathbf{F}) = xy(\mathbf{ab}).\mathbf{F} \qquad (6.1.9a)$$

Similarly

$$\{\mathbf{a}(\mathbf{b}+\mathbf{c})\}.\mathbf{F} = \mathbf{a}\{(\mathbf{b}+\mathbf{c}).\mathbf{F}\} = \mathbf{a}(\mathbf{b}.\mathbf{F}) + \mathbf{a}(\mathbf{c}.\mathbf{F})$$
$$= (\mathbf{ab}).\mathbf{F} + (\mathbf{ac}).\mathbf{F}. \qquad (6.1.9b)$$

A sum of n dyads

$$\mathbf{\Phi} = \sum_{i=1}^{n} \mathbf{a}_i \mathbf{b}_i$$

is termed a dyadic and the dyadic

$$\mathbf{\Phi}_c = \sum_{i=1}^{n} \mathbf{b}_i \mathbf{a}_i$$

is termed the conjugate dyadic. Clearly the indefinite product is not commutative.

If $n = 3$, equation (6.1.7) may be written either as

$$\mathbf{f} = \mathbf{F}.(\mathbf{a}_1\mathbf{b}_1 + \mathbf{a}_2\mathbf{b}_2 + \mathbf{a}_3\mathbf{b}_3) \qquad (6.1.10a)$$
$$= \mathbf{F}.\mathbf{\Phi}$$

or

$$\mathbf{f} = (\mathbf{b}_1\mathbf{a}_1 + \mathbf{b}_2\mathbf{a}_2 + \mathbf{b}_3\mathbf{a}_3).\mathbf{F}$$
$$= \mathbf{\Phi}_c.\mathbf{F} \qquad (6.1.10b)$$

In equation (6.1.10a) **F** is a prefactor, in equation (6.1.10b) **F** is a postfactor.

Dyads have a connection with matrices. This is discussed briefly in section 6.43.

It may be noted in passing that, if **u** is a unit vector

$$\mathbf{F}.(\mathbf{uu}) = (\mathbf{F}.\mathbf{u})\mathbf{u} = F_u\mathbf{u}(= \mathbf{u}\times(\mathbf{u}\times\mathbf{F}) - \mathbf{F}) \qquad (6.1.11a)$$

is the component of **F** in the u direction.

Also, if $\mathbf{u}_1$, $\mathbf{u}_2$ are two unit vectors,

$$\mathbf{F}(\mathbf{u}_1\mathbf{u}_2) = (\mathbf{F}.\mathbf{u}_1)\mathbf{u}_2 = F_1\mathbf{u}_2 \qquad (6.1.11b)$$

F_1 being the resolute of **F** in the $\mathbf{u}_1$ direction, e.g. $\mathbf{F}(\mathbf{ij}) = F_x\mathbf{j} = (\mathbf{ji}).\mathbf{F}$. In the first of these **F** is a prefactor, in the second a postfactor.

It is not difficult to see that the following relations are obeyed.

$\mathbf{\Phi}$ and $\mathbf{\Psi}$ are equal if

$$\mathbf{\Phi}.\mathbf{F} = \mathbf{\Psi}.\mathbf{F} \qquad (6.1.12)$$

for arbitrary **F**.

If $\mathbf{\Phi}.\mathbf{F} = \mathbf{0}$ for all $\mathbf{F}$, (6.1.13)

$\mathbf{\Phi} = \mathbf{0}$, the zero dyadic.

A cross-product of a dyad with a vector can also be defined.

$$(\mathbf{ab}) \times \mathbf{F} = \mathbf{a}(\mathbf{b} \times \mathbf{F}) \tag{6.1.14a}$$

and

$$\mathbf{F} \times (\mathbf{ab}) = (\mathbf{F} \times \mathbf{a})\mathbf{b}. \tag{6.1.14b}$$

These are both dyads. The cross-product of a dyadic with a vector follows immediately. If the antecedents and the consequents can be expressed in terms of three non-coplanar vectors, the dyadic is complete. If, on the other hand, either the antecedents or the consequents can be expressed in terms of less than three vectors only, the dyadic is singular. Thus dyadics containing only one or two dyads are, by definition, singular. Any complete dyadic of n dyads $n > 3$ can be reduced to a dyadic of three dyads so that either the antecedents or consequents are three arbitrary non-coplanar vectors $\mathbf{d}_1$, $\mathbf{d}_2$, $\mathbf{d}_3$.

$$\begin{aligned}\mathbf{\Phi} &= \sum_{i=1}^{n} \mathbf{a}_i \mathbf{b}_i = \sum_{i=1}^{n} \mathbf{a}_i(\alpha_i \mathbf{d}_1 + \beta_i \mathbf{d}_2 + \gamma_i \mathbf{d}_3) \\ &= \sum_{i=1}^{n} \{(\alpha_i \mathbf{a}_i)\mathbf{d}_1 + (\beta_i \mathbf{a}_i)\mathbf{d}_2 + (\gamma_i \mathbf{a}_i)\mathbf{d}_3\} \\ &= \mathbf{c}_1 \mathbf{d}_1 + \mathbf{c}_2 \mathbf{d}_2 + \mathbf{c}_3 \mathbf{d}_3 \end{aligned} \tag{6.1.15}$$

where

$$\mathbf{c}_1 = \sum_{i=1}^{n} \alpha_i \mathbf{a}_i, \text{ etc.}$$

A dyadic is symmetrical if $\mathbf{\Phi} = \mathbf{\Phi}_c$ and antisymmetrical if $\mathbf{\Phi} = -\mathbf{\Phi}_c$. Writing

$$\mathbf{\Phi} = \tfrac{1}{2}(\mathbf{\Phi} + \mathbf{\Phi}_c) + \tfrac{1}{2}(\mathbf{\Phi} - \mathbf{\Phi}_c) \tag{6.1.16}$$

it follows that any dyadic can be split up into the sum of a symmetrical dyadic and an antisymmetrical dyadic. A symmetrical dyadic is self-conjugate.

It is possible to define a scalar product of two dyadics. Let

$$\mathbf{\Phi} = \sum_m \mathbf{a}_m \mathbf{b}_m, \qquad \mathbf{\Psi} = \sum_n \mathbf{c}_n \mathbf{d}_n$$

$$\begin{aligned}\mathbf{\Phi}.\mathbf{\Psi} &= \sum_m \sum_n \mathbf{a}_m(\mathbf{b}_m . \mathbf{c}_n)\mathbf{d}_n \\ &= \sum_m \sum_n \mathbf{a}_m \mathbf{d}_n(\mathbf{b}_m . \mathbf{c}_n). \end{aligned} \tag{6.1.17}$$

The commutative law is not obeyed as

$$\mathbf{\Psi}.\mathbf{\Phi} = \sum_m \sum_n \mathbf{c}_n(\mathbf{d}_n.\mathbf{a}_m)\,\mathbf{b}_m$$

$$= \sum_m \sum_n \mathbf{c}_n\,\mathbf{b}_m(\mathbf{d}_n.\mathbf{a}_m) \neq \mathbf{\Phi}.\mathbf{\Psi}. \qquad (6.1.18)$$

It is easily seen that

$$(\mathbf{\Phi}.\mathbf{\Psi})_c = \mathbf{\Psi}_c\,\mathbf{\Phi}_c \qquad (6.1.19)$$

$$\mathbf{F}.(\mathbf{\Phi}.\mathbf{\Psi}) = (\mathbf{F}.\mathbf{\Phi}).\mathbf{\Psi} \qquad (6.1.20)$$

and

$$(\mathbf{\Phi}.\mathbf{\Psi}).\mathbf{\Gamma} = \mathbf{\Phi}.(\mathbf{\Psi}.\mathbf{\Gamma}). \qquad (6.1.21)$$

Powers of $\mathbf{\Phi}$ may thus be defined

$$\mathbf{\Phi}^2 = \mathbf{\Phi}.\mathbf{\Phi} \qquad (6.1.22)$$

and

$$\mathbf{\Phi}^{n+1} = \mathbf{\Phi}.\mathbf{\Phi}^n$$

(n a positive integer). By the associative law this is unique.

Let $\mathbf{a}_1$, $\mathbf{a}_2$, $\mathbf{a}_3$, be a set of non-coplanar vectors and $\mathbf{b}_1$, $\mathbf{b}_2$, $\mathbf{b}_3$ the reciprocal set.

Let

$$\mathbf{I} = \mathbf{a}_1\,\mathbf{b}_1 + \mathbf{a}_2\,\mathbf{b}_2 + \mathbf{a}_3\,\mathbf{b}_3. \qquad (6.1.23)$$

If $\mathbf{\Phi}$ is the dyadic

$$\sum_i \mathbf{c}_i\,\mathbf{d}_i$$

$$\mathbf{I}.\mathbf{\Phi} = (\mathbf{a}_1\,\mathbf{b}_1 + \mathbf{a}_2\,\mathbf{b}_2 + \mathbf{a}_3\,\mathbf{b}_3).\mathbf{\Phi}$$

$$= \sum_i \{\mathbf{a}_1(\mathbf{b}_1.\mathbf{c}_i)\,\mathbf{d}_i + \mathbf{a}_2(\mathbf{b}_2.\mathbf{c}_i)\,\mathbf{d}_i + \mathbf{a}_3(\mathbf{b}_3.\mathbf{c}_i)\,\mathbf{d}_i\}.$$

Now

$$\mathbf{a}_1(\mathbf{b}_1.\mathbf{c}_i) + \mathbf{a}_2(\mathbf{b}_2.\mathbf{c}_i) + \mathbf{a}_3.(\mathbf{b}_3.\mathbf{c}_i) = \mathbf{c}_i$$

and so

$$\mathbf{I}.\mathbf{\Phi} = \sum_i \mathbf{c}_i\,\mathbf{d}_i = \mathbf{\Phi}. \qquad (6.1.24a)$$

Similarly

$$\mathbf{\Phi}.\mathbf{I} = \mathbf{\Phi}. \qquad (6.1.24b)$$

Thus $\mathbf{I}$ is an idemfactor and may be referred to as the identical dyadic. It is possible to write $\mathbf{\Phi}^{\circ} = \mathbf{I}$.

For any non-singular dyadic $\mathbf{\Phi}$, a reciprocal dyadic may be defined in a unique manner.

Let $\mathbf{\Phi}$ be reduced to the form

$$\mathbf{a}_1\mathbf{b}_1 + \mathbf{a}_2\mathbf{b}_2 + \mathbf{a}_3\mathbf{b}_3 \tag{6.1.25}$$

where $\mathbf{a}_1$, $\mathbf{a}_2$, $\mathbf{a}_3$; $\mathbf{b}_1$, $\mathbf{b}_2$, $\mathbf{b}_3$ are sets of non-coplanar vectors. Let $\mathbf{a}_1'$, $\mathbf{a}_2'$, $\mathbf{a}_3'$; $\mathbf{b}_1'$, $\mathbf{b}_2'$, $\mathbf{b}_3'$ be the reciprocal sets of vectors.

Let

$$\mathbf{\Phi}' = \mathbf{b}_1'\mathbf{a}_1' + \mathbf{b}_2'\mathbf{a}_2' + \mathbf{b}_3'\mathbf{a}_3' \tag{6.1.26}$$

$$\mathbf{\Phi}.\mathbf{\Phi}' = (\mathbf{a}_1\mathbf{b}_1 + \mathbf{a}_2\mathbf{b}_2 + \mathbf{a}_3\mathbf{b}_3).(\mathbf{b}_1'\mathbf{a}_1' + \mathbf{b}_2'\mathbf{a}_2' + \mathbf{b}_3'\mathbf{a}_3').$$

By virtue of the reciprocal property

$$\mathbf{b}_r.\mathbf{b}_s' = \delta_{rs}$$

and so

$$\mathbf{\Phi}.\mathbf{\Phi}' = \mathbf{a}_1\mathbf{a}_1' + \mathbf{a}_2\mathbf{a}_2' + \mathbf{a}_3\mathbf{a}_3' = \mathbf{I}. \tag{6.1.27a}$$

Similarly

$$\mathbf{\Phi}'.\mathbf{\Phi} = \mathbf{I}. \tag{6.1.27b}$$

Thus $\mathbf{\Phi}'$ may be regarded as the reciprocal of $\mathbf{\Phi}$ and written $\mathbf{\Phi}^{-1}$. If $\mathbf{\Phi}$ is singular, it is not possible to find the reciprocal sets of vectors as there is no reciprocal set for two vectors or one vector only. Thus the reciprocal is not then defined. It is possible, following what has been done, to define the dyadic $\mathbf{\Phi}^n$ for n integral, positive zero or negative if $\mathbf{\Phi}$ is non-singular.

6.11. *Scalar and vector of a dyadic*

The scalar of the dyadic

$$\mathbf{\Phi} = \sum_i \mathbf{a}_i\mathbf{b}_i \tag{6.11.1}$$

is defined by

$$\Phi_s = \sum_i \mathbf{a}_i.\mathbf{b}_i \tag{6.11.2}$$

and the vector is defined by

$$\mathbf{\Phi}_v = \sum_i \mathbf{a}_i \times \mathbf{b}_i. \tag{6.11.3}$$

It follows that

$$\Phi_{cs} = \Phi_s \tag{6.11.4}$$

and

$$\mathbf{\Phi}_{cv} = -\mathbf{\Phi}_v. \tag{6.11.5}$$

Writing

$$\mathbf{\Phi} = \tfrac{1}{2}(\mathbf{\Phi} + \mathbf{\Phi}_c) + \tfrac{1}{2}(\mathbf{\Phi} - \mathbf{\Phi}_c)$$

it follows that, for an antisymmetric dyadic

$$\begin{aligned}\mathbf{F}.\boldsymbol{\Phi} &= \tfrac{1}{2}\mathbf{F}.(\boldsymbol{\Phi}-\boldsymbol{\Phi}_c)\\ &= \tfrac{1}{2}\mathbf{F}.\sum_i(\mathbf{a}_i\mathbf{b}_i-\mathbf{b}_i\mathbf{a}_i)\\ &= \tfrac{1}{2}\sum_i(\mathbf{F}.\mathbf{a}_i)\mathbf{b}_i-(\mathbf{F}.\mathbf{b}_i)\mathbf{a}_i\\ &= \tfrac{1}{2}\sum_i(\mathbf{a}_i\times\mathbf{b}_i)\times\mathbf{F}=\tfrac{1}{2}\boldsymbol{\Phi}_v\times\mathbf{F}.\end{aligned} \tag{6.11.6}$$

If $\boldsymbol{\Phi}_v$ vanishes, $\boldsymbol{\Phi}$ is symmetric.

6.12. *Dyadics in rectangular cartesian coordinates*

If the antecedents and consequents of a dyadic $\boldsymbol{\Phi}$ are expressed in terms of the unit vectors **i**, **j**, **k** and the dyads involved are expanded according to the distributive law

$$\begin{aligned}\boldsymbol{\Phi} = \Phi_{11}\,\mathbf{ii}+\Phi_{12}\,\mathbf{ij}+\Phi_{13}\,\mathbf{ik}+\Phi_{21}\,\mathbf{ji}+\Phi_{22}\,\mathbf{jj}+\Phi_{23}\,\mathbf{jk}+\Phi_{31}\,\mathbf{ki}\\ +\Phi_{32}\,\mathbf{kj}+\Phi_{33}\,\mathbf{kk}\end{aligned} \tag{6.12.1}$$

and

$$\begin{aligned}\boldsymbol{\Phi}_c = \Phi_{11}\,\mathbf{ii}+\Phi_{21}\,\mathbf{ij}+\Phi_{31}\,\mathbf{ik}+\Phi_{12}\,\mathbf{ji}+\Phi_{22}\,\mathbf{jj}+\Phi_{32}\,\mathbf{jk}+\Phi_{13}\,\mathbf{ki}\\ +\Phi_{23}\,\mathbf{kj}+\Phi_{33}\,\mathbf{kk}.\end{aligned} \tag{6.12.2}$$

These are referred to as the nonion forms.

For symmetric dyadics

$$\Phi_{rs}=\Phi_{sr} \tag{6.12.3a}$$

and for antisymmetric dyadics

$$\Phi_{rs}+\Phi_{sr}=0 \qquad (r\neq s), \qquad \Phi_{ss}=0. \tag{6.12.3b}$$

Now

$$\begin{aligned}\boldsymbol{\Phi} &= (\Phi_{11}\,\mathbf{i}+\Phi_{21}\,\mathbf{j}+\Phi_{31}\,\mathbf{k})\,\mathbf{i}\\ &\quad+(\Phi_{12}\,\mathbf{i}+\Phi_{22}\,\mathbf{j}+\Phi_{32}\,\mathbf{k})\,\mathbf{j}\\ &\quad+(\Phi_{13}\,\mathbf{i}+\Phi_{23}\,\mathbf{j}+\Phi_{33}\,\mathbf{k})\,\mathbf{k}\\ &= \boldsymbol{\psi}_1\,\mathbf{i}+\boldsymbol{\psi}_2\,\mathbf{j}+\boldsymbol{\psi}_3\,\mathbf{k}, \text{ say.}\end{aligned} \tag{6.12.4}$$

It follows that

$$\begin{aligned}\phi_3\,\mathbf{i} &= \alpha_{11}\,\boldsymbol{\psi}_1+\alpha_{12}\,\boldsymbol{\psi}_2+\alpha_{13}\,\boldsymbol{\psi}_3\\ \phi_3\,\mathbf{j} &= \alpha_{21}\,\boldsymbol{\psi}_1+\alpha_{22}\,\boldsymbol{\psi}_2+\alpha_{23}\,\boldsymbol{\psi}_3\\ \phi_3\,\mathbf{k} &= \alpha_{31}\,\boldsymbol{\psi}_1+\alpha_{32}\,\boldsymbol{\psi}_2+\alpha_{33}\,\boldsymbol{\psi}_3\end{aligned} \tag{6.12.5}$$

where α_{rs} is the cofactor of Φ_{rs} in the determinant

$$\begin{vmatrix} \Phi_{11} & \Phi_{12} & \Phi_{13} \\ \Phi_{21} & \Phi_{22} & \Phi_{23} \\ \Phi_{31} & \Phi_{32} & \Phi_{33} \end{vmatrix}$$

and ϕ_3 is the value of the determinant.

Consider now the dyadic Ψ defined by

$$\begin{aligned} \phi_3 \Psi = {} & \alpha_{11}\,\mathbf{ii} + \alpha_{22}\,\mathbf{ij} + \alpha_{31}\,\mathbf{ik} \\ & + \alpha_{12}\,\mathbf{ji} + \alpha_{22}\,\mathbf{jj} + \alpha_{32}\,\mathbf{jk} \\ & + \alpha_{13}\,\mathbf{ki} + \alpha_{23}\,\mathbf{kj} + \alpha_{33}\,\mathbf{kk}. \end{aligned} \qquad (6.12.6)$$

Using the properties of cofactors in a determinant, it follows that

$$\mathbf{\Psi} . \mathbf{\Phi} = \mathbf{ii} + \mathbf{jj} + \mathbf{kk}. \qquad (6.12.7)$$

Now

$$\mathbf{F} . (\mathbf{ii} + \mathbf{jj} + \mathbf{kk}) = (\mathbf{F} . \mathbf{i})\,\mathbf{i} + (\mathbf{F} . \mathbf{j})\,\mathbf{j} + (\mathbf{f} . \mathbf{k})\,\mathbf{k} = \mathbf{F}. \qquad (6.12.8)$$

Thus

$$\mathbf{ii} + \mathbf{jj} + \mathbf{kk} = \mathbf{I} \qquad (6.12.9)$$

and

$$\mathbf{\Psi} = \mathbf{\Phi}^{-1}. \qquad (6.12.10)$$

6.13. *Dyadic invariants*

Consider now the possibility of finding a scalar λ and vector $\mathbf{F}$ for a dyadic $\mathbf{\Phi}$ such that

$$\mathbf{\Phi} . \mathbf{F} = \lambda \mathbf{F}. \qquad (6.13.1)$$

Expanding the terms, it follows that the condition is given by

$$\begin{aligned} (\Phi_{11} - \lambda)\, F_x + \Phi_{12}\, F_y + \Phi_{13}\, F_z &= 0 \\ \Phi_{21}\, F_x + (\Phi_{22} - \lambda)\, F_y + \Phi_{23}\, F_z &= 0 \\ \Phi_{31}\, F_x + \Phi_{32}\, F_y + (\Phi_{33} - \lambda)\, F_z &= 0 \end{aligned} \qquad (6.13.2)$$

This is, of course, simply the process of finding the eigenvalues of the matrix $|\Phi_{rs}|$ and the corresponding eigenvectors. The moduli of these are clearly not important. The condition that (6.13.2) has non-zero solutions for F_x, F_y, F_z is

$$\begin{vmatrix} \Phi_{11} - \lambda & \Phi_{12} & \Phi_{13} \\ \Phi_{21} & \Phi_{22} - \lambda & \Phi_{23} \\ \Phi_{31} & \Phi_{32} & \Phi_{33} - \lambda \end{vmatrix} = 0 \qquad (6.13.3)$$

This is a cubic equation which may be rewritten as

$$\lambda^3 - \phi_1 \lambda^2 + \phi_2 \lambda - \phi_3 = 0. \tag{6.13.4}$$

ϕ_3 has been defined previously

$$\phi_1 = \Phi_{11} + \Phi_{22} + \Phi_{33} \tag{6.13.5a}$$

$$\phi_2 = \Phi_{11}\Phi_{22} + \Phi_{22}\Phi_{33} + \Phi_{33}\Phi_{11} - \Phi_{12}\Phi_{21} - \Phi_{13}\Phi_{31} - \Phi_{23}\Phi_{32}. \tag{6.13.5b}$$

The values of λ are obviously independent of the coordinate system. This must be true therefore for ϕ_1, ϕ_2, ϕ_3. These are thus invariants of the dyadic.

Because ϕ_1, ϕ_2, ϕ_3 are all real, at least one of the eigenvalues is real. If the dyadic is symmetric, more may be said. In the first place all eigenvalues are real. The proof is as follows. Let $\mathbf{F}_r$ be the eigenvector which corresponds to the eigenvalue λ_r. Then

$$\mathbf{\Phi} . \mathbf{F}_r = \lambda_r \mathbf{F}_r. \tag{6.13.6}$$

If $\mathbf{F}_r^*$ is the complex conjugate of $\mathbf{F}$

$$\mathbf{F}_r^* . \mathbf{\Phi} . \mathbf{F}_r = \lambda \mathbf{F}_r^* . \mathbf{F}_r. \tag{6.13.7}$$

Expanding and using the symmetry property it can be seen that the left-hand side of (6.13.7) is real. Also $\mathbf{F}_r^* . \mathbf{F}_r$ is real and so λ_r is real.

Suppose now that the three eigenvalues λ_r, $r = 1, 2, 3$ are unequal. Then

$$\mathbf{F}_r . \mathbf{\Phi} . \mathbf{F}_s = \mathbf{F}_r . \lambda_s \mathbf{F}_s \tag{6.13.8a}$$

$$\mathbf{F}_s . \mathbf{\Phi} . \mathbf{F}_r = \mathbf{F}_s . \lambda_r \mathbf{F}_r. \tag{6.13.8b}$$

As $\mathbf{\Phi}$ is symmetric, the two left-hand sides are equal and so

$$0 = (\lambda_r - \lambda_s) \mathbf{F}_r . \mathbf{F}_s \tag{6.13.9a}$$

and so

$$\mathbf{F}_r . \mathbf{\Phi} . \mathbf{F}_s = 0. \tag{6.13.9b}$$

Thus if the eigenvalues are unequal, the eigenvectors are normal to one another.

The moduli of the eigenvectors are unimportant. For convenience they may be taken as unity. If they are $\mathbf{i}_1$, $\mathbf{i}_2$, $\mathbf{i}_3$ it can be seen that

$$\mathbf{i}_s . \mathbf{\Phi} . \mathbf{i}_r = 0 \qquad (r \neq s) \tag{6.13.10a}$$

$$\mathbf{i}_s . \mathbf{\Phi} . \mathbf{i}_s = \lambda_s. \tag{6.13.10b}$$

It is not difficult to see that, if two roots coincide, it is possible to construct out of two vectors satisfying $\mathbf{\Phi}.\mathbf{F} = \lambda\mathbf{F}$, two unit vectors obeying these relations.

From the above, it follows that the dyadic may be written in the nonion form.

$$\mathbf{\Phi} = \lambda_1 \mathbf{i}_1 \mathbf{i}_1 + \lambda_2 \mathbf{i}_2 \mathbf{i}_2 + \lambda_3 \mathbf{i}_3 \mathbf{i}_3 \qquad (6.13.11)$$

$$\mathbf{F}.\mathbf{\Phi}.\mathbf{G} = \lambda_1 \mathbf{F}_1 \mathbf{G}_1 + \lambda_2 \mathbf{F}_2 \mathbf{G}_2 + \lambda_3 \mathbf{F}_3 \mathbf{G}_3 \qquad (6.13.12)$$

and

$$\mathbf{F}.\mathbf{\Phi}.\mathbf{F} = \lambda_1 F_1^2 + \lambda_2 F_2^2 + \lambda_3 F_3^2. \qquad (6.13.13)$$

If all the λ are positive, $\mathbf{F}.\mathbf{\Phi}.\mathbf{F} > 0$ except when $\mathbf{F} = \mathbf{0}$. In this case the dyadic may be said to be positive definite.

6.14. *Dyadics and the vector operator* $\mathbf{\nabla}$.

Dyadics may be formed in which the vector operator $\mathbf{\nabla}$ is a component part. By analogy with the relation

$$\delta V = \delta\mathbf{r}.\mathbf{\nabla} V$$

there exists a dyadic $\mathbf{\nabla F}$ defined by

$$\delta\mathbf{F} = \delta\mathbf{r}.\mathbf{\nabla F}. \qquad (6.14.1)$$

It is not difficult to see that

$$\mathbf{\nabla F} = \mathbf{i}\,\frac{\partial \mathbf{F}}{\partial x} + \mathbf{j}\,\frac{\partial \mathbf{F}}{\partial x} + \mathbf{k}\,\frac{\partial \mathbf{F}}{\partial z} \qquad (6.14.2a)$$

the conjugate dyadic being defined by

$$\mathbf{F\nabla} = \frac{\partial \mathbf{F}}{\partial x}\,\mathbf{i} + \frac{\partial \mathbf{F}}{\partial y}\,\mathbf{j} + \frac{\partial \mathbf{F}}{\partial z}\,\mathbf{k}. \qquad (6.14.2b)$$

The scalar of these is $\mathbf{\nabla}.\mathbf{F}$ and the vector of $\mathbf{\nabla F}$ and $-\mathbf{F\nabla}$ is $\mathbf{\nabla} \times \mathbf{F}$.

If $\mathbf{\nabla} \times \mathbf{F}$ is zero, the dyadic is symmetric and $\mathbf{F} = \mathbf{\nabla}\psi$. In this case

$$\mathbf{\nabla F} = \mathbf{\nabla\nabla}\psi. \qquad (6.14.3)$$

The divergence and curl may be defined in a similar manner.

$$\mathbf{\nabla}.\mathbf{\Phi} = \mathbf{i}.\frac{\partial \mathbf{\Phi}}{\partial x} + \mathbf{j}.\frac{\partial \mathbf{\Phi}}{\partial y} + \mathbf{k}.\frac{\partial \mathbf{\Phi}}{\partial z} \qquad (6.14.4)$$

and

$$\mathbf{\nabla} \times \mathbf{\Phi} = \mathbf{i} \times \frac{\partial \mathbf{\Phi}}{\partial x} + \mathbf{j} \times \frac{\partial \mathbf{\Phi}}{\partial y} + \mathbf{k} \times \frac{\partial \mathbf{\Phi}}{\partial z}\cdot \qquad (6.14.5)$$

Similarly

$$\mathbf{\Phi}.\mathbf{\nabla} = \frac{\partial \mathbf{\Phi}}{\partial x}.\mathbf{i} + \frac{\partial \mathbf{\Phi}}{\partial y}.\mathbf{j} + \frac{\partial \mathbf{\Phi}}{\partial z}.\mathbf{k} \qquad (6.14.6)$$

and

$$\mathbf{\Phi} \times \nabla = \frac{\partial \mathbf{\Phi}}{\partial x} \times \mathbf{i} + \frac{\partial \mathbf{\Phi}}{\partial y} \times \mathbf{j} + \frac{\partial \mathbf{\Phi}}{\partial z} \times \mathbf{k}. \tag{6.14.7}$$

Many formulae analogous to those involving the application of ∇ to scalars and vectors may be proved, for example

$$\nabla . f\mathbf{\Phi} = (\nabla f) . \mathbf{\Phi} + f\nabla . \mathbf{\Phi}. \tag{6.14.8}$$

Others will be mentioned in the examples and exercises at the end of the chapter.

6.15. *Integral theorems involving dyadics*

There exist dyadic analogues of Stokes's theorem and the Gauss divergence theorem. Let $\mathbf{F} = \mathbf{\Phi} . \mathbf{A}$ where $\mathbf{A}$ is an arbitrary constant vector

$$\int_{\mathscr{C}} \mathbf{F} . d\mathbf{r} = \int_S (d\mathscr{A} \times \nabla) . \mathbf{F}$$

by Stokes's theorem, $\mathscr{C}$ being the rim of the surface S.

$$\int_{\mathscr{C}} d\mathbf{r} . \mathbf{\Phi} . \mathbf{A} = \int_S (d\mathscr{A} \times \nabla) . \mathbf{\Phi} . \mathbf{A}$$

$\mathbf{A}$ although constant is arbitrary and so

$$\int_{\mathscr{C}} d\mathbf{r} . \mathbf{\Phi} = \int_S (d\mathscr{A} \times \nabla) . \mathbf{\Phi}. \tag{6.15.1}$$

Similarly

$$\int_{\mathscr{C}} d\mathbf{r} \times \mathbf{\Phi} = \int_S (d\mathscr{A} \times \nabla) \times \mathbf{\Phi}. \tag{6.15.2}$$

By the Gauss divergence theorem

$$\int_S d\mathscr{A} . \mathbf{F} = \int_V d\tau \nabla . \mathbf{F}$$

V being a volume surrounded by the surface S, whence

$$\int_S d\mathscr{A} . \mathbf{\Phi} . \mathbf{A} = \int_V d\tau \nabla . \mathbf{\Phi} . \mathbf{A}.$$

As $\mathbf{A}$, although constant is arbitrary

$$\int_S d\mathscr{A} . \mathbf{\Phi} = \int_V d\tau \nabla . \mathbf{\Phi}. \tag{6.15.3}$$

Similarly

$$\int_S d\mathscr{A} \times \mathbf{\Phi} = \int_V d\tau \nabla \times \mathbf{\Phi} \tag{6.15.4}$$

6.2. Vector spaces

So far the concept of a vector has been discussed only in the familiar three-dimensional space. A more extended discussion of a vector may be given which extends that of section 1.32, and a short account follows. This is of necessity incomplete. (See further, for example, P. R. Halmos, *Finite Dimensional Vector Spaces*, Van Nostrand, (1958).) Abstract vector spaces comprise quantities called vectors which obey the following axioms.

There exists a sum $\mathbf{x} + \mathbf{y}$ of two vectors $\mathbf{x}$ and $\mathbf{y}$ such that

$$\mathbf{x} + \mathbf{y} = \mathbf{y} + \mathbf{x} \tag{6.2.1a}$$
$$\mathbf{x} + (\mathbf{y} + \mathbf{z}) = (\mathbf{x} + \mathbf{y}) + \mathbf{z}. \tag{6.2.1b}$$

A unique vector $\mathbf{0}$ exists such that

$$\mathbf{x} + \mathbf{0} = \mathbf{x}. \tag{6.2.1c}$$

To each vector $\mathbf{x}$ there corresponds a unique vector $-\mathbf{x}$ such that

$$\mathbf{x} + (-\mathbf{x}) = \mathbf{0}. \tag{6.2.1d}$$

If α is a number (or scalar) and $\mathbf{x}$ a vector, there exists a vector $\alpha\mathbf{x}$ such that

$$\alpha(\beta\mathbf{x}) = (\alpha\beta)\mathbf{x} \tag{6.2.2a}$$
$$\alpha(\mathbf{x} + \mathbf{y}) = \alpha\mathbf{x} + \alpha\mathbf{y} \tag{6.2.2b}$$
$$(\alpha + \beta)\mathbf{x} = \alpha\mathbf{x} + \beta\mathbf{x} \tag{6.2.2c}$$

and

$$1\mathbf{x} = \mathbf{x}. \tag{6.2.2d}$$

The inner product (or scalar product) $(\mathbf{x}, \mathbf{y})$ is a number associated with two vectors $\mathbf{x}$ and $\mathbf{y}$ such that

$$(\mathbf{x}, \mathbf{y}) = (\mathbf{y}, \mathbf{x}) \tag{6.2.3a}$$
$$(\alpha_1 \mathbf{x}_1 + \alpha_2 \mathbf{x}_2, \mathbf{y}) = \alpha_1(\mathbf{x}_1, \mathbf{y}) + \alpha_2(\mathbf{x}_2, \mathbf{y}). \tag{6.2.3b}$$

If the scalar product of two vectors is zero, the vectors are said to be normal to one another. When $\mathbf{x} = \mathbf{y}$, the self scalar product is defined similarly, with properties as follows

$$(\mathbf{x}, \mathbf{x}) > 0 \tag{6.2.4a}$$
$$(\mathbf{x}, \mathbf{x}) = 0 \text{ only if } \mathbf{x} = \mathbf{0}. \tag{6.2.4b}$$

The positive root of the self scalar product is termed the norm.

$$\|\mathbf{x}\| = \sqrt{(\mathbf{x}, \mathbf{x})} > 0. \tag{6.2.4c}$$

Clearly

$$\|\alpha\mathbf{x}\| = |\alpha|\|\mathbf{x}\| \tag{6.2.4d}$$

when α is a number.

There exist vector spaces for which there is no inner product but these will not be considered here.

A set of vectors $\mathbf{i}_r$ is orthonormal if $(\mathbf{i}_r, \mathbf{i}_s) = \delta_{rs}$. It is complete if it is not contained in a larger orthonormal set. For example, in three-dimensional Euclidean space the set **i**, **j**, **k** is a complete orthonormal set, but the set **i**, **j** is merely orthonormal, being incomplete.

Vectors such as $\mathbf{i}_r$ are termed unit vectors, and for any vector **x** a unit vector $\hat{\mathbf{x}}$ may be defined by

$$\hat{\mathbf{x}} = \mathbf{x}/\|\mathbf{x}\| \qquad \text{for} \qquad \|\mathbf{x}\|^2 (\hat{\mathbf{x}}, \hat{\mathbf{x}}) = (\mathbf{x}, \mathbf{x}) = \|\mathbf{x}\|^2.$$

Suppose that there are n vectors in the complete orthonormal set. The set of vectors

$$\mathbf{x}^* = \sum_{j=1}^{k \leq n} x_j \mathbf{i}_j \tag{6.2.5}$$

for all possible x_j is said to be a linear manifold or subspace of the vector space. The whole space is a particular subspace of itself. If $k < n$, there may be said to be a strict subspace. $\mathbf{i}_j (1 \leqslant j \leqslant k < n)$ is an incomplete orthonormal set.

Let $\mathbf{i}_j$ $1 \leqslant j \leqslant l \leqslant n$ be an orthonormal set which may or may not be complete.

Let

$$x_j = (\mathbf{x}, \mathbf{i}_j) \tag{6.2.6}$$

Let

$$\mathbf{x}' = \mathbf{x} - \sum_{j=1}^{l} x_j \mathbf{i}_j$$

$$\begin{aligned}(\mathbf{x}' . \mathbf{i}_k) &= (\mathbf{x}, \mathbf{i}_k) - \sum_{j=1}^{l} x_j (\mathbf{i}_j, \mathbf{i}_k) \\ &= x_k - \sum_{j=1}^{l} x_j \delta_{jk} \\ &= 0. \end{aligned} \tag{6.2.7}$$

Also

$$
\begin{aligned}
0 < \|\mathbf{x}'\|^2 &= (\mathbf{x}', \mathbf{x}') \\
&= (\mathbf{x} - \sum_{j=1}^{l} x_j \mathbf{i}_j, x - \sum_{j=1}^{l} x_j \mathbf{i}_j) \\
&= (\mathbf{x}, \mathbf{x}) - 2 \sum_{j=1}^{l} x_j (\mathbf{x}, \mathbf{i}_j) + \sum_{j=1}^{l} \sum_{k=1}^{l} x_j x_k \delta_{jk} \\
&= (\mathbf{x}, \mathbf{x}) - \sum_{j=1}^{l} x_j^2.
\end{aligned}
$$

Thus

$$
(\mathbf{x}, \mathbf{x}) - \sum_{j=1}^{l} x_j^2 \geqslant 0. \tag{6.2.8}
$$

This is termed Bessel's inequality.

If $\mathbf{u}$ is any unit vector, it follows that

$$
(\mathbf{x}, \mathbf{x}) - \{(\mathbf{x}, \mathbf{u})\}^2 \geqslant 0
$$

and if $\mathbf{u}\|\mathbf{y}\| = \mathbf{y}$, it follows that

$$
\|\mathbf{x}\|^2 \|\mathbf{y}\|^2 - (\mathbf{x}, \mathbf{y})^2 \geqslant 0
$$

whence

$$
\|\mathbf{x}\| \|\mathbf{y}\| \geqslant (\mathbf{x}, \mathbf{y}) \tag{6.2.9}
$$

the so-called Schwarz inequality.

There are many equivalent conditions for an orthonormal set I to be complete. They are as follows.

(*a*) If $(\mathbf{x}, \mathbf{i}_j) = 0 \; 1 \leqslant j \leqslant n$, then $\mathbf{x} = \mathbf{0}$. (6.2.10a)

(*b*) The subspace spanned by I is the whole space. (6.2.10b)

(*c*)
$$
\mathbf{x} = \sum_{j=1}^{n} (\mathbf{x}, \mathbf{i}_j) \mathbf{i}_j. \tag{6.2.10c}
$$

(*d*)
$$
(\mathbf{x}, \mathbf{y}) = \sum_{j=1}^{n} (\mathbf{x}, \mathbf{i}_j)(\mathbf{y}, \mathbf{i}_j). \tag{6.2.10d}
$$

(*e*)
$$
\|\mathbf{x}\|^2 = \sum_{j=1}^{n} (\mathbf{x}, \mathbf{i}_j)^2. \tag{6.2.10e}
$$

The proof is as follows.

If $(\mathbf{x}, \mathbf{i}_j) = 0$ for all j and $\mathbf{x} \neq \mathbf{0}$, then the unit vector $\hat{\mathbf{x}}$ may be adjoined to the set of unit vectors. There would then be an orthonormal set greater than I. Thus if I is complete (6.2.10a) holds.

If there exists an $\mathbf{x}$ which is not a linear combination of the $\mathbf{i}_j$, it follows from (6.2.8) that

$$\mathbf{x}' = \mathbf{x} - \sum_{j=1}^{n} (\mathbf{x}, \mathbf{i}_j)\mathbf{i}_j \neq \mathbf{0}. \tag{6.2.11}$$

Also

$$(\mathbf{x}', \mathbf{i}_k) = x_k - \sum_{j=1}^{n} x_j \delta_{jk} = 0. \tag{6.2.12}$$

It follows therefore that (6.2.10a) implies (6.2.10b). As the subspace spans the whole space

$$\mathbf{x} = \sum_{j=1}^{n} x_j \mathbf{i}_j$$

and

$$(\mathbf{x}, \mathbf{i}_k) = \sum_{j=1}^{n} x_j(\mathbf{i}_j, \mathbf{i}_k) = \sum_{j=1}^{n} x_j \delta_{jk} = x_k. \tag{6.2.13}$$

Thus (6.2.10c) follows from (6.2.10b).

From (6.2.13)

$$\begin{aligned}(\mathbf{x}, \mathbf{y}) &= \left(\sum_{j=1}^{n} x_j \mathbf{i}_j, \sum_{k=1}^{n} y_k \mathbf{i}_k \right) \\ &= \sum_{j=1}^{n} \sum_{k=1}^{n} x_j y_k (\mathbf{i}_j, \mathbf{i}_k) \\ &= \sum_{j=1}^{n} \sum_{k=1}^{n} x_j y_k \delta_{jk} \\ &= \sum_{j=1}^{n} x_j y_j. \end{aligned} \tag{6.2.14}$$

Thus (6.2.10d) follows from (6.2.10c). (6.2.10e) follows by putting $\mathbf{y} = \mathbf{x}$ in (6.2.10d).

Suppose now that I is incomplete and that there exists a unit vector $\mathbf{u}$ orthogonal to each $\mathbf{i}_j$. Then

$$1 = \|\mathbf{u}\|^2 = \sum_{j=1}^{n} (\mathbf{u}, \mathbf{i}_j)^2 = 0.$$

Thus there is a contradiction and so (6.2.10e) implies completeness, and the following logical circle has been set up.

Completeness implies (6.2.10a) which implies (6.2.10b) which implies (6.2.10c) which implies (6.2.10d) which implies (6.2.10e) which implies completeness. That is completeness and the five conditions (6.2.10) are equivalent.

The concept of a distance is also associated with vector spaces. For a distance between the end-points of two vectors $\mathbf{x}$, $\mathbf{y}$, it is reasonable to expect the following properties.

$$d(\mathbf{x},\mathbf{y}) = d(\mathbf{y},\mathbf{x}) \tag{6.2.15a}$$
$$d(\mathbf{x},\mathbf{y}) \geqslant 0 \tag{6.2.15b}$$
$$d(\mathbf{x},\mathbf{y}) = 0 \text{ if and only if } \mathbf{x} = \mathbf{y} \tag{6.2.15c}$$
$$d(\mathbf{x},\mathbf{y}) = d(\mathbf{x}+\boldsymbol{\alpha},\mathbf{y}+\boldsymbol{\alpha}) \tag{6.2.15d}$$

(a change of origin being immaterial), and the so-called triangle in equality

$$d(\mathbf{x},\mathbf{y}) \leqslant d(\mathbf{x},\mathbf{z}) + d(\mathbf{y},\mathbf{z}). \tag{6.2.15e}$$

A suitable distance function is $\|\mathbf{x}-\mathbf{y}\|$. This clearly obeys all the properties (6.2.15a, b, c, d) and it remains to be shown that it obeys (6.2.15e) also.

$$\begin{aligned}
\|\mathbf{x}-\mathbf{y}\|^2 &= \|\mathbf{x}-\mathbf{z}-(\mathbf{y}-\mathbf{z})\|^2 \\
&= (\mathbf{x}-\mathbf{z}-(\mathbf{y}-\mathbf{z}), \mathbf{x}-\mathbf{z}-(\mathbf{y}-\mathbf{z})) \\
&= (\mathbf{x}-\mathbf{z}, \mathbf{x}-\mathbf{z}) + (\mathbf{y}-\mathbf{z}, \mathbf{y}-\mathbf{z}) + 2(\mathbf{x}-\mathbf{z}, \mathbf{z}-\mathbf{y}) \\
&= \|\mathbf{x}-\mathbf{z}\|^2 + \|\mathbf{y}-\mathbf{z}\|^2 + 2(\mathbf{x}-\mathbf{z}, \mathbf{z}-\mathbf{y}) \\
&\leqslant \|\mathbf{x}-\mathbf{z}\|^2 + \|\mathbf{y}-\mathbf{z}\|^2 + 2\|\mathbf{x}-\mathbf{z}\|\,\|\mathbf{z}-\mathbf{y}\|
\end{aligned}$$

by the Schwarz inequality (6.2.9).

Thus

$$\|\mathbf{x}-\mathbf{y}\| \leqslant \|\mathbf{x}-\mathbf{z}\| + \|\mathbf{y}-\mathbf{z}\|. \tag{6.2.16}$$

The angle θ between two vectors may be defined by

$$0 \leqslant \theta \leqslant \pi, \cos\theta = \frac{(\mathbf{x},\mathbf{y})}{\|\mathbf{x}\|\,\|\mathbf{y}\|} \tag{6.2.17}$$

and the Schwarz inequality is equivalent to

$$-1 \leqslant \cos\theta \leqslant 1.$$

It may be remarked that it is possible to consider more general vector spaces with complex components. In this case the inner produce property (6.2.3a) is replaced by $(\mathbf{x},\mathbf{y}) = (\mathbf{y},\mathbf{x})^*$, the asterisk denoting complex conjugate, $x_j y_j$ by $x_j y_j^*$ and x_j^2 by $|x_j|^2$. These spaces will not, however, be considered here.

6.21. *n dimensional Euclidean space*

By analogy with three-dimensional space, a vector $\mathbf{F}$ is defined as a set of numbers $(F_1, \dots F_n)$, $\mathbf{F}+\mathbf{G}$ is the set $(F_1+G_1, \dots, F_n+G_n)$ and $\alpha\mathbf{F}$ is the set $(\alpha F_1, \dots, \alpha F_n)$.

The set of vectors $\mathbf{i}_j$ $1 \leqslant j \leqslant n$ such that $\mathbf{i}_j . \mathbf{i}_k = \delta_{jk}$ is complete, and

$$\mathbf{F} = \sum_{j=1}^{n} F_j \mathbf{i}_j. \tag{6.21.1}$$

The inner product is

$$\mathbf{F}.\mathbf{G} = \sum_{j=1}^{n} F_j G_j \tag{6.21.2}$$

and the norm is

$$\|\mathbf{F}\| = \sqrt{\left(\sum_{j=1}^{n} F_j^2\right)}. \tag{6.21.3}$$

The angle between two vectors being given by

$$\cos\theta = \frac{\sum_{j=1}^{n} F_j G_j}{\sqrt{\left(\sum_{j=1}^{n} F_j^2 \sum_{k=1}^{n} G_k^2\right)}}. \tag{6.21.4}$$

A point $\mathbf{x}$ is given by

$$\mathbf{x} = \sum_{j=1}^{n} x_j \mathbf{i}_j.$$

The x_j may be said to be the coordinates of x.

A curve in n dimensions is given by the set of functions $x_j = x_j(u)$ where u is some parameter.

The elementary arc length along the curve is given by

$$(\delta s)^2 = \sum_{j=1}^{n} (\delta x_j)^2 = \sum_{j=1}^{n} \{x_j'(u)\}^2 (\delta u)^2. \tag{6.21.5}$$

The element of hypervolume is given by

$$d\tau_n = dx_1 \dots dx_n \tag{6.21.6}$$

and in a general system of coordinates by

$$d\tau_n = \frac{\partial(x_1, \dots x_n)}{\partial(u_1, \dots u_n)} (du_1 \dots du_n). \tag{6.21.7}$$

A gradient of a function $V(\mathbf{x}) \equiv V(x_1, \ldots x_n)$ may be defined as follows

$$\frac{\partial V}{\partial \mathbf{x}} = \sum_{j=1}^{n} \frac{\partial V}{\partial x_j} \mathbf{i}_j \tag{6.21.8}$$

and the divergence of a vector by

$$\frac{\partial}{\partial \mathbf{x}} . \mathbf{F} = \sum_{j=1}^{n} \frac{\partial F_j}{\partial x_j} . \tag{6.21.9}$$

A hypersurface is defined by a relation of the form

$$V(\mathbf{x}) = V(x_1, \ldots, x_n) = \text{constant.} \tag{6.21.10}$$

The unit normal is defined by

$$\mathbf{n} = \frac{\dfrac{\partial V}{\partial \mathbf{x}}}{\left\| \dfrac{\partial V}{\partial \mathbf{x}} \right\|} . \tag{6.21.11}$$

An element of the hypersurface has components

$$d\sigma_i = dx_1 \ldots dx_{i-1}\, dx_{i+1} \ldots dx_n \tag{6.21.12}$$

$$= \frac{d\tau_n}{dx_i} . \tag{6.21.13}$$

If now the scalar value of the element of surface is required

$$d\sigma^2 = \sum_{i=1}^{n} d\sigma_i^2 = \sum_{i=1}^{n} \left(\frac{d\tau_n}{dx_i} \right)^2 . \tag{6.21.14}$$

It is often convenient to write the equation of a hypersurface in the form

$$x_i = x_i(u_i, \ldots u_{n-1}) \tag{6.21.15}$$

where the u's are parameters. In this case

$$\begin{aligned} d\sigma_i &= \frac{\partial(x_1, x_2, \ldots x_{i-1}, x_{i+1}, \ldots x_n)}{\partial(u_1, \ldots, u_{n-1})} \\ &= A_i\, du_1 \ldots du_{n-1}. \end{aligned} \tag{6.21.16}$$

In this case

$$\begin{aligned} d\sigma^2 &= \sum_{i=1}^{n} d\sigma_i^2 \\ &= \sum_{i=1}^{n} A_i^2 (du_1 \ldots du_{n-1})^2 \end{aligned}$$

and

$$d\sigma = +\sqrt{\left(\sum_{i=1}^{n} A_i^2\right)}\, du_1 \ldots du_{n-1}. \qquad (6.21.17)$$

(Further reference may be made to D. M. Y. Sommerville, *Introduction to the Geometry of N Dimensions*, Dover, (1958).)

6.22. *The space L^2*

Let R be some range of integration over t. This may be for example part of a curve, a volume or a hypervolume. Let $x(t)$ be a function such that

$$\int_R |x(t)|^2\, dt$$

is finite.

In this section, x may be complex.

It may be noted that if $x_1(t)$, $x_2(t)$ differ at only a finite number of points by a finite amount, the integrals involving them will be the same, that is

$$\int_R x_1(t)\, y^*(t)\, dt = \int_R x_2(t)\, y^*(t)\, dt$$

and

$$\int_R |x_1(t)|^2\, dt = \int_R |x_2(t)|^2\, dt.$$

This possibility will, however, be excluded in what follows.

It is possible to consider a 'function vector' $\mathbf{x}$ which corresponds to the function $x(t)$. Axioms (6.2.1) and (6.2.2) will hold. For example if $x(t)$, $y(t)$ have the required properties, so also do $x(t) + y(t)$ and $\alpha x(t)$.

The inner product is defined by

$$(\mathbf{x}, \mathbf{y}) = \int_R x(t)\, y^*(t)\, dt = (\mathbf{y}, \mathbf{x})^* \qquad (6.22.1)$$

$$\|\mathbf{x}\| = (\mathbf{x}, \mathbf{x})^{1/2} = \int_R |x(t)|^2\, dt \qquad (6.22.2)$$

L^2 possesses an infinite number of dimensions because, in general, functions in L^2 cannot be represented by a linear combination of a finite number of functions. As a result convergence is involved.

Without going into great detail, if ϕ_n is a complete set of orthonormal functions such that

$$\int_R \phi_m(t)\,\phi_n^*(t)\,dt = \delta_{mn} \tag{6.22.3}$$

it can be shown that

$$x(t) = \sum_{n=1}^{\infty} (\mathbf{x}, \boldsymbol{\varphi}_n^*)\,\phi_n. \tag{6.22.4a}$$

This corresponds to (6.22.10c).

Similarly

$$(\mathbf{x}, \mathbf{y}) = \sum_{n=1}^{\infty} \int_R x(t)\,\phi_n^*(t)\,dt \int_R y^*(t)\,\phi_n(t)\,dt \tag{6.22.4b}$$

corresponding to (6.2.10d) and

$$\|\mathbf{x}\|^2 = \sum_{n=1}^{\infty} \left|\int_R x(t)\,\phi_n^*(t)\,dt\right|^2 \tag{6.22.4c}$$

corresponding to (6.2.10c), and provided that the series converge the formulae derived for finite spaces may be used. If $\mathbf{x}_n$ is a sequence of function vectors and if

$$\lim_{n\to\infty} \|\mathbf{x}_n - \mathbf{x}\| = 0. \tag{6.22.5a}$$

$\mathbf{x}_n$ is said to converge strongly, or converge in mean to $\mathbf{x}$. This is written $\mathbf{x}_n \to x$. This condition can be shown, by use of the triangle equality, to be equivalent to the so-called Riesz-Fischer condition

$$\lim_{n,\,m\to\infty} \|\mathbf{x}_m - \mathbf{x}_n\| = 0. \tag{6.22.5b}$$

A weaker form of convergence is possible.

If $\mathbf{x}_n \to \mathbf{x}$, then by the Schwarz equality

$$|(\mathbf{x}_n - \mathbf{x}, \mathbf{y})| \leqslant \|\mathbf{x} - \mathbf{x}_n\|\,\|\mathbf{y}\|$$

and so

$$\lim_{n\to\infty} (\mathbf{x}_n - \mathbf{x}, \mathbf{y}) = 0. \tag{6.22.6}$$

If this holds for all $\mathbf{y}$, $\mathbf{x}_n$ converges weakly to $\mathbf{x}$. Weak convergence does not, however, imply strong convergence. The expansion of a function $x(t)$ in L^2 in terms of a complete orthonormal set may be looked at in another way. The coefficients $x_n = (x, \phi_n^*)$ of $x(t)$ can be thought of as coordinates of a point in an infinite dimensional

Euclidean space. Such a space is termed a Hilbert space. The completeness of such a space is defined through conditions such as (6.2.10), the summation to n being replaced by a summation to infinity. (See further, J. L. Synge, *The Hypercircle in Mathematical Physics*, Cambridge Univ. Press, (1957), and J. von Neumann, *Mathematical Principles of Quantum Mechanics*, Princeton Univ. Press, (1955).)

6.3. **Orthogonal transformations and tensors**

Consider a Euclidean space of n dimensions with unit vectors $\mathbf{i}_\alpha$ along the axes ($\mathbf{i}_\alpha . \mathbf{i}_\beta = \delta_{\alpha\beta}$). ($n$ is 3 when ordinary space is under consideration.)

If $\mathbf{C}$ is any vector

$$\mathbf{C} = \sum_{\alpha=1}^{n} C_\alpha \mathbf{i}_\alpha. \tag{6.3.1}$$

C_α are the components of $\mathbf{C}$.

It will be convenient from now on to leave out the summation sign and to use the summation convention that repeated small Greek suffices imply a summation from 1 to n. Thus

$$\mathbf{C} = C_\alpha \mathbf{i}_\alpha. \tag{6.3.2}$$

Suppose now that there is a rotation of axes and that the new unit vectors are $\mathbf{i}'_\beta$

$$\mathbf{C} = C'_\beta \mathbf{i}'_\beta. \tag{6.3.3}$$

C'_β are the components of $\mathbf{C}$ in the new frame of reference.

$$C'_\beta = \mathbf{C} . \mathbf{i}'_\beta = C_\alpha \mathbf{i}_\alpha . \mathbf{i}'_\beta = C_\alpha a_{\alpha\beta} \tag{6.3.4}$$

where

$$a_{\alpha\beta} = \mathbf{i}_\alpha . \mathbf{i}'_\beta. \tag{6.3.5}$$

(6.3.4) is a linear transformation of C_α.

Also

$$C_\alpha = \mathbf{C} . \mathbf{i}_\alpha = C_\gamma \mathbf{i}'_\gamma . \mathbf{i}_\alpha = C'_\gamma a_{\alpha\gamma} \tag{6.3.6}$$

and so

$$C'_\beta = C_\alpha a_{\alpha\beta} = C'_\gamma a_{\alpha\gamma} a_{\alpha\beta} = C'_\gamma \delta_{\beta\gamma}.$$

Thus

$$a_{\alpha\gamma} a_{\alpha\beta} = \delta_{\beta\gamma} \tag{6.3.7}$$

and so

$$|a_{ij}|^2 = 1. \tag{6.3.8a}$$

The transformation corresponds to a rotation if

$$|a_{ij}| = -1. \tag{6.3.8b}$$

Thus if n scalars $C_1, \ldots C_n$ represent a vector, it is necessary that they transform in this manner. In particular this holds for $\mathbf{x}$ the vector of the representative point. $x_\alpha = a_{\alpha\beta} x'_\beta$.

The inner produce is an invariant or scalar for

$$\begin{aligned}(\mathbf{C}, \mathbf{D})' &= C'_\alpha D'_\alpha = C_\beta a_{\beta\alpha} D_\lambda a_{\lambda\alpha} \\ &= C_\beta D_\lambda \delta_{\beta\lambda} = C_\beta D_\beta = (\mathbf{C}, \mathbf{D}).\end{aligned} \tag{6.3.9}$$

The same property holds for $C^2 = (\mathbf{C}, \mathbf{C})$.

Let U be some scalar function of $\mathbf{x}$. That is, U is a scalar field defined within the space.

Let

$$C_\alpha = \frac{\partial U}{\partial x_\alpha} \tag{6.3.10}$$

$$C'_\beta = \frac{\partial U}{\partial x'_\beta} = \frac{\partial U}{\partial x_\alpha}\frac{\partial x_\alpha}{\partial x'_\alpha} = \frac{\partial U}{\partial x_\alpha} a_{\alpha\beta} = C_\alpha a_{\alpha\beta}. \tag{6.3.11}$$

Thus

$$C_\alpha = \frac{\partial U}{\partial x_\alpha}$$

is a vector, the transformation law being obeyed.

It is sometimes convenient to indicate differentiation with respect x_α by a comma

$$\frac{\partial U}{\partial x_\alpha} = U_{,\alpha}.$$

The quantity $\partial C_\alpha / \partial x_\alpha$ is a scalar for

$$\begin{aligned}U' &= \frac{\partial C'_\beta}{\partial x'_\beta} = \frac{\partial C_\alpha}{\partial x'_\beta} a_{\alpha\beta} = \frac{\partial C_\alpha}{\partial x_\gamma}\frac{\partial x_\gamma}{\partial x'_\beta} a_{\alpha\beta} \\ &= \frac{\partial C_\alpha}{\partial x_\gamma} a_{\gamma\beta} a_{\alpha\beta} = \frac{\partial C_\alpha}{\partial x_\gamma}\delta_{\gamma\alpha} = \frac{\partial C_\alpha}{\partial x_\alpha} = U\end{aligned} \tag{6.3.12}$$

Similarly

$$\frac{\partial^2 U}{\partial x_\alpha^2} = \frac{\partial^2 U}{\partial x_\alpha \partial x_\alpha} \tag{6.3.13}$$

is also an invariant. (6.3.13)

Let $\mathbf{B}$ be a linear vector function of $\mathbf{A}$.

$$B_\alpha = T_{\alpha\beta} A_\beta.$$

The $T_{\alpha\beta}$ are associated in three dimensions with a dyadic. If they are such that on a transformation of coordinates they transform so that **A** continues to transform into **B**, they are said to form a tensor of rank 2, there being two indices. (A vector is a tensor of rank 1 and a scalar a tensor of rank zero.)

$$\begin{aligned} B'_\gamma &= a_{\gamma\mu} B_\mu = a_{\gamma\mu} T_{\delta\lambda} A_\lambda \\ &= a_{\gamma\mu} T_{\mu\lambda} a_{\lambda\delta} A'_\delta \end{aligned} \tag{6.3.14}$$

whence

$$T'_{\gamma\delta} = a_{\gamma\mu} T_{\mu\lambda} a_{\lambda\delta}. \tag{6.3.15}$$

If $T_{\alpha\beta} = T_{\beta\alpha}$ the tensor is symmetric and if $T_{\alpha\beta} + T_{\beta\alpha} = 0$, it is antisymmetric. Also

$$T_{\alpha\beta} = \tfrac{1}{2}(T_{\alpha\beta} + T_{\beta\alpha}) + \tfrac{1}{2}(T_{\alpha\beta} - T_{\beta\alpha}) \tag{6.3.16}$$

and so an arbitrary tensor of rank 2 may be split into a symmetric and an antisymmetric part.

The concept of tensor may be extended to a set of n^s quantities with s suffices, appropriate transformation laws being obeyed. A tensor may be symmetric or antisymmetric with respect to two of its indices.

It is possible to define a generalization of a vector product as follows.

Let $\epsilon_{\alpha\beta\gamma}$ be a function such that

$$\begin{aligned} &\epsilon_{\alpha\beta\gamma} = 0 \text{ if any of } \alpha, \beta, \gamma \text{ are equal} \\ &\epsilon_{\alpha\beta\gamma} = 1 \text{ if } \alpha, \beta, \gamma \text{ form an even permutation of } p, q, r \\ &\epsilon_{\alpha\beta\gamma} = -1 \text{ if } \alpha, \beta, \gamma \text{ form an odd permutation of } p, q, r \end{aligned} \tag{6.3.17}$$

Then

$$\epsilon_{\alpha\beta\gamma} = -\epsilon_{\alpha\gamma\beta}.$$

where p, q, r are integers such that $1 \leqslant p < q < r \leqslant n$.

Consider the quantity

$$C_\alpha = \epsilon_{\alpha\beta\gamma} A_\beta B_\gamma. \tag{6.3.18}$$

In three dimensions the suffices run from 1 to 3 only and (6.3.18) is equivalent to

$$\begin{aligned} C_1 &= A_2 B_3 - A_3 B_2 \\ C_2 &= A_3 B_1 - A_1 B_3 \\ C_3 &= A_1 B_2 - B_2 A_1. \end{aligned} \tag{6.3.19}$$

If only right-handed frames are used, C_α behaves as a vector. If a transformation to a left-handed frame is made, a reflection is involved and C_α does not behave as a true vector, and is in fact an axial vector. A generalization of the curl is given by

$$\epsilon_{\alpha\beta\gamma}\frac{\partial C_\gamma}{\partial x_\beta}. \tag{6.3.20}$$

This also is an axial vector.

Consider now the number of terms involved in a vector product. The quantity

$$T_{\beta\gamma} = A_\beta B_\gamma - A_\gamma B_\beta \tag{6.3.21}$$

has $\frac{1}{2}(n^2 - n)$ independent terms.

On the other hand

$$C_\alpha = \epsilon_{\alpha\beta\gamma} T_{\beta\gamma} \tag{6.3.22}$$

has n independent terms.

Both $T_{\beta\gamma}$ and C_α are different generalizations of the vector product. It is only when $n = 3$ that $\frac{1}{2}(n^2 - n) = n$ and the number of terms is the same.

6.31. *Stokes's theorem and the Gauss divergence theorem*

Stokes's theorem (5.1.1) may be written in tensor form as

$$\int_{\mathscr{C}} dx_\alpha F_\alpha = \int_S d\mathscr{A}\, n_\alpha \epsilon_{\alpha\beta\gamma}\frac{\partial F_\gamma}{\partial x_\beta}. \tag{6.31.1}$$

n_α is the component of the unit vector.

The Gauss divergence theorem (5.2.1) may be written as

$$\int_V \frac{\partial F_\alpha}{\partial x_\alpha}\, d\tau = \int_S F_\alpha n_\alpha\, d\mathscr{A} \tag{6.31.2}$$

These may be extended to deal with tensors of rank higher than unity in the same way as the extensions to dyadics were proved.

For example

$$\int_S n_\alpha G_{\alpha\beta\gamma}\, d\mathscr{A} = \int_V \frac{\partial G_{\alpha\beta\gamma}}{\partial x_\alpha}\, d\tau. \tag{6.31.3}$$

Let $G_{\alpha\beta\gamma} a_\beta b_\gamma = F_\alpha a_\beta$, b_γ be arbitrary constants. Then (6.32.2) becomes

$$\int_S a_\beta b_\gamma G_{\alpha\beta\gamma} n_\alpha\, d\mathscr{A} = \int_V a_\beta b_\gamma \frac{\partial G_{\alpha\beta\gamma}}{\partial x_\alpha}\, d\tau$$

and as a_β, b_γ are arbitrary, the result follows.

The theorems may also be extended to a higher number of dimensions.

6.32. *Covariance and contravariance*

The element of distance in Euclidean space is given by

$$ds^2 = dx_\mu \, dx_\mu. \tag{6.32.1}$$

Suppose now that, instead of using cartesian coordinates, an arbitrary system is used involving n coordinates u^α

$$ds^2 = \frac{\partial x_\mu}{\partial u^\alpha} du^\alpha \frac{\partial x_\mu}{\partial u^\beta} du^\beta$$

$$= g_{\alpha\beta} \, du^\alpha \, du^\beta \tag{6.32.2}$$

where

$$g_{\alpha\beta} = \frac{\partial x_\mu}{\partial u^\alpha} \frac{\partial x_\mu}{\partial u^\beta}. \tag{6.32.3}$$

The $_{\alpha\beta}$ are dependent upon the u. By definition $g_{\alpha\beta} = g_{\beta\alpha}$. Systems of coordinates such that $g_{\alpha\beta} = 0$ $(\alpha \neq \beta)$ are said to be orthogonal.

Suppose now that the set u^α represent the coordinates of a point and that $\bar{u}^\beta$ form another set of coordinates

$$d\bar{u}^\beta = \frac{\partial \bar{u}^\beta}{\partial u^\alpha} du^\alpha. \tag{6.32.4}$$

The n quantities du^α are taken to be the components of the elementary displacement vector in the first frame of reference. The components in the second frame are $d\bar{u}_\beta$. This elementary displacement vector is the prototype of contravariant vectors. A^α is a contravariant vector if it transforms according to the law

$$\bar{A}^\beta = \frac{\partial \bar{u}^\beta}{\partial u^\alpha} A^\alpha. \tag{6.32.5}$$

Now $\partial \bar{u}^\beta / \partial u^\alpha$ depends upon the values of the u (or $\bar{u}$) and so the position of A^α must be known before its law of transformation may be determined.

Consider now a scalar or invariant ψ. For this the transformation law is obviously

$$\bar{\psi} = \psi. \tag{6.32.6}$$

Now

$$\frac{\partial \bar{\psi}}{\partial \bar{u}^\alpha} = \frac{\partial \bar{\psi}}{\partial u_\beta} \frac{\partial u^\beta}{\partial \bar{u}^\alpha} = \frac{\partial \psi}{\partial u_\beta} \frac{\partial u^\beta}{\partial \bar{u}_\alpha} \tag{6.32.7}$$

The transformation law of $\partial\psi/\partial u^\beta$ is thus not the same as that of a contravariant vector. Vectors such as B_α which transform according to the law

$$\bar{B}_\alpha = \frac{\partial u^\beta}{\partial u_\alpha} B_\beta \tag{6.32.8}$$

are termed covariant vectors.

Covariant vectors are written with a suffix and contravariant vectors are written with a superfix. When rectangular cartesian coordinates are used, there is no difference, as

$$\bar{x}_\alpha = a_{\alpha\beta} x_\beta \qquad \text{and} \qquad x_\beta = a_{\alpha\beta} \bar{x}_\alpha$$

and both types of vector transform in a similar manner.

It is possible to construct tensors such as

$$A_{\alpha\mu} = a_\mu b_\alpha, \qquad B_\nu^{\mu\lambda} = c_\nu d^\mu e^\lambda$$

and so on. Tensors such as $B_\gamma^{\mu\lambda}$ are said to be mixed.

Contraction is said to take place when there is summation over a superscript and a subscript at the same time, e.g.

$$A_\alpha^\alpha = \sum_{\alpha=1}^{n} A_\alpha^\alpha$$

(Further details of tensor theory are given, for example, in D. F. Lawden *Introduction to Tensor Calculus and Relativity*, Methuen London (1967).)

6.4. Relations with matrices

Many of the formulae of this book, may be rewritten as formulae involving matrices. (For the details and definitions of matrix algebra see e.g. W. G. Bickley and R. S. H. G. Thomson, English Universities Press, 1964.)

6.41. *Vectors and matrices*

An N dimensional vector **a** (including $N = 3$ as a particular case) with components a_r $1 \leqslant r \leqslant N$, may be regarded either as a column matrix

$$a^* = \begin{bmatrix} a_1 \\ \cdot \\ \cdot \\ \cdot \\ a_N \end{bmatrix}$$

or as a row matrix

$$a = [a_1, \dots a_N].$$

The matrix relation

$$a + b = c \tag{6.41.1}$$

is, written in full,

$$[a_1, \dots a_N] + [b_1, \dots b_N] = [c_1, \dots c_N] \tag{6.41.2}$$

where

$$c_r = a_r + b_r \qquad (1 \leqslant r \leqslant N). \tag{6.41.3}$$

Similarly $\alpha\mathbf{a}$ is equivalent to the matrix $\alpha a = [\alpha a_1, \dots \alpha a_N]$.

If a, and b are row matrices, then the scalar product $\mathbf{a} \cdot \mathbf{b}$ is the value of the single element in the 1×1 matrix

$$\left[\sum_{r=1}^{N} a_r b_r\right]. \tag{6.41.4}$$

This being the matrix which is the product either of ab^* or ba^*, the asterisk denoting the transpose. Similar remarks apply for the column matrix representation.

6.42. *Complex numbers and matrices*

It may easily be verified that, if x, y are real numbers, the algebra of square matrices of the form

$$\begin{bmatrix} x, & y \\ -y, & x \end{bmatrix}$$

is identical with the algebra of complex numbers $x + iy$. This is because

$$\begin{bmatrix} x, & y \\ -y, & x \end{bmatrix} = x\begin{bmatrix} 1, & 0 \\ 0, & 1 \end{bmatrix} + y\begin{bmatrix} 0, & 1 \\ -1, & 0 \end{bmatrix}, \tag{6.42.1}$$

$$\begin{bmatrix} 1, & 0 \\ 0, & 1 \end{bmatrix} \begin{bmatrix} 1, & 0 \\ 0, & 1 \end{bmatrix} = \begin{bmatrix} 1, & 0 \\ 0, & 1 \end{bmatrix} \tag{6.42.2}$$

and

$$\begin{bmatrix} 0, & 1 \\ -1, & 0 \end{bmatrix} \begin{bmatrix} 0, & 1 \\ -1 & 0 \end{bmatrix} = \begin{bmatrix} -1, & 0 \\ 0, & -1 \end{bmatrix} = -\begin{bmatrix} 1, & 0 \\ 0, & 1 \end{bmatrix}. \tag{6.42.3}$$

The matrix

$$\begin{bmatrix} 1, & 0 \\ 0, & 1 \end{bmatrix}$$

corresponds to 1, and the matrix

$$\begin{bmatrix} 0, & 1 \\ -1, & 0 \end{bmatrix}$$

to i.

6.43. *Matrices and dyads*

Consider the 3×3 matrix defined by

$$C = \begin{bmatrix} a_x b_x, & a_x b_y, & a_x b_z \\ a_y b_x, & a_y b_y, & a_y b_z \\ a_z b_x, & a_z b_y, & a_z b_z \end{bmatrix} \tag{6.43.1}$$

Let F be a column matrix with components F_x, F_y, F_z. Then

$$CF = \begin{bmatrix} a_x b_x, & a_x b_y, & a_x b_z \\ a_y b_x, & a_y b_y, & a_y b_z \\ a_z b_x, & a_y b_z, & a_z b_z \end{bmatrix} \begin{bmatrix} F_x \\ F_y \\ F_z \end{bmatrix}$$

$$= \begin{bmatrix} a_x(b_x F_x + b_y F_y + b_z F_z) \\ a_y(b_x F_x + b_y F_y + b_z F_z) \\ a_z(b_x F_x + b_y F_y + b_z F_z) \end{bmatrix}$$

$$= \begin{bmatrix} a_x \\ a_y \\ a_z \end{bmatrix} (\mathbf{b}.\mathbf{F}) \tag{6.43.2}$$

This is analogous with the property of dyads $(\mathbf{ab}).\mathbf{F} = \mathbf{a}(\mathbf{b}.\mathbf{F})$.

Similarly

$$F^* C = [b_x, b_y, b_z](\mathbf{a}.\mathbf{F}) \tag{6.43.3}$$

which corresponds to

$$\mathbf{F}.(\mathbf{ab}) = (\mathbf{F}.\mathbf{a})\mathbf{b}$$

The process discussed in section 6.13 for finding a scalar λ and vector $\mathbf{F}$ for a dyadic $\boldsymbol{\Phi}$ such that

$$\boldsymbol{\Phi}.\mathbf{F} = \lambda \mathbf{F} \tag{6.43.4}$$

is clearly the same as that of finding the eigenvalues of the matrix $[\Phi_{rs}]$ which is associated with the dyadic $\boldsymbol{\Phi}$ and section 6.13 may be translated into matrix notation without any difficulty.

6.44. *Tensors and matrices*

Consider a space in n dimensions, and suppose that a vector $\mathbf{C}$ can be written as $C_\alpha \mathbf{i}_\alpha$ with the usual notation. Suppose that there are two successive rotations. After the first rotation

$$\mathbf{C} = C''_\beta \mathbf{i}'_\beta \tag{6.44.1}$$

and after the second rotation

$$\mathbf{C} = C''_\gamma \mathbf{i}''_\beta. \tag{6.44.2}$$

Let

$$a_{\alpha\beta} = \mathbf{i}_\alpha . \mathbf{i}'_\beta, \qquad b_{\beta\gamma} = \mathbf{i}'_\beta . \mathbf{i}''_\beta.$$

Then

$$C'_\beta = C_\alpha a_{\alpha\beta} \tag{6.44.3}$$

$$C''_\gamma = C'_\beta b_{\beta\gamma} \tag{6.44.4}$$

as has been indicated in section 6.3.

Now

$$C''_\gamma = C'_\beta b_{\beta\gamma} = C_\alpha a_{\alpha\beta} b_{\beta\gamma} = C_\alpha f_{\alpha\gamma} \tag{6.44.5}$$

where

$$f_{\alpha\gamma} = a_{\alpha\beta} b_{\beta\gamma}. \tag{6.44.6}$$

$a_{\alpha\beta} b_{\beta\gamma}$ is the coefficient in the α'th column and the γ'th row of the matrix which is the product of $[a_{rs}][b_{rs}]$. Thus, a tensor of rank 2 is analogous with a square matrix. Equations (6.44.3), (6.44.4), (6.44.5), and (6.44.6) may be written in matrix form as

$$C'^* = C^* A \qquad C''^* = C'^* B$$

$$C''^* = C^* AB = C^* F$$

where

$$F = AB. \tag{6.44.7}$$

Thus a tensor arises from generalizing the ideas associated with a square matrix. A scalar is a tensor of rank zero, a vector is a tensor of rank unity, and a square matrix is a tensor of rank two.

Worked examples

1. Express the formula for the expansion of the vector triple product in terms of dyadics:

$$(\mathbf{P} \times \mathbf{Q}) \times \mathbf{R} = \mathbf{Q}(\mathbf{P}.\mathbf{R}) - \mathbf{P}(\mathbf{Q}.\mathbf{R}).$$
$$= \mathbf{Q}.\mathbf{\Psi} - \mathbf{P}.\mathbf{\Phi}$$

where $\mathbf{\Psi} = \mathbf{PR}$, $\mathbf{\Phi} = \mathbf{QR}$.

2. Express the general equation of the quadric in dyadic form. The general quadric is

$$ax^2 + by^2 + cz^2 + 2fyz + 2gzx + 2hxy + px + qy + rz + d = 0.$$

The quadratic terms may be written as

$$(x\mathbf{i} + y\mathbf{j} + z\mathbf{k}).[a\mathbf{ii} + b\mathbf{jj} + c\mathbf{kk} + f(\mathbf{jk} + \mathbf{kj}) + g(\mathbf{ki} + \mathbf{ik}) + h(\mathbf{ij} + \mathbf{ji})].(x\mathbf{i}$$

$$+ y\mathbf{j} + z\mathbf{k}) = \mathbf{r}.\boldsymbol{\Omega}_0.\mathbf{r}.$$

The linear term may be written as

$$2(p\mathbf{i} + q\mathbf{j} + r\mathbf{k}).(x\mathbf{i} + y\mathbf{j} + z\mathbf{k}) = 2\mathbf{S}.\mathbf{r}.$$

The general quadric may thus be written in the form

$$\mathbf{r}.\boldsymbol{\Omega}_0.\mathbf{r} + 2\mathbf{S}.\mathbf{r} + d = 0.$$

It is not in fact essential for the dyadic to be symmetric, for if $\boldsymbol{\Omega}^*$ is any antisymmetric dyadic

$$\mathbf{r}.\boldsymbol{\Omega}^*.\mathbf{r} = \mathbf{r}.(\tfrac{1}{2}\boldsymbol{\Omega}_v\mathbf{r}) = 0$$

and so the general quadric may be written as

$$\mathbf{r}.\boldsymbol{\Omega}_0.\mathbf{r} + 2\mathbf{S}.\mathbf{r} + d = 0$$

where $\boldsymbol{\Omega} = \boldsymbol{\Omega}_0 + \boldsymbol{\Omega}^*$.

3. The body force on a medium is such that the force across any element of surface $\delta\mathscr{A}$ is $\mathbf{M}.\delta\mathscr{A}$. Show that this is equivalent to a certain distribution of force per unit volume in the medium.

The total force acting on a volume V within a surface S is

$$-\int_S \mathbf{M}.d\mathscr{A}.$$

By an extension of the Gauss divergence theorem this is

$$\int_V -(\nabla.\mathbf{M})\,d\tau.$$

Thus the equivalent force per unit volume is $-\nabla.\mathbf{M}$.

It will be seen that if the stress dyadic $\mathbf{M}$ is antisymmetric $\nabla.\mathbf{M}$ is zero, and so there is no equivalent force per unit volume. Conversely, for a given force per unit volume, it is always possible to assume that the equivalent stress tensor is symmetric.

4. Prove that, the moment of inertia about the line through O with direction specified by the unit vector $\mathbf{u}$ may be written in the form $\mathbf{u}.\boldsymbol{\Psi}.\mathbf{u}$.

The perpendicular from a point with position vector $\mathbf{r}_i$ to the line through

O with direction specified by unit vector $\mathbf{u}$ is given by $p_i = |\mathbf{u} \times \mathbf{r}_i|$. The moment of inertia about the line is given by

$$\sum_i m_i p_i^2 = \sum_i m_i(\mathbf{u} \times \mathbf{r}_i).(\mathbf{u} \times \mathbf{r}_i)$$

$$= \mathbf{u}.\left\{\sum_i m_i \mathbf{r}_i \times (\mathbf{u} \times \mathbf{r}_i)\right\}$$

$$= \mathbf{u}.\sum_i m_i[r_i^2 \mathbf{u} - \mathbf{r}_i(\mathbf{r}_i.\mathbf{u})]$$

$$= \mathbf{u}.\psi(\mathbf{u})$$

where $\psi(\mathbf{u})$ is a linear vector function of $\mathbf{u}$. Now by section 6.1, a linear vector function of $\mathbf{u}$ can be written as $\boldsymbol{\Psi}.\mathbf{u}$. In this case

$$\boldsymbol{\Psi} = A\mathbf{ii} + B\mathbf{jj} + C\mathbf{kk} - D(\mathbf{jk} + \mathbf{kj}) - E(\mathbf{ki} + \mathbf{ik}) - F(\mathbf{ij} + \mathbf{ji})$$

where

$$A = \sum_i m_i(y_i^2 + z_i^2)$$

$$D = \sum_i m y_i z_i, \text{ etc.}$$

By section 6.13, it is possible by a change of axes to write $\boldsymbol{\Psi}$ in the form

$$\boldsymbol{\Psi} = A\mathbf{ii} + B\mathbf{jj} + C\mathbf{kk}.$$

In this case A, B, C are the principal moments of inertia and the new axes are the principal axes.

5. Prove that

$$\nabla.(\mathbf{PQ}) = (\nabla.\mathbf{P})\mathbf{Q} + (\mathbf{P}.\nabla)\mathbf{Q}.$$

Let $\mathbf{A}$ be a constant vector

$$\nabla.(\mathbf{PQ}).\mathbf{A} = \nabla.(\mathbf{P}\psi)$$

where

$$\begin{aligned}\psi &= \mathbf{Q}.\mathbf{A}\\ &= (\nabla.\mathbf{P})\psi + (\mathbf{P}.\nabla)\psi\\ &= (\nabla.\mathbf{P})(\mathbf{Q}.\mathbf{A}) + (\mathbf{P}.\nabla)\mathbf{Q}.\mathbf{A}.\end{aligned}$$

$\mathbf{A}$ is arbitrary and so

$$\nabla.(\mathbf{PQ}) = (\nabla.\mathbf{P})\mathbf{Q} + (\mathbf{P}.\nabla)\mathbf{Q}.$$

6. The force per unit volume on an electric charge distribution is given by

$$\rho\mathbf{E} + \mathbf{J} \times \mathbf{B}$$

where ρ is the electric charge density, $\mathbf{E}$ the electric field, $\mathbf{J}$ the electric current density and $\mathbf{B}$ the magnetic induction. Find the equivalent stress dyadic.

The force over a volume V is given by

$$\int_V (\rho\mathbf{E} + \mathbf{J} \times \mathbf{B})\, d\tau$$

and using Maxwell's equations (chapter 3, example 12), this becomes

$$\int_V \left\{(\nabla.\mathbf{D})\mathbf{E} + \left(\nabla \times \mathbf{H} - \frac{\partial \mathbf{D}}{\partial t}\right) \times \mathbf{B}\right\} d\tau = \int_V \left\{(\nabla.\mathbf{D})\mathbf{E} + (\nabla \times \mathbf{H}) \times \mathbf{B} + \mathbf{D} \times \frac{\partial \mathbf{B}}{\partial t} - \frac{\partial}{\partial t}(\mathbf{D} \times \mathbf{B})\right\} d\tau$$

where $\mathbf{D}$ is the electric displacement and $\mathbf{H}$ is the magnetic field. Now

$$\frac{\partial \mathbf{B}}{\partial t} = -\nabla \times \mathbf{E}$$

and so the integral becomes

$$\int_V \left\{(\nabla.\mathbf{D})\mathbf{E} + (\nabla \times \mathbf{H}) \times \mathbf{B} - \mathbf{D} \times (\nabla \times \mathbf{E}) - \frac{\partial}{\partial t}(\mathbf{D} \times \mathbf{B})\right\} d\tau.$$

Now $\mathbf{D} = \epsilon\mathbf{E}$, $\mathbf{B} = \mu\mathbf{H}$ and also $\nabla.\mathbf{B} = 0$, so this can be written

$$\int_V \epsilon\{(\nabla.\mathbf{E})\mathbf{E} - \mathbf{E} \times (\nabla \times \mathbf{E})\} + \mu\{(\nabla.\mathbf{H})\mathbf{H} - \mathbf{H} \times (\nabla \times \mathbf{H})\}\, d\tau - \frac{\partial}{\partial t}\int_V (\mathbf{D} \times \mathbf{B})\, d\tau.$$

The first integral can be written as

$$\int_V [\epsilon\{(\nabla.\mathbf{E})\mathbf{E} + (\mathbf{E}.\nabla)\mathbf{E} - \tfrac{1}{2}\nabla E^2\} + \mu\{(\nabla.\mathbf{H})\mathbf{H} + (\mathbf{H}.\nabla)\mathbf{H} - \tfrac{1}{2}\nabla H^2\}]\, d\tau$$

$$= \int_V \epsilon\nabla(\mathbf{EE}) + \mu\nabla(\mathbf{HH}) - \tfrac{1}{2}\nabla(\epsilon E^2 + \mu H^2)\, d\tau$$

$$= \int_V \nabla\{\epsilon\mathbf{EE} + \mu\mathbf{HH} - \tfrac{1}{2}\mathbf{I}(\epsilon E^2 + \mu H^2)\}\, d\tau$$

$$= \int_S \mathbf{\Psi}.d\mathscr{A}$$

where $\mathbf{\Psi} = \mathbf{DE} + \mathbf{BH} - \tfrac{1}{2}\mathbf{I}(\mathbf{D}.\mathbf{E} + \mathbf{B}.\mathbf{H})$.

This stress dyadic is symmetric. It does not, however, represent an actual stress. All that can be said is that the force system is consistent with this stress.

The quantity

$$\int_V (\mathbf{D} \times \mathbf{B})\, d\tau$$

is termed the electromagnetic momentum and may be regarded as the momentum associated with the electromagnetic field. If this interpretation be adopted, the electromagnetic stresses over the surface equal the electromagnetic forces on the charges and currents and the rate of increase of the electromagnetic momentum within the surface.

7. The equation of a quadric is

$$\mathbf{r}.\mathbf{\Omega}.\mathbf{r} + 2\mathbf{s}.\mathbf{r} + d = 0.$$

Find the equation of the tangent plane at the point with position vector $\mathbf{r}_0$.

Let $\mathbf{r}_0 + \mathbf{a}$ be the position vector of a neighbouring point on the quadric.

$$(\mathbf{r}_0 + \mathbf{a}).\mathbf{\Omega}.(\mathbf{r}_0 + \mathbf{a}) + 2\mathbf{s}.(\mathbf{r}_0 + \mathbf{a}) + d = 0.$$

Now as $\mathbf{a}$ is small this becomes

$$\mathbf{r}_0.\mathbf{\Omega}.\mathbf{r}_0 + \mathbf{r}_0.\mathbf{\Omega}.\mathbf{a} + \mathbf{a}.\mathbf{\Omega}.\mathbf{r}_0 + 2\mathbf{s}.\mathbf{r}_0 + 2\mathbf{s}.\mathbf{a} + d = 0.$$

$\mathbf{r}_0$ is on the quadric and $\mathbf{\Omega}$ is symmetric. It follows that

$$\mathbf{a}.(\mathbf{\Omega}.\mathbf{r}_0 + s) = 0.$$

If now $\mathbf{r}$ is the position vector of a point on the tangent plane, it follows that $(\mathbf{r} - \mathbf{r}_0).(\mathbf{\Omega}.\mathbf{r}_0 + \mathbf{s}) = 0$ is the equation of the largest plane. This is $\mathbf{r}.\mathbf{\Omega}.\mathbf{r}_0 - \mathbf{s}.\mathbf{r}_0 - \mathbf{r}_0.\mathbf{\Omega}.\mathbf{r}_0 + \mathbf{s}.\mathbf{r}_0 = 0$, and as $\mathbf{r}_0$ lies on the quadric, this is

$$\mathbf{r}.\mathbf{\Omega}.\mathbf{r}_0 + \mathbf{s}.(\mathbf{r} + \mathbf{r}_0) + d = 0.$$

8. $\mathbf{\Psi} = \mathbf{ii} - \mathbf{jj} + 2\mathbf{kk} + 2\mathbf{ji} + \mathbf{ki} + 4\mathbf{ij} - 3\mathbf{kj} + 3\mathbf{jk} + \mathbf{ik}$. Find the eigenvalues, eigenvectors and invariants.

To obtain the eigenvalues and eigenvectors, $\mathbf{\Psi}.\mathbf{r} = \lambda\mathbf{r}$, that is

$$(x + 2y + z)\mathbf{i} + (4x - y - 3z)\mathbf{j} + (x + 3y + 2z)\mathbf{k} = \lambda(x\mathbf{i} + y\mathbf{j} + z\mathbf{k}).$$

Thus the eigenvalue equations are given by

$$\begin{aligned} (1 - \lambda)x + 2y + z &= 0 \\ 4x - (1 + \lambda)y - 3z &= 0 \\ x + 3y + (2 - \lambda)z &= 0. \end{aligned}$$

The invariants are given by

$$\phi_1 = 1 - 1 + 2 = 2$$

$$\phi_2 = (1.-1) + (2.-1) + 1.2 - (2.4 + 1.1 + 3.-3) = -1$$

$$\phi_3 = \begin{vmatrix} 1 & 2 & 1 \\ 4 & -1 & -3 \\ 1 & 3 & 2 \end{vmatrix} = -2$$

The eigenvalue equation is

$$0 = \lambda^3 - 2\lambda^2 - \lambda + 2 = (\lambda^2 - 1)(\lambda - 2).$$

The eigenvalues are thus −1, 1, 2.

If $\lambda = 1$,

$$\begin{aligned} 2y + z &= 0 \\ 4x - 2y - 3z &= 0 \\ x + 3y + z &= 0. \end{aligned}$$

The first two of these equations are satisfied by

$$\frac{x}{-1} = \frac{y}{1} = \frac{z}{-2}$$

and this also satisfies the third.

Thus the eigenvector is proportional to (−1, 1, −2) and the associated unit vector is

$$\left(-\frac{1}{\sqrt{6}}, \frac{1}{\sqrt{6}}, \frac{2}{\sqrt{6}}\right).$$

The other eigenvalues and eigenvectors may be determined in a similar manner and the orthogonality relation may be verified.

The canonical form of the dyadic will be

$$-\mathbf{ii} + \mathbf{jj} + 2\mathbf{kk}.$$

9. Find the function vector $\boldsymbol{\delta}$ which corresponds to the delta function in the space L^2 discussed in section 6.22. This is rather a contradiction, as

$$\int_R |\delta(t - t')|^2 \, dt$$

does not converge and so strictly the problem cannot be solved. However, an expression can be obtained by the following formal analysis. Using (6.22.4a)

$$\delta(t - t') = \sum_{n=1}^{\infty} \delta_n \phi_n,$$

where

$$\begin{aligned} \delta_n &= \int_R \delta(t - t') \phi_n^*(t) \, dt \\ &= \phi_n(t'). \end{aligned}$$

Thus

$$\delta(t - t') = \sum_{n=1}^{\infty} \phi_n(t') \phi_n(t)$$

and

$$\boldsymbol{\delta} = \{\phi_n(t')\}.$$

10. Prove that a vector space is defined for a set of scalar point functions $\phi(\mathbf{r})$ defined over a volume V and vanishing on its boundary S when the inner product is defined by

$$(\phi, \psi) = -\int_V \phi \nabla^2 \psi \, d\tau.$$

It is easily verified that axioms (6.2.1) and (6.2.2) hold. The axiom (6.2.3b) also holds.

Now

$$-\int_V \phi \nabla^2 \psi \, d\tau = -\int_V \nabla.(\phi \nabla \psi) \, d\tau + \int_V (\nabla \phi).(\nabla \psi) \, d\tau$$

$$= -\int_S \phi \frac{\partial \psi}{\partial n} d\mathscr{A} + \int_V (\nabla \phi).(\nabla \psi) \, d\tau$$

$$= \int_V (\nabla \phi).(\nabla \psi) \, d\tau$$

by the boundary condition.

And so

$$(\phi, \psi) = \int_V (\nabla \phi).(\nabla \psi) \, d\tau.$$

It follows that (6.2.3a) holds

$$(\phi, \phi) = \int_V (\nabla \phi)^2 \, d\tau$$

$$> 0 \text{ if } \phi = 0$$

$$= 0 \text{ if } \phi = 0.$$

($\nabla \phi$ vanishes if ϕ is a constant, but because of the boundary condition the only ϕ possible for this is $\phi = 0$.)

Thus the norm axioms also hold, and a vector space is fully defined.

11. Let Mu denote the operator

$$Mu = -\sum_i \sum_j \frac{\partial}{\partial x_i} \left\{ A_{ij}(\mathbf{x}) \frac{\partial u}{\partial x_j} \right\}$$

$A_{ij} = A_{ji}$ in an N dimensional Euclidean space, and let V be a hypervolume in it bounded by a hypersurface S. Prove that

$$\int_V vMu \, d\tau = \int_V \sum_i \sum_j A_{ij} \frac{\partial v}{\partial x_i} \frac{\partial u}{\partial x_j} d\tau - \int_S v \sum_i \sum_j A_{ij} \frac{\partial u}{\partial x_j} d\mathscr{A}_i$$

all summations being between 1 and N.

$$\int_V vMu\,d\tau = -\sum_i \sum_j \int_V v \frac{\partial}{\partial x_i}\left(A_{ij}\frac{\partial u}{\partial x_j}\right) d\tau$$

$$= -\sum_i \sum_j \int_V \frac{\partial}{\partial x_i}\left(A_{ij} v \frac{\partial u}{\partial x_j}\right) d\tau$$

$$+ \sum_i \sum_j \int_V A_{ij} \frac{\partial u}{\partial x_i}\frac{\partial v}{\partial x_j}\, d\tau.$$

The first term on the left-hand side of the equation becomes, by the generalized divergence theorem

$$-\sum_i \sum_j \int_S A_{ij} v \frac{\partial u}{\partial x_j}\, d\mathscr{A}_i$$

and the result follows.

It will be observed that this problem could be regarded as one in tensor analysis and the summation signs could be omitted.

12. $s(x)$ is a function of x such that

$$\int_{-1}^{+1} \{s(x)\}^2\,dx = \int_{-1}^{+1} s(x)\,s_0(x)\,dx$$

where $s_0(x)$ is a known function of x.

Prove that

$$\int_{-1}^{+1} \{s(x)\}^2\,dx < \int_{-1}^{+1} \{s_0(x)\}^2\,dx$$

unless $s(x) = s_0(x)$, and provide a geometrical interpretation.

Corresponding to $f(x)$ defined in $-1 < x < 1$ there is a function vector **f**. If $f(x)$ is such that

$$\int_{-1}^{+1} \{f(x)\}^2\,dx$$

is finite, and an inner product is defined by

$$(\;,\mathbf{g}) = \int_{-1}^{+1} f(x)g(x)\,dx,$$

the system is associated with an L^2 space.

The relation for **s** may be written in abstract form

$$(\mathbf{s}, \mathbf{s}) - (\mathbf{s}, \mathbf{s}_0) = 0.$$

This can be rewritten as

$$(\mathbf{s},\mathbf{s})+(\mathbf{s}-\mathbf{s}_0,\mathbf{s}-\mathbf{s}_0)=(\mathbf{s}_0,\mathbf{s}_0)$$

whence

$$(\mathbf{s},\mathbf{s})<(\mathbf{s}_0,\mathbf{s}_0)\,\mathbf{s}\neq\mathbf{s}_0.$$

Alternatively

$$(\mathbf{s}-\tfrac{1}{2}\mathbf{s}_0,\mathbf{s}-\tfrac{1}{2}\mathbf{s}_0)=\tfrac{1}{4}(\mathbf{s}_0,\mathbf{s}_0).$$

This represents a hypersphere, centre $\frac{1}{2}\mathbf{s}_0$ and radius $\frac{1}{2}\|\mathbf{s}_0\|$.

13. Prove that the relation

$$\int_R \alpha(t)\,x(t)\,dt=\text{constant}=k$$

is equivalent to a hyperplane in function space.

It is possible to consider always that

$$\int_R \{\alpha(t)\}^2\,dt=1$$

and so the equation becomes

$$\int_R \alpha(t)\,x(t)\,dt=k\int_R \{\alpha(t)\}^2\,dt.$$

If now an inner product be defined by

$$(\mathbf{x},\mathbf{y})=\int_R x(t)\,y(t)\,dt$$

the equation in function space is

$$(\mathbf{x},\boldsymbol{\alpha})=(k\boldsymbol{\alpha},\boldsymbol{\alpha})$$

or

$$(\mathbf{x}-k\boldsymbol{\alpha},\boldsymbol{\alpha})=0.$$

This is the equation of a hyperplane whose normal is $\boldsymbol{\alpha}$ ($\boldsymbol{\alpha}$ is a unit vector). k is the perpendicular from the origin to the plane.

14. If $T_{ij\ldots u}$ is a tensor of rank n, prove that

$$\frac{\partial T_{ij\ldots u}}{\partial x_v}$$

is a tensor of rank $n+1$. It is not necessary to bother about the indices $ij\ldots u$.

$\partial\phi/\partial x_v$ has been shown to transform as a tensor of rank 1 and so the result follows.

15. Prove that $\epsilon_{ijk}\epsilon_{lmn}$ is unity if ijk are all different and lmn is a cyclic permutation of ijk, is minus unit if ijk are all different and lmn is a non-cyclic permutation of ijk and is zero otherwise.

The third result follows immediately if any two of ijk, or lmn are the same as in this case are, or both of ϵ_{ijk}, ϵ_{lmn} are zero.

If ijk and lmn are all different

$$\epsilon_{ijk} = \pm 1 \qquad \text{and} \qquad \epsilon_{lmn} = \pm 1.$$

If ijk and lmn are cyclic permutations of each other, then either

$$\epsilon_{ijk} = \epsilon_{lmn} = +1$$

or

$$\epsilon_{ijk} = \epsilon_{lmn} = -1.$$

In both cases

$$\epsilon_{ijk}\,\epsilon_{lmn} = 1.$$

If ijk and lmn are non-cyclic permutations of one another

$$\epsilon_{ijk} = -\epsilon_{lmn}$$

and the result follows.

16. Suppose that a set of cartesian coordinates is defined by $x_1 = x$, $x_2 = y$, $x_3 = z$, $x_4 = ict$ where c is the velocity of light *in vacuo*. If a signal be emitted from the point with position vector $\mathbf{r}_0$ at time t_0 and at time t it has reached the set of points with position vector $\mathbf{r}$, prove that, with an obvious notation

$$(x_\alpha - x_{\alpha_0})(x_\alpha - x_{\alpha_0}) = 0.$$

At any time, the relation between r and t is given by

$$|\mathbf{r} - \mathbf{r}_0| = c(t - t_0).$$

This may be rewritten as

$$(\mathbf{r} - \mathbf{r}_0)^2 - c^2(t - t_0)^2 = 0$$

or

$$(x - x_0)^2 + (y - y_0)^2 + (z - z_0)^2 + i^2 c^2(t - t_0)^2 = 0$$

whence

$$(x_\alpha - x_{\alpha_0})(x_\alpha - x_{\alpha_0}) = 0.$$

The 'four vector' $\mathbf{x}$ may be regarded as representing the coordinates of an event which happens at a particular place at a particular time. The interval between two points in space time is given by $(x_\alpha - y_\alpha)(x_\alpha - y_\alpha)$ and the elementary interval by

$$\begin{aligned} ds^2 &= dx_\alpha\, dx_\alpha \\ &= dx^2 + dy^2 + dz^2 - c^2\, dt^2. \end{aligned}$$

It will be seen that ds^2 is not of necessity positive. This is different from what has been seen previously, when the (interval)^2 between two points has always been positive. If ds^2 is positive the interval is space-like and if ds^2 is negative the interval is time-like. If ds^2 is zero, the interval between two different events is zero and this occurs when a signal is propagated from one to the other with the speed of light.

17. With the same notation as in example 16, show that the transformation $(-1 < \beta < 1)$

$$x_1' = x_1, \qquad x_2' = x_2,$$

$$x_3' = \frac{1}{(1-\beta^2)^{1/2}}(x_3 + i\beta x_4), \qquad x_4' = \frac{1}{(1-\beta^2)^{1/2}}(x_4 - i\beta x_3)$$

leaves $x_\alpha x_\alpha$ unchanged, and interpret this transformation

$$x_1'^2 + x_2'^2 + x_3'^2 + x_4'^2 = x_1^2 + x_2^2 + \frac{1}{(1-\beta^2)}(x_3^2 + 2i\beta x_3 x_4 - \beta^2 x_4^2)$$

$$+ \frac{1}{(1-\beta^2)}(x_4^2 - 2i\beta x_3 x_4 - \beta^2 x_3^2) = x_1^2 + x_2^2 + x_3^2 + x_4^2.$$

In space time coordinates this becomes

$$x' = x, \qquad y' = y, \qquad z' = \frac{1}{(1-\beta^2)^{1/2}}(z - \beta ct),$$

$$t' = \frac{1}{\sqrt{(1-\beta^2)}}\left(t - \frac{\beta}{c}z\right).$$

Suppose that a point is fixed in S', the frame of reference associated with (x', y', z'). The coordinates in S are (x, y, z). As z' is fixed $dz/dt = \beta c$ and so the transformation is between two frames of reference. S' is moving with constant velocity $v\mathbf{k}$ relative to S where $\beta c = v$. This transformation is an example of the transformation class known as the Lorentz transformations, and the transformation tensor is given by

$$a_{rs} = \begin{bmatrix} 1 & 0 & 0 & 0 \\ 0 & 1 & 0 & 0 \\ 0 & 0 & \dfrac{1}{(1-\beta^2)^{1/2}} & \dfrac{i\beta}{(1-\beta^2)^{1/2}} \\ 0 & 0 & \dfrac{-i\beta}{(1-\beta^2)^{1/2}} & \dfrac{1}{(1-\beta^2)^{1/2}} \end{bmatrix}$$

It is sometimes convenient to write

$$dx^2 + dy^2 + dz^2 - c^2 dt^2 = -c^2 d\tau^2$$

and

$$\frac{d\tau}{dt} = \left[1 - \frac{1}{c^2}\left\{\left(\frac{dx}{dt}\right)^2 + \left(\frac{dy}{dt}\right)^2 + \left(\frac{dz}{dt}\right)^2\right\}\right]^{1/2}$$

$$= \left(1 - \frac{v^2}{c^2}\right)^{1/2} = (1-\beta^2)^{1/2}.$$

τ is termed the proper time.

18. Express the differential equations for the electric and vector potentials in terms of four-vectors.

The differential equations of interest are

$$\nabla^2 \mathbf{A} - \frac{1}{c^2}\frac{\partial^2 \mathbf{A}}{\partial t^2} = -\mu_0 \mathbf{J}$$

$$\nabla^2 V - \frac{1}{c^2}\frac{\partial^2 V}{\partial t^2} = -\frac{\rho}{\epsilon_0}$$

$$\nabla . \mathbf{A} + \frac{1}{c^2}\frac{\partial V}{\partial t} = 0$$

and

$$\nabla . \mathbf{J} + \frac{\partial \rho}{\partial t} = 0.$$

If $J_1 = J_x$, $J_2 = J_y$, $J_3 = J_z$, $J_4 = ic\rho$, the continuity equation can be written as

$$\frac{\partial J_\alpha}{\partial x_\alpha} = 0.$$

If $A_1 = A_x$, $A_2 = A_y$, $A_3 = A_z$, $A_4 = iV/c$,

$$\nabla . \mathbf{A} + \frac{1}{c^2}\frac{\partial V}{\partial t} = 0,$$

the gauge condition, becomes

$$\frac{\partial A_\alpha}{\partial x_\alpha} = 0.$$

The operator

$$\nabla^2 - \frac{1}{c^2}\frac{\partial^2}{\partial t^2} = \frac{\partial^2}{\partial x_\alpha \partial x_\alpha}$$

and so the differential equations for $\mathbf{A}$ and V become

$$\frac{\partial^2 \mathbf{A}}{\partial x_\alpha \partial x_\alpha} = -\mu_0 \mathbf{J}$$

$$\frac{\partial^2 V}{\partial x_\alpha \partial x_\alpha} = -\frac{\rho}{\epsilon_0} = -\mu_0 c^2 \rho.$$

These can both be taken from the equation

$$\frac{\partial^2 A_\beta}{\partial x_\alpha \partial x_\alpha} = -\mu_0 J_\beta.$$

Under the Lorentz transformation

$$J_x' = J_x, \qquad J_y' = J_y,$$

$$J_z' = \frac{1}{\sqrt{(1-\beta^2)}}(J_z - v\rho), \qquad \rho' = \frac{1}{\sqrt{(1-\beta^2)}}\left(\rho - \frac{v}{c^2}J_z\right).$$

The quantities $J_\alpha J_\alpha$, $A_\alpha A_\alpha$, $J_\alpha A_\alpha$, that is $J^2 - c^2\rho^2$, $A^2 - V^2/c^2$, $\mathbf{J}.\mathbf{A} - \rho V$, are invariant.

19. Prove that if $T_{\mu\gamma}$ is a tensor, $T_{\alpha\alpha}$ is an invariant or scalar.

$$T'_{\alpha\alpha} = a_{\alpha\beta}\, a_{\alpha\gamma}\, T_{\beta\gamma} = \delta_{\beta\gamma}\, T_{\beta\gamma} = T_{\beta\beta}.$$

This is termed the trace.

20. A medium is given a small continuous deformation by external forces. Show that the deformation consists of a rigid body motion and a straining. Let x_α be the coordinates of a point and let $x_\alpha + \xi_\alpha$ be those of a neighbouring point.

After deformation, these have changed instead to

$$x_\alpha + \delta x_\alpha \qquad \text{and} \qquad x_\alpha + \delta x_\alpha + \xi_\alpha + \delta\xi_\alpha.$$

The displacement of the second point is given by $\delta x_\alpha + \Delta\xi_\alpha$

$$\delta x_\alpha + \frac{\partial \xi_\alpha}{\partial x_\beta}\,\delta x_\beta = \left(\delta_{\alpha\beta} + \frac{\partial \xi_\alpha}{\partial x_\beta}\right)\delta x_\beta.$$

The motion of the second point relative to the first is given by

$$\frac{\partial \xi_\alpha}{\partial x_\beta}\,\delta x_\beta = \frac{1}{2}\left(\frac{\partial \xi_\alpha}{\partial x_\beta} + \frac{\partial \xi_\alpha}{\partial x_\beta}\right)\delta x_\beta + \frac{1}{2}\left(\frac{\partial \xi_\alpha}{\partial x_\beta} - \frac{\partial \xi_\beta}{\partial x_\alpha}\right)\delta x_\beta.$$

The second term can be written as

$$-\tfrac{1}{2}\delta\mathbf{r} \times (\nabla \times \boldsymbol{\xi})$$

which by section 2.71 represents a rigid body rotation.

The tensor

$$e_{\alpha\beta} = \frac{1}{2}\left(\frac{\partial \xi_\alpha}{\partial x_\beta} + \frac{\partial \xi_\beta}{\partial x_\alpha}\right) - e_{\beta\alpha}$$

is called the strain tensor. Thus, ignoring the rigid body motion, the displacement of the second point is given by

$$\delta x'_\alpha = (\delta_{\alpha\beta} + e_{\alpha\beta})\,\delta x_\beta.$$

The change in elementary volume will be given by

$$\delta x'_1\, \delta x'_2\, \delta x'_3 = \frac{\partial(x'_1, x'_2, x'_3)}{\partial(x_1, x_2, x_3)} \cdot \delta x_1\, \delta x_2\, \delta x_3$$

$$= |\delta_{\alpha\beta} + e_{\alpha\beta}|\; \delta x_1\, \delta x_2\, \delta x_3.$$

It is always possible to choose axes so that

$$e_{\alpha\beta} = 0 \quad (\alpha \neq \beta).$$

and so the determinant becomes

$$(1+e_{11})(1+e_{22})(1+e_{33}) = 1+e_{11}+e_{22}+e_{33}$$

if the e_{rr} are all small.

Thus

$$\frac{\delta\tau' - \delta\tau}{\delta\tau} = e_{11}+e_{22}+e_{33} = e_{\alpha\alpha}$$

which is the dilation. By example 19, $e_{\alpha\alpha}$ is a scalar and so the same formula would hold for any rectangular axes.

Exercises

1. Prove that

$$\nabla\nabla\left(\frac{1}{R}\right) = \frac{3}{R^5}\mathbf{RR} - \frac{1}{R^3}\mathbf{I}.$$

2. The equation of a quadric is given by

$$\mathbf{r}.\boldsymbol{\Omega}.\mathbf{r} + 2\mathbf{s}.\mathbf{r} + d = 0$$

where Ω is symmetric. Find the equation of the polar plane of the point with position vector $\mathbf{r}_0$ and the equation of the tangent cone with its vertex at this point.

3. Show that $\mathbf{a}.(\mathbf{r}\times\boldsymbol{\Gamma}) = (\mathbf{a}\times\mathbf{r}).\boldsymbol{\Gamma}$.

4. Prove that $\nabla(\mathbf{P}\times\mathbf{Q}) = (\nabla\mathbf{P})\times\mathbf{Q} - (\nabla\mathbf{Q})\times\mathbf{P}$.

5. Prove that the angular momentum of a rigid body rotating about a fixed point is given by

$$\mathbf{H} = \boldsymbol{\Psi}.\boldsymbol{\omega}$$

where $\boldsymbol{\Psi}$ is the inertia dyadic discussed in example 4 and $\boldsymbol{\omega}$ is the angular velocity.

What is the kinetic energy?

6. If the elements of each of the dyads comprising the dyadics $\boldsymbol{\Phi}_r$ are small compared with unity, prove that, to the first order of small quantities

$$(\mathbf{I}+\boldsymbol{\Phi}_1).(\mathbf{I}+\boldsymbol{\Phi}_2)(\ldots)(\mathbf{I}+\boldsymbol{\Phi}_n) = \mathbf{I}+\boldsymbol{\Phi}_1+\boldsymbol{\Phi}_2+\ldots\boldsymbol{\Phi}_n$$

and that to the first order

$$(\mathbf{I}+\boldsymbol{\Phi})^n = \mathbf{I}+n\boldsymbol{\Phi}$$

n being a positive negative or zero integer.

7. Prove that the set of quadrics defined by for varying λ by

$$\mathbf{r}.\boldsymbol{\Psi}_\lambda.\mathbf{r} = 1$$

where $\boldsymbol{\Psi}_\lambda^{-1} = \boldsymbol{\Phi} + \lambda\mathbf{I}$ is confocal.

8. If a vector space is defined as in example 10, verify that the Schwarz inequality holds when the volume is the sphere $r < a$ and

$$\psi(\mathbf{r}) = \left(1 - \frac{r^2}{a^2}\right)\cos\theta, \qquad \phi(\mathbf{r}) = \left(1 - \frac{r}{a}\right)\sin\theta.$$

9. With the system defined in example 11 and writing

$$Nu = \sum_i \sum_j A_{ij} \frac{\partial u}{\partial x_j}(\mathbf{n}.\mathbf{i}_i)$$

show that

$$\int_V (vMu - uMv)\,d\tau = \int_S (uNv - vNu)\,d\mathscr{A}.$$

10. $f(x)$ is the set of functions such that

$$f(x + 2\pi) = f(x).$$

If

$$\phi_0(x) = \frac{1}{\sqrt{2\pi}}, \qquad \phi_{2r-1}(x) = \frac{\sin x}{\sqrt{\pi}}.\phi_2(x) = \frac{\cos x}{\sqrt{\pi}} \qquad (r > 0),$$

verify if $\mathbf{f}$ is the function vector corresponding to

$$f(x) = \sum_{\xi=0}^{\infty} f_r \phi_r(x),$$

and the inner product is defined by

$$(\mathbf{f}, \mathbf{g}) = \int_{-\pi}^{\pi} f(x) g(x)\,dx$$

that

$$(\mathbf{f}, \mathbf{g}) = \sum_{\xi=0}^{\infty} f_n g_n.$$

It may be assumed that the set $\phi_s(x)$ is complete.

11. Consider the class of differentiable functions $u(x)$ defined over $0 < x < 1$ such that $u'(0) - \alpha u(0) = 0$ and $u'(1) + \beta u(1) = 0$ $\alpha,\ \beta > 0$. Prove that there is an abstract vector space associated with the inner product

$$(\mathbf{u}, \mathbf{v}) = -\int_0^1 u \frac{d^2 v}{dx^2}\,dx.$$

12. Prove that for any vector space

$$||\mathbf{u}-\mathbf{v}||^2+||\mathbf{u}+\mathbf{v}||^2=2(||\mathbf{u}||^2+||\mathbf{v}||^2).$$

13. If $u(x)$ is a function which is differentiable over $0<x<1$ such that $u(0)=u(1)=0$, show that if

$$v=-\frac{d^2u}{dx^2},$$

then

$$(\mathbf{u},\mathbf{v})>2||\mathbf{u}||^2$$

where

$$(\mathbf{u},\mathbf{v})=\int_0^1 u(x)\,v(x)\,dx.$$

14. If $T_{ij\ldots u}$ is a tensor of rank n, prove that

$$\frac{T_{ij\ldots\alpha}}{\partial x_\alpha}$$

is a tensor of rank $n-1$.

15. Prove that $\epsilon_{\alpha ij}\epsilon_{\alpha mn}=\delta_{im}\delta_{jn}-\delta_{in}\delta_{jm}$.

16. Show that a four-vector, which has as three of its components the momentum of a body, when the velocity is very small, is given by

$$m_0\left\{\frac{v_x}{\sqrt{(1-\beta^2)}},\frac{v_y}{\sqrt{(1-\beta^2)}},\frac{v_z}{\sqrt{(1-\beta^2)}},\frac{ic}{\sqrt{(1-\beta^2)}}\right\}$$

and find the four-vector which has acceleration forming three of its components under the same condition.

17. Show that the equations

$$\nabla\times\mathbf{H}-\epsilon\frac{\partial\mathbf{D}}{\partial t}=\mathbf{J},\qquad \nabla.\mathbf{D}=\rho$$

can be rewritten in the form

$$\frac{\partial f_{m\alpha}}{\partial x_\alpha}=J_m$$

and that the equations

$$\nabla\times\mathbf{B}+\frac{\partial\mathbf{E}}{\partial t}=\mathbf{0},\qquad \nabla.\mathbf{B}=0$$

can be written in the form

$$\frac{\partial F_{ij}}{\partial x_k}+\frac{\partial F_{ki}}{\partial x_j}+\frac{\partial F_{jk}}{\partial x_i}=0$$

where i, j, k are any three of the four numbers, and verify that if $\mathbf{D} = \epsilon_0 \mathbf{E}$, $\mathbf{B} = \mu_0 \mathbf{H}$ the equations are satisfied by

$$F_{jk} = \frac{\partial A_k}{\partial x_j} - \frac{\partial A_j}{\partial x_k}$$

A_k being the k component of the four potential mentioned in example 18.

f_{rs}, F_{rs} are both antisymmetric, and

$$f_{14} = -icD_x, \qquad f_{24} = -icD_y, \qquad f_{34} = -icD_z$$

$$f_{23} = H_x, \qquad f_{31} = H_y, \qquad f_{12} = H_z$$

$$F_{14} = -\frac{i}{c}E_x, \qquad F_{24} = -\frac{i}{c}E_y, \qquad F_{34} = -\frac{i}{c}E_z$$

$$F_{23} = B_x, \qquad F_{31} = B_y, \qquad F_{12} = B_z.$$

18. Prove that, under the Lorentz transformation of example 17

$$B_z' = B_z, \qquad E_z' = E_z$$

$$\mathbf{B}_\perp' = \frac{1}{\sqrt{(1-\beta^2)}}\left(\mathbf{B}_\perp - \frac{1}{c^2}\mathbf{v} \times \mathbf{E}_\perp\right)$$

$$\mathbf{E}_\perp' = \frac{1}{\sqrt{(1-\beta^2)}}(\mathbf{E}_\perp + \mathbf{v} \times \mathbf{B}_\perp)$$

and verify that Maxwell's equations are unchanged.

19. Assuming the generalization of Hooke's law

$$T_{rs} = c_{rs\alpha\beta}\, e_{\alpha\beta}$$

where $e_{\alpha\beta}$ is the strain tensor and T_{rs} is the stress tensor, show that the energy of strain per unit volume is

$$\tfrac{1}{2} c_{\alpha\beta\gamma\delta}\, e_{\alpha\beta}\, e_{\gamma\delta}.$$

20. If T_{rs} is the stress tensor, α_r is the acceleration of a point in a medium with density ρ and f_r the force per unit mass, show that

$$T_{r\alpha,\,\alpha} + \rho f_r = \rho\alpha_r.$$

Appendix: The Delta Function

The delta function and its derivatives are examples of what are termed generalized functions. In this book only a short treatment is given, sufficient for use in vector analysis.

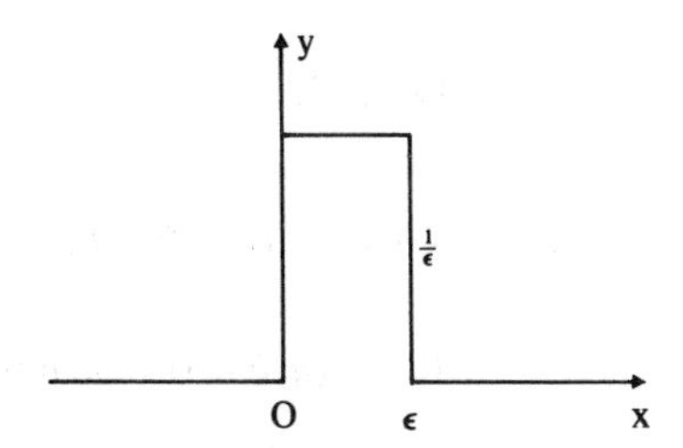

Consider the function $\delta_\epsilon(x)$ which is defined by

$$\delta_\epsilon(x) = \frac{1}{\epsilon\sqrt{\pi}} e^{-x^2/\epsilon^2} \qquad (\epsilon > 0). \tag{A.1}$$

Then

$$\int_{-\infty}^{\infty} \delta_\epsilon(x)\, dx = 1 \qquad \text{for all } \epsilon. \tag{A.2}$$

Let $f(x)$ be such that $|f(x) - f(0)| < M|x|$ where $M > 0$ and M is independent of x. Then it may be shown (e.g., D. S. Jones, *Generalized Functions*, McGraw-Hill (1966), p. 51) that

$$\lim_{\epsilon \to 0} \int_{-\infty}^{\infty} \delta_\epsilon(x) f(x)\, dx = f(0). \tag{A.3}$$

Define now a generalized function $\delta(x)$, defined for real x, such that

$$\delta(x) = 0 \qquad (x \neq 0) \tag{A.4}$$

$$\int_{-\infty}^{\infty} f(x)\, \delta(x)\, dx = f(0). \tag{A.5}$$

Comparing equations (A.1) and (A.3) with (A.4) and (A.5) respectively, it follows in a certain sense that

$$\delta(x) = \lim_{\epsilon \to 0} \delta_\epsilon(x). \tag{A.6}$$

This would imply, were it taken literally, that

$$\delta(x) = 0 \qquad (x \neq 0) \tag{A.7a}$$
$$\delta(x) = \infty \qquad (x = 0) \tag{A.7b}$$

so to speak, and it is for this reason that $\delta(x)$ is referred to as a generalized function rather than as a function.

The following properties of $\delta(x)$ follow immediately

$$\int_{-\infty}^{\infty} \delta(x)\,dx = 1 \tag{A.8a}$$

$$\int_a^b f(x)\,\delta(x)\,dx = f(0) \qquad (a < 0 < b) \tag{A.8b}$$

$$= 0 \qquad (a, b < 0 \text{ or } a, b > 0). \tag{A.8c}$$

If $H(x)$ (the Heaviside Function or the unit function) is defined by

$$H(x) = 1 \qquad (x > 0) \tag{A.9a}$$
$$= \tfrac{1}{2} \qquad (x = 0) \tag{A.9b}$$
$$= 0 \qquad (x < 0), \tag{A.9c}$$

in a sense

$$H'(x) = \delta(x). \tag{A.10}$$

The derivatives of the delta function are interpreted as follows.

Clearly

$$\delta'(x) = 0 \qquad (x \neq 0).$$

Now

$$\int_{-\infty}^{\infty} f(x)\,\delta_\epsilon'(x)\,dx = \Big[f(x)\,\delta_\epsilon(x)\Big]_{-\infty}^{\infty} - \int_{-\infty}^{\infty} f'(x)\,\delta_\epsilon(x)\,dx$$
$$= -\int_{-\infty}^{\infty} f'(x)\,\delta_\epsilon(x)\,dx.$$

Consequently we interpret

$$\int_{-\infty}^{\infty} f(x)\,\delta'(x)\,dx = -\int_{-\infty}^{\infty} f'(x)\,\delta(x)\,dx = -f'(0) \tag{A.11}$$

an appropriate continuity condition being imposed on $f'(x)$. Similarly

$$\int_{-\infty}^{\infty} f(x)\,\delta^n(x)\,dx = (-1)^n f^n(0). \tag{A.12}$$

A number of other properties follow by formal manipulation which can be justified by the theory of generalized functions.

$$\int_{-\infty}^{\infty} \delta(x-x')\,f(x)\,dx = \int_{-\infty}^{\infty} \delta(y)\,f(x'+y)\,dy = f(x'). \tag{A.13}$$

If $\alpha > 0$,

$$\delta(\alpha x) = \frac{1}{\alpha}\,\delta(x) \tag{A.14}$$

$$\int_{x_1}^{x_2} f(x)\,\delta(g(x))\,dx = f(x_0)/g'(x_0). \tag{A.15}$$

This holds when $g(x)$ is monotonically increasing from $-\infty$ to ∞ over the range $x_1 < x < x_2$, and $g(x)$ has a simple zero at $x = x_0$. $x_1 < x_0 < x_2$. Extensions to two and three dimensions follow easily. The two-dimensional function is defined by

$$\delta(\mathbf{r}-\mathbf{r}') = \delta(x-x')\,\delta(y-y') \tag{A.16}$$

and if D is some domain in the xy plane

$$\iint_D f(\mathbf{r}_\perp)\,\delta(\mathbf{r}_\perp - \mathbf{r}'_\perp)\,dx\,dy = f(\mathbf{r}'_\perp) \tag{A.17a}$$

or

$$= 0 \tag{A.17b}$$

according as to whether $\mathbf{r}'_\perp$ is the position vector of a point within or without D.

Similarly, the three-dimensional delta function is defined by

$$\delta(\mathbf{r}-\mathbf{r}') = \delta(x-x')\,\delta(y-y')\,\delta(z-z') \tag{A.18}$$

and

$$\int_V f(\mathbf{r})\,\delta(\mathbf{r}-\mathbf{r}')\,d\tau = f(\mathbf{r}') \tag{A.19a}$$

or

$$= 0 \tag{A.19b}$$

according as $\mathbf{r}'$ is the position vector of a point within or without V. A useful interpretation of the three-dimensional delta function is as follows.

The charge density distribution associated with a charge Q, situated at a point with position vector $\mathbf{r}'$ is given by

$$\rho(\mathbf{r}) = Q\,\delta(\mathbf{r}-\mathbf{r}') \qquad \text{(A.20)}$$

for $\rho(\mathbf{r})$ vanishes everywhere except at $\mathbf{r}=\mathbf{r}'$ and

$$\int_V Q\,\delta(\mathbf{r}-\mathbf{r}')\,d\tau = Q \qquad \text{(A.21)}$$

where V is any volume including the point with position vector $\mathbf{r}'$. The charge density associated with a point electric dipole may be calculated in a similar manner.

Consider two point charges. $-Q$ at the point with position vector $\mathbf{r}'$ and $+Q$ at the point with position vector $\mathbf{r}'+\mathbf{n}l$ where $\mathbf{n}$ is a unit vector. Then

$$\rho - -Q\,\delta(\mathbf{r}-\mathbf{r}') + Q\,\delta(\mathbf{r}-\mathbf{r}'-\mathbf{n}l). \qquad \text{(A.22)}$$

If l is small, this becomes to the first order

$$\rho = Q\mathbf{n}l\,.\,\boldsymbol{\nabla}'\{-\delta(\mathbf{r}-\mathbf{r}')\} \qquad \text{(A.23)}$$

The dipole moment $\mathbf{p} = Q\mathbf{n}l$ and making l tend to zero with Q tending to infinity so that Ql remains finite

$$\rho = -\mathbf{p}\,.\,\boldsymbol{\nabla}'\,\delta(\mathbf{r}-\mathbf{r}') = \mathbf{p}\,.\,\boldsymbol{\nabla}\,\delta(\mathbf{r}-\mathbf{r}'). \qquad \text{(A.24)}$$

For general orthogonal coordinate systems the element of volume is

$$\delta\tau = \delta s_1\,\delta s_2\,\delta s_3 = h_1\,h_2\,h_3\,\delta u_1\,\delta u_2\,\delta u_3 \qquad \text{(A.25)}$$

and

$$\delta(\mathbf{r}) = (h_1\,h_2\,h_3)^{-1}\,\delta(u_1)\,\delta(u_2)\,\delta(u_3) \qquad \text{(A.26)}$$

for

$$\int \delta(\mathbf{r})\,d\tau = \iiint (h_1\,h_2\,h_3)^{-1}\,\delta(u_1)\,\delta(u_2)\,\delta(u_3)\,h_1\,h_2\,h_3\,du_1\,du_2\,du_3 \qquad \text{(A.27)}$$

where the integration is over the whole of space

$$= 1. \qquad \text{(A.28)}$$

More generally

$$\delta(\mathbf{r}-\mathbf{r}') = (h_1\,h_2\,h_3)^{-1}\,\delta(u_1-u_1')\,\delta(u_2-u_2')\,\delta(u_3-u_3'). \qquad \text{(A.29)}$$

For example, in spherical coordinates

$$\delta(\mathbf{r}-\mathbf{r}') = \frac{1}{r^2\sin\theta}\,\delta(r-r')\,\delta(\theta-\theta')\,\delta(\phi-\phi'). \qquad \text{(A.30)}$$

Index

Abstract vector space, 12
Acceleration, 19
Acceleration, Coriolis, 72
Acceleration, relative, 20
Addition of dyadics, 182
Addition of vectors, 3
Angular velocity, 57
Antecedents of dyadic, 181
Area, 47
Argand diagram, 62
Associative law for dyadics, 184
Associative law for vectors, 7
Axial vector, 58

Beltrami field, 96
Bessels inequality, 193
Bound vector, 4

Central axis, 67
Centroid, 8
Circulation, 172
Commutative law, scalar product, 40
Commutative law, vector product, 45
Complete dyadic, 183
Complete set of vectors, 193
Complex number, 62
Complex numbers and matrices, 206
Component of vector, 8
Conjugate dyadic, 182
Consequents of dyadic, 181
Conservation, equation of, 170
Conservative field, 119
Contravariant tensor, 204
Covariant tensor, 204
Convergence, strong, 199
Convergence, weak, 199
Coordinates, curvilinear, 60
Coordinates, cylindrical, 23
Coordinates, oblate spheroidal, 114
Coordinates, parabolic cylinder, 79
Coordinates, rectangular cartesian, 10
Coordinates, spherical, 24
Coplanar vectors, 13
Couple, 67
Curl, 91
Curl in curvilinear coordinates, 99
Curl in cylindrical coordinates, 101
Curl in rectangular cartesian coordinates, 91
Curl in spherical coordinates, 101
Current density vector, 107

Del, 83
Delta function, 225
Dipole, 228
Directional derivative, 86
Differentiation of vectors, 17
Differentiation with respect to vector, 87
Dirichlet condition, 152
Dirichlet integral, 161
Distributive law for vectors, 7
Divergence, 90
Divergence in curvilinear coordinates, 99
Divergence in cylindrical coordinates, 101
Divergence in rectangular cartesian coordinates, 90
Divergence in spherical coordinates, 101

Dyad, 181
Dyads and matrices, 207
Dyadic, 182
Dyadic invariants, 187

Eigenvalue, 187
Eigenvector, 187
Electric displacement, 107
Electric intensity, 107
Element of arc, 117
Element of area, 124
Elementary rotation, 56
Elliptic harmonic motion, 31
Equations involving vectors, 52

Free vector, 4
Field, 81
Frame of reference, 9
Function space, 198

Gauss divergence theorem, 148
Gradient, 83
Gradient in curvilinear coordinates, 97
Gradient in cylindrical coordinates, 85
Gradient in rectangular cartesian coordinates, 85
Gradient in spherical coordinates, 85
Gram-Schmidt orthogonalization process, 77
Green's function, 156
Green's first theorem, 151
Green's second theorem, 151
Green's vector theorems, 152

Harmonic function, 91
Helmoltz's theorem, 160
Hilbert space, 200
Hodograph, 17

Incomplete set of vectors, 192
Infinitesimal rotation, 56
Inner product, 191
Integral along curve, 116
Integral over surface, 122
Integral over volume, 127
Invariants of dyadic, 187
Irrotational vector, 92

L^2, 198
Lagranges identity, 51
Lami's theorem, 2
Laplacian operator, 91
Laplacian operator in curvilinear coordinates, 100
Laplacian operator in cylindrical coordinates, 101
Laplacian operator in rectangular cartesian coordinates, 91
Laplacian operator in spherical coordinates, 101
Line, equation of, 14
Line vector, 4
Linear dependence, 13
Linear manifold, 192
Linear vector function, 181

Magnetic field vector, 107
Magnetic induction, 107
Matrices and complex numbers, 206
Matrices and dyads, 207
Matrices and tensors, 208
Matrices and vectors, 205
Maxwell's equations, 107
Modulus of vector, 5
Moment of force, 46
Moment of momentum, 46

n dimensional vector, 196
Neumann condition, 152
Nonion form of dyadic, 186
Norm, 192
Normal, unit, 84

Orthogonal curvilinear coordinates, 60
Orthogonal frame of reference, 9
Orthonormal sets of vectors, 192

Pitch of wrench, 67
Plane electromagnetic wave, 115
Plane, equation of, 42
Poisson's equation, 155

Polar vector, 58
Poloidal vector field, 108
Position vector, 4
Positive direction of curve, 117
Positive direction of surface, 122
Positive definite dyadic, 189
Post factor of dyadic, 182
Power of dyadic, 184
Poynting vector, 173
Poynting's theorem, 172
Prefactor of dyadic, 182
Product of dyadics, 183
Product, scalar, 39
Product, scalar triple, 47
Product, vector, 44
Product, vector triple, 50
Pseudo-scalar, 59
Pseudo-vector, 59

Radiation condition, 167
Reciprocal dyadic, 185
Reciprocal set of vectors, 49
Rectangular cartesian coordinates, 10
Relative acceleration, 19
Relative velocity, 19
Resolution of vector, 8
Resultant of forces, 20
Rigid body motion, 53
Right-handed frame, 10
Ritz method, 161
Rotation, finite, 54
Rotation, elementary, 56

Scalar, 2
Scalar field, 81
Scalar of dyadic, 185
Scalar potential, 160
Scalar product of vectors, 39
Schwarz inequality, 193
Singular dyadic, 183
Sliding vector, 4
Solenoidal vector, 90
Solid angle, 126
Space, abstract vector, 191
Space, function, 198
Space, Hilbert, 200
Speed, 19
Stokes' theorem, 145
Strong convergence, 199
Subspace, 192
Summation convention, 200
Surface area, 124
Symmetric dyadic, 183

Tensor, 200
Tensors and matrices, 208
Toroidal vector field, 108
Trefftz method, 161
Triple scalar product, 47
Triple vector product, 50

Uniqueness theorem, 159
Unit vector, 5

Vector, 2
Vector, axial, 58
Vector, bound, 4
Vector, components of, 8
Vector, curl of, 91
Vector, derivative of, 17
Vector, divergence of, 90
Vector field, 81
Vector, free, 4
Vector function, 81
Vector, irrotational, 92
Vector, modulus, 5
Vector, *n* dimensional, 196
Vector, polar, 58
Vector, potential, 160
Vector product, 44
Vector, scalar product, 39
Vector, scalar triple product, 47
Vector, solenoidal, 90
Vector sum, 2
Vector, unit, 5
Vector, vector triple product, 50
Vectors and matrices, 205
Velocity, 19
Velocity, angular, 57
Velocity relative, 19

Weak convergence, 199
Wrench, 67
Work, 41